Vibrations and waves in physics

Vibrations and waves in physics

IAIN G. MAIN

Senior Lecturer in Physics, University of Liverpool

SECOND EDITION

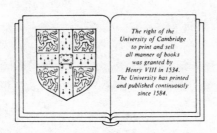

The right of the
University of Cambridge
to print and sell
all manner of books
was granted by
Henry VIII in 1534.
The University has printed
and published continuously
since 1584.

CAMBRIDGE UNIVERSITY PRESS

CAMBRIDGE

NEW YORK · PORT CHESTER · MELBOURNE · SYDNEY

Published by the Press Syndicate of the University of Cambridge
The Pitt Building, Trumpington Street, Cambridge CB2 1RP
40 West 20th Street, New York, NY 10011, USA
10 Stamford Road, Oakleigh, Melbourne 3166, Australia

First published 1978
Reprinted 1979, 1980
Second edition 1984
Reprinted 1985, 1987, 1988, 1990

Printed in Great Britain at
the University Press, Cambridge

Library of Congress catalogue card number: 83-23999

British Library cataloguing in publication data

Main, Iain G.
Vibrations and waves in physics. – 2nd ed

1. Waves 2. Vibrations
I. Title
531'.1133   QC157

ISBN 0 521 26124 4 hard covers
ISBN 0 521 27846 5 paperback
(First edition ISBN 0 521 21662 1 hard covers
ISBN 0 521 29220 4 paperback)

AS

For my parents

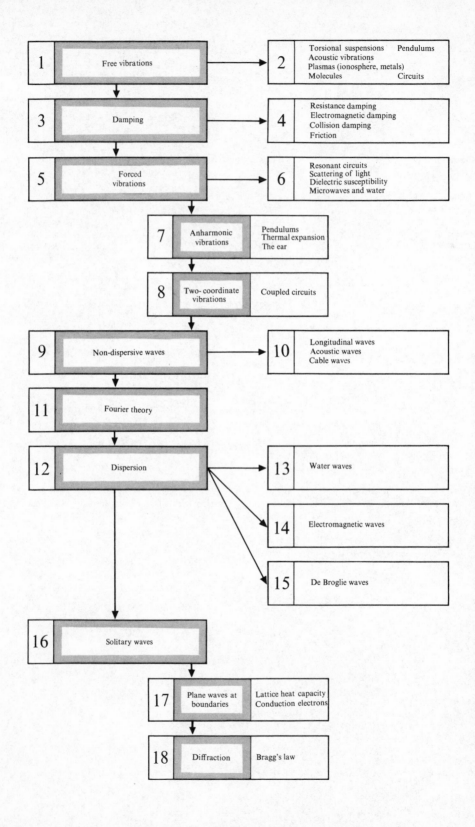

Contents

Preface to the first edition

It would be generally agreed that an undergraduate course on vibrations and waves should lead the student to a thorough understanding of the basic concepts, demonstrate how these concepts unify a wide variety of familiar physics, and open doors to some more advanced topics on which they shed light. The fundamental ideas can be introduced with reference to mechanical systems which are easy to visualize and to illustrate; less tangible phenomena can then be treated with the same mathematics, leaving one free to concentrate on more interesting problems of formulation and interpretation.

In such a course there is always a risk that springs and strings will take over completely, submerging the real physics. The theory must be developed with some care and thoroughness if the student is to comprehend it in sufficient depth to be able to apply it readily, and all too often the physics have to be left half baked. This textbook is an attempt to provide, in a volume of moderate size, an account which is systematic and coherent, but which also treats the physical examples in some depth.

The diagram on p. vi shows the plan of the book, and indicates the relative weights attached to fundamental and illustrative material. The theoretical development (indicated by the downward flow through the toned areas on the left) is continuous, but is punctuated by regular excursions into physics. These excursions are made as soon as the necessary theoretical ground has been cleared, but there is no reason why the order should not be varied by any reader who wishes to press on with the theory first.

The two kinds of material are treated in distinct ways. The prototype systems of the expository chapters are true models, endowed with the properties that we choose for them, and equations predominate over numbers. The physics sections, by contrast, give due emphasis to the sizes of the quantities involved. One of the things we find here is that a great deal of physics is 'extreme', showing much simpler behaviour than the general equations would suggest. A few order-of-magnitude estimates may reveal that we are dealing with a very heavily damped system, a mass controlled vibration or a shallow-water wave, and in such a case we can often achieve a fairly advanced understanding of the phenomenon without detailed analysis. This approach is particularly valuable in the case of those quantum phenomena, such as the scattering of light and microwave dispersion in water, which can be illuminated by a semi-classical discussion.

Although my hope is that the book can be read with profit at any time during a first-degree course, the mathematics background of a typical entrant to a physics-based course has been kept in mind. A working knowledge of calculus and trigonometry is assumed, and the reader is presumed to be learning, if he has not already learnt, how to tackle the solution of the simpler differential equations and the most elementary facts of complex number algebra. The notation of vector algebra (but not of vector calculus) is called upon in a few places.

Central to any discussion of wave motion is the problem of what to call the propagation vector k and its magnitude $2\pi/\lambda$. The best solution seems to be to use as few names as possible. I have therefore called k the wavevector, using the same name for the signed scalar k which replaces it in one-dimensional problems. I avoid terms such as wavenumber, which in different hands can mean $1/\lambda$ (sometimes denoted by k) or $2\pi/\lambda$, and merely draw these pitfalls to the attention of any reader for whom this is only one of several textbooks in use at once.

Any attempt to bring together different branches of physics must lead to the multiple use of symbols. In this book R, for example, has to do duty for resistances, reflection coefficients, the molar gas constant and the bond length of a diatomic molecule, not to mention a response function $R(\omega)$. In most cases there is no overlap and, to avoid introducing a host of strange characters or cluttering the landscape with suffixes, I have usually adopted the most familiar symbol. An exception occurs in chapter 17, where k (the wavevector) and k (the Boltzmann constant) meet; at that point the latter becomes k_B. I have not provided a table of symbols and their various uses, which could only instil unnecessary anxiety.

Numerical answers are provided for all problems requiring them. They are quoted with a precision (usually two significant figures) which is consistent with that of the data. At first the student will repeatedly find that his answers are 'more accurate' than the ones at the back of the book.

But considerations of precision should be in the back of a physicist's mind at all times, not just in the laboratory. Numerical problems provide an opportunity, which the pocket-calculator revolution has enhanced, for the student to develop these reflexes.

The suggestion that I write this book came from Professor J. M. Cassels. At one time it was intended that he should be a co-author, and we worked very closely on most of the material which now forms the first half of the book. That material has been re-worked twice since other calls upon his time obliged him to withdraw from the enterprise, but many of the ideas, and the more incisive turns of phrase, are his. He taught me much, not least about the elimination of what Thomas Young's anonymous critic in the *Edinburgh Review* for 1803 called a 'vibratory and undulatory mode of reasoning'.

I am deeply indebted to my colleague Dr M. F. Thomas, who prepared detailed comments on two drafts of the book and worked all the problems. I also received valuable guidance from Professor M. M. Woolfson. My family gave cheerful encouragement, and advice on how to draw springs, and the professionals of Cambridge University Press displayed generous open-mindedness in the face of my amateur ideas on book design.

I am grateful to the University of Liverpool for an eight-month period of Study Leave which enabled me to accelerate the completion of a project which had been under way for the preceding six years.

Liverpool   I.G.M.
April 1977

Preface to the second edition

Encouraged by the friendly reception given to the first edition, I have preserved its basic form and most of the details. The new and revised material occurs mainly in the latter half, on waves. There is one completely new chapter, intended to provide an elementary introduction to the so-called solitary wave, or soliton, which has become such a pervasive feature of physical science. It seemed to me that it should be possible to base an explanation of the solitary wave on the simplest notions of non-linear waves, and chapter 16 is my modest attempt. Consequential changes were necessary in chapter 12, but I believe the outcome is a more helpful treatment of dispersion, even for those who have no immediate need of the solitary wave material. Apart from these and other, smaller, changes, I have added some 30 new problems, the majority of which introduce new physical examples of vibrations and waves.

Most of the work was done during a further period of Study Leave from the University of Liverpool. Of the many people who made valuable comments on the first edition, I am particularly grateful to my Liverpool colleagues Professor J. R. Holt and Dr A. N. James, and their students.

A number of new diagrams were drawn for this edition by Roderick Main.

Liverpool   I.G.M.
September 1983

Notes for reference

Four kinds of equality symbol ($=$, $\equiv$, $\approx$ and $\sim$) are used in this book. The first is the ordinary 'equals' sign. The second will always be used to define something; thus $\gamma \equiv b/m$ is the definition of γ.

The symbol $\approx$ means 'is approximately equal to'. In relating the ideal behaviour of theoretical model systems to the behaviour of real physical systems, we frequently have to make approximations. One of the physicist's favourite approximation techniques is expansion in a power series. The *binomial series*

$$(1+x)^n = 1 + nx + \frac{n(n-1)x^2}{2!} + \ldots$$

where $|x| < 1$, is the most popular. Usually we take only the first two terms and write

$$(1+x)^n \approx 1 + nx$$

The size of the third term can be used as an estimate of the error which neglect of the higher terms has introduced.

In this book the symbol $\sim$ is used to indicate a coarser kind of approximation in which the linked quantities are said to be 'of the same order (of magnitude)'. This phrase means that they do not differ by more than about a factor of 10.

See page 348 for notes on constants and units.

1

Free vibrations

Most physical systems possess certain properties which enable them, under suitable conditions, to vibrate; we shall examine a few examples in chapter 2. But in order to discover the essential features of vibrational behaviour, we first consider a 'model system': an imaginary prototype system which possesses those properties which are necessary for vibrational behaviour, and no others. This is a well-tried procedure in physics. The basic idea is that, after examining in detail the behaviour of the model, we shall be able to recognize, in real, complicated systems, features which can lead to vibrational behaviour.

The model system in this case is a mass m attached to one end of a light helical spring whose other end is fixed (fig. 1.1). In equilibrium we suppose that adjacent coils of the spring are not in contact, so that there is scope for it to be compressed as well as stretched. We choose to ignore all forces not due to the elasticity of the spring: gravity, friction and viscosity are all 'switched off'.

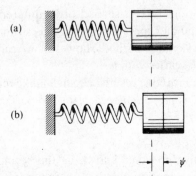

Fig. 1.1 (a) The prototype vibrator in equilibrium. (b) The mass is instantaneously displaced a distance ψ to the right of its equilibrium position.

If the system has been disturbed at some earlier time, it may not be in equilibrium. The mass may be at some position a distance ψ to the right of its equilibrium position, as in fig. 1.1(b). In that case the spring will exert a force towards the left. If the displacement is to the left, the force will act to the right. In either case the magnitude of the force will increase as the size of the displacement increases: the mass always experiences a *return force* which tends to change the displacement ψ towards the value zero.

It is easy to see that the free motion of this system takes the form of a vibration. The mass is given an acceleration

$$\ddot{\psi} \equiv \frac{d^2\psi}{dt^2}$$

which is determined by Newton's second law

$$m\ddot{\psi} = F_s \tag{1.1}$$

where F_s is the spring force. It will thus have acquired a certain velocity, and a corresponding momentum, by the time it reaches its equilibrium position, and so it will overshoot. Now the mass is acted on by a return force in the opposite direction; it is decelerated, brought to rest, and accelerated back to its equilibrium position where it overshoots again. The direction of the displacement continually alternates.

It is clear that both the elasticity or 'stiffness' of the spring and the inertial property of the mass are necessary for vibrational motion: the stiffness ensures that the mass tries to return to its equilibrium position, whereas the inertia makes it overshoot. We shall find that all vibrational phenomena depend on the existence of a pair of quantities analogous to stiffness and inertia.

1.1. Harmonic motion

The equation of motion (1.1) is a second-order differential equation from which we wish to find an expression giving ψ as a function of the time t. The equation is too vague as it stands, however: we can solve it only if we know exactly how F_s varies with ψ.

In order to get quantitative results, we shall make the simplest possible assumption, that F_s is *proportional to* ψ for the particular spring we are dealing with. We write

$$F_s = -s\psi \tag{1.2}$$

where s is a positive constant known as the spring constant, or the *stiffness*. With this assumption, our system now possesses all the properties of the imaginary object known to physics as the Harmonic Oscillator.

The equation of motion (1.1) now becomes

$$m\ddot{\psi} = -s\psi \tag{1.3}$$

In order to make the forthcoming results easier to carry over to other vibrating systems, we shall write this equation in a standard form

$$\ddot{\psi} + \omega_0^2 \psi = 0 \tag{1.4}$$

which contains a new, positive quantity

$$\omega_0 \equiv |(s/m)^{1/2}| \tag{1.5}$$

It is easy to verify, by differentiation and substitution, that (1.4) is satisfied by an expression of the form

$$\psi(t) = A \cos(\omega_0 t + \phi) \tag{1.6}$$

where A is any constant length and ϕ is any constant angle. When a quantity depends on time in this way it is said to vary harmonically. A vibration in which ψ varies harmonically is known as *harmonic motion*; it occurs for a mass on a spring only when the spring obeys (1.2).

The controlling quantity in (1.6) is the *phase angle* $\omega_0 t + \phi$, sometimes called simply the phase. The phase angle increases uniformly with time; values of any angle differing by an integral multiple of 2π are physically indistinguishable, however. Thus harmonic motion is *periodic*, repeating itself endlessly in a sequence of identical *cycles*. All measurable quantities, such as the displacement, speed, direction of travel and acceleration of the mass, recur whenever the phase angle increases by 2π. This will happen at times separated by the interval τ given by $\omega_0 \tau = 2\pi$ (fig. 1.2). This characteristic time interval is called the *period* of the vibration.

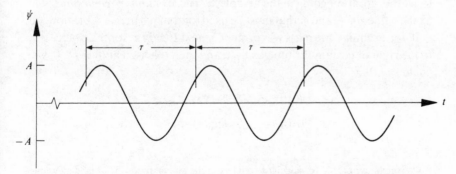

Fig. 1.2 How ψ varies with time during harmonic vibration. Identical events are separated by time intervals τ. The origin $t = 0$ can be placed at any convenient stage in the cycle of events.

Of the two constants in the phase angle, ω_0 is fixed by (1.5) but any value of ϕ, the *phase constant*, will give an acceptable solution of (1.4). Changing the value of ϕ merely makes all events in the cycle happen earlier or later by the same amount, without affecting the sequence of events within any cycle. Again, retarding or advancing the action by any whole number of periods will produce no observable difference. Thus we can increase or decrease ϕ by any whole multiple of 2π without changing anything physically.

During a cycle of vibration, ψ takes on all values between the limits $\pm A$; we call A the *amplitude*. The number of cycles per unit time,

$$\nu_0 = \omega_0/2\pi$$

is the *frequency*. If the time is measured in seconds, ν_0 is quoted in hertz (Hz); a vibration at 5000 cycles per second, for example, has a frequency of 5 kHz.

The quantity ω_0 clearly has the same dimensions as ν_0. Since $\omega_0 t$ appears in (1.6) as an angle, we shall usually call ω_0 the *angular frequency* to distinguish it from ν_0. Although ν_0 is easier to measure, ω_0 makes for tidier formulas containing fewer factors of 2π. To distinguish the two quantities further, we shall measure ω_0 in inverse seconds (s^{-1}) rather than hertz.[†]

Boundary conditions. Equation (1.6), being the solution of a second-order differential equation (1.4), correctly contains two arbitrary constants. Any pair of values of A and ϕ will describe a vibration which *can* be executed by the mass and spring provided. In practice, however, we shall be dealing with a particular vibration whose details have been determined by some other physical conditions: usually the method used to set the vibration going in the first place. These *boundary conditions* will fix the values of A and ϕ that must be used for that particular vibration.

If, for example, the mass was originally held steady at some distance A_1 to the right of its equilibrium position, and then released at time $t = 0$, we could say that

$$\psi(0) = A \cos \phi = A_1$$
$$\dot{\psi}(0) = -\omega_0 A \sin \phi = 0$$

$$(1.7)$$

[†] One might use rad s^{-1} for angular frequencies. The radian, however, being dimensionless, is not so much a unit as a signal saying 'angle'. It is more convenient to omit the radian when discussing vibrations, as we shall be cancelling angular frequencies with other quantities measured in s^{-1} and having nothing to do with angles.

where $\dot{\psi}$ means $d\psi/dt$. These two boundary conditions (which are actually *initial conditions* in this case) are sufficient to fix A and ϕ. The second condition tells us that ϕ is either 0 or π; we reject the latter alternative because the first condition makes $\cos \phi$ positive. Thus $A = A_1$ and (1.6) becomes

$$\psi(t) = A_1 \cos \omega_0 t \qquad (1.8)$$

Different starting arrangements would lead to different answers for A and ϕ.

Phase differences. In the example above we chose to label the moment at which the vibration was started as $t = 0$. This choice assumes that we are measuring time with a stopwatch which we start at the instant when the mass is released. There is no reason why we should not use an ordinary clock and start the vibration at some different time t_0; but by choosing the starting time to be zero we have been able to arrange that ϕ is zero also. As we saw above, having a different value for ϕ would merely advance or retard all the action by the same amount in time.

The real significance of the phase constant becomes apparent when we are dealing with two or more vibrations of the same frequency. Some examples are shown in fig. 1.3. In each case one of the vibrations is of the kind described by (1.8); that is, we have started it in the way described and have chosen $t = 0$ as the starting time.

The second vibration in fig. 1.3(a) has a different amplitude A_2, but it again has zero phase constant. The two displacements vary exactly in step with each other, and are always in the ratio A_2/A_1. We say that these two vibrations are *in phase*.

Under all other circumstances we say that the vibrations are *out of phase*. In fig. 1.3(b) the second vibration has $\phi > 0$, and all events such as passing the equilibrium point from left to right, or reaching the point of maximum positive ψ, happen earlier for this vibration than for the other. In this case we say that the second vibration leads in phase by ϕ, or has a *phase advance* of ϕ relative to the first. When the second vibration has $\phi < 0$ as in fig. 1.3(c), it is said to lag in phase by $|\phi|$, or to have a *phase lag* of $|\phi|$.

Clearly a phase lag greater than $180°$ is equivalent to a phase lead of less than $180°$, and it is usually more convenient to quote the smaller value. The two special cases $\phi = \pm\pi$ are of course equivalent; vibrations such as those in fig. 1.3(d) with a $180°$ phase difference are said to be *in antiphase*. If $|\phi| = \frac{1}{2}\pi$ they are *in quadrature*.

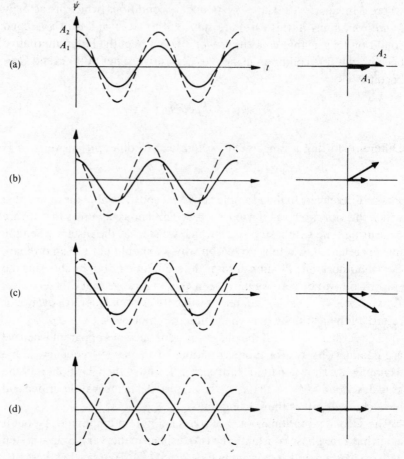

Fig. 1.3 Phase differences. In each diagram the vibration with amplitude A_1 has a phase constant of zero. The vibration with amplitude A_2 has (a) $\phi = 0$; (b) $\phi > 0$; (c) $\phi < 0$; (d) $\phi = \pm \pi$. Vector diagrams representing these vibrations are shown on the right.

Choosing our zero of time in a different way would have led to a non-zero phase constant for the first vibration in these examples. The phase constant of the other vibration would have been increased or decreased by the same amount, however, and the *phase difference* would have been unaffected.

Vector diagrams. For handling two or more harmonic vibrations of the same frequency, a geometrical method is helpful. We have already noted that the phase angle $\omega_0 t + \phi$ increases uniformly with time as the vibration takes place. The displacement at any moment t is proportional to the cosine of this angle. We can therefore generate $\psi(t)$ by letting a radius vector of length A rotate anticlockwise, as in fig. 1.4, and projecting it on to some fixed axis. An axis through the origin is the most convenient. The radius vector should make an angle ϕ with the axis at time $t = 0$, and should rotate with angular speed ω_0. The rotating vector is sometimes called a *phasor*.

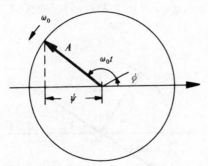

Fig. 1.4 The rotating vector which generates $\psi(t)$. At time $t = 0$ the vector makes an angle ϕ (anticlockwise) with the reference axis. At other times it makes an angle $\omega_0 t + \phi$. Its projection on the reference axis is $\psi(t)$.

A second vibration, with a different amplitude and phase constant, is represented by a vector of a different length, rotating at a fixed angle (the phase difference) to the first. This angle is measured in an anticlockwise sense from the first vector to the second if the second vibration leads the first; conversely, a phase lag is measured clockwise.

Since the vibrations have a common frequency, we shall usually be more interested in amplitudes and phase constants. Their values can be specified by means of a static diagram showing the rotating vectors in their $t = 0$ positions. Vector diagrams for the four examples in fig. 1.3 are shown beside the $\psi - t$ plots.

Velocity and acceleration. As an example of the use of vector diagrams, we illustrate the phase relationships between the displacement, the velocity and the acceleration of the mass during harmonic vibration. The velocity can be obtained as a function of time by differentiating (1.6) with respect to t; we find

$$\dot{\psi}(t) = -\omega_0 A \sin(\omega_0 t + \phi)$$
$$= \omega_0 A \cos(\omega_0 t + \phi + \tfrac{1}{2}\pi) \tag{1.9}$$

A second differentiation gives the acceleration

$$\ddot{\psi}(t) = -\omega_0^2 A \cos(\omega_0 t + \phi)$$
$$= \omega_0^2 A \cos(\omega_0 t + \phi + \pi)$$

These expressions show that $\dot{\psi}$ and $\ddot{\psi}$, like ψ, vary harmonically, and that the frequency is the same in all three cases. The amplitude of $\dot{\psi}$ (the maximum speed reached by the mass) is $\omega_0 A$, and that of $\ddot{\psi}$ is $\omega_0^2 A$. The velocity, however, leads the displacement in phase by 90°; the acceleration leads the velocity by a further 90°, bringing it into antiphase with the displacement, as is necessary if (1.3) is to be obeyed.

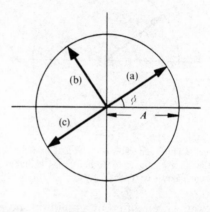

Fig. 1.5 Vectors showing the relative phases of (a) the displacement, (b) the velocity, and (c) the acceleration, for a vibration with amplitude A and phase constant ϕ. The velocity amplitude is $\omega_0 A$ and the acceleration amplitude is $\omega_0^2 A$.

In vector diagram terms (fig. 1.5) each differentiation with respect to t can be seen as a multiplication of the length of the rotating vector by ω_0, together with an anticlockwise rotation through 90°. The vectors drawn in fig. 1.5 are all lengths. On other occasions it may be more convenient to choose vectors which represent velocities, accelerations or forces; these quantities should not be mixed in the same diagram, however.

Energy. As usual in mechanical systems, two kinds of energy are present. When the mass is moving with speed $|\dot{\psi}|$ in either direction, its kinetic energy is

$$T = \tfrac{1}{2}m\dot{\psi}^2 \tag{1.10}$$

When the spring is stretched or compressed by an amount $|\psi|$, it stores potential energy

$$V = \tfrac{1}{2}s\psi^2 \tag{1.11}$$

The total energy

$$W = T + V = \tfrac{1}{2}m\dot{\psi}^2 + \tfrac{1}{2}s\psi^2 \tag{1.12}$$

remains constant during the vibration, since all dissipative forces like friction and viscosity are assumed to be absent. Thus

$$\frac{dW}{dt} = 0 \tag{1.13}$$

Using (1.12) we obtain

$$m\dot{\psi}\ddot{\psi} + s\psi\dot{\psi} = 0$$

$$m\ddot{\psi} + s\psi = 0$$

The last equation is just (1.3), reached by a new route.

To discover how T and V vary individually with time, we substitute (1.9) into (1.10), and (1.6) into (1.11), to find

$$T = \tfrac{1}{2}m\omega_0^2 A^2 \sin^2(\omega_0 t + \phi)$$

$$V = \tfrac{1}{2}sA^2 \cos^2(\omega_0 t + \phi)$$

These are plotted in fig. 1.6. The value of their constant sum is

$$W = \tfrac{1}{2}m(\omega_0 A)^2 = \tfrac{1}{2}sA^2$$

since $s = m\omega_0^2$ by (1.5). For a given mass and spring, the total energy is proportional to the square of the amplitude, but does not depend on the phase constant.

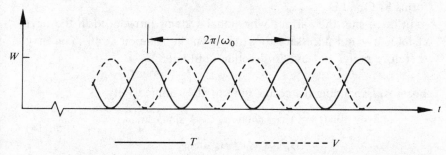

Fig. 1.6 Variation of kinetic energy T and potential energy V during harmonic vibration. The total energy W is constant.

We can think of the vibration as the repeated transfer of a fixed amount of energy from the mass to the spring and back again, twice in each cycle. When the spring is stretched or compressed its maximum distance ($\psi = \pm A$) the mass comes momentarily to rest and the kinetic energy vanishes. At that moment all the energy of the system is stored in the spring as potential energy. When the mass is passing through $\psi = 0$ (in either direction) it has its maximum speed $\omega_0 A$ and contains the entire energy of the system, since the spring is neither stretched nor compressed. At other points in the cycle there is a variable mixture of kinetic and potential energy, but their sum never changes.

Summary. The free motion of the model system is a vibration because the system possesses the two essential properties of stiffness and inertia. In the special case of a spring which exerts a return force proportional to the displacement of the mass, the vibration is harmonic. For a given mass and a given spring the frequency is fixed, but the amplitude and the phase constant depend on boundary conditions.

The velocity and the acceleration of the mass also vary harmonically, with the same frequency as the displacement. Because displacement and velocity are in quadrature, the energy of the system flows back and forth between the mass and the spring twice in each cycle.

1.2. Alternative mathematics for harmonic motion

There are several alternative ways of writing the solution (1.6). Each has its own special advantages, and in the rest of the book we shall adopt whichever one is most suitable for the particular purpose in hand. From now on we shall refer to (1.6) as form A; here we derive, with little further comment, three other versions which we identify for convenience as forms B, C and D.

In discussing the various vibrational systems introduced in the next chapter it is not necessary to write down the solution at all. You may therefore prefer to leave this section until later.

Form B. Expanding the cosine in (1.6) gives immediately

$$\psi(t) = A \cos \phi \cos \omega_0 t - A \sin \phi \sin \omega_0 t$$
$$= B_p \cos \omega_0 t + B_q \sin \omega_0 t \tag{1.14}$$

We call this form B. Information about the amplitude and the phase constant of the vibration is conveyed by means of two new constants

$$B_p \equiv A \cos \phi$$

$$B_q \equiv -A \sin \phi$$

each of which can be either positive or negative. These constants will be fixed by the boundary conditions, as were A and ϕ. Thus, in the example considered in the previous section we would find $B_p = A_1$ and $B_q = 0$: comparison of (1.8) with (1.14) gives the same result directly.

It is worth noting here that a second vibration of amplitude A_2, in quadrature with the previous one, has $B_p = 0$ and $B_q = \pm A_2$, and so takes the simple form

$$\psi(t) = \pm A_2 \sin \omega_0 t$$

Here the plus sign means that the second vibration lags the first, and *vice versa*. By adopting a vibration with $\phi = 0$ as a standard of phase (as in fig. 1.3) we can use the suffixes p ('phase') and q ('quadrature') as memory aids.

Form C. We can always try to solve a differential equation like (1.4) by substituting a trial expression of the form

$$\psi = C\, e^{pt}$$

In the present case we find that such a solution is acceptable if the relation

$$p^2 = -\omega_0^2$$

is satisfied. This will be so for two values of p, given by $\pm i\omega_0$. The general solution may therefore be written as a linear combination

$$\psi(t) = C \exp(i\omega_0 t) + C' \exp(-i\omega_0 t) \tag{1.15}$$

in which the constants C and C' are both complex.

This solution is actually too general for our purpose. It has four arbitrary constants (Re C, Im C, Re C' and Im C') whereas the boundary conditions can only cope with two. It is easy to see what is wrong with (1.15) as it stands. In order to represent a single physical displacement, ψ may be either real or imaginary, but must not be complex. The expression on the right of (1.15) can be made real by insisting that the two terms are complex conjugates of each other, that is

$$[C' \exp(-i\omega_0 t)]^* = C \exp(i\omega_0 t)$$

The complex conjugate of a product of two quantities is the product of their complex conjugates, and so

$$C'^* \exp(i\omega_0 t) = C \exp(i\omega_0 t)$$

$$C' = C^*$$

Form C now appears as

$$\psi(t) = C \exp(i\omega_0 t) + C^* \exp(-i\omega_0 t) \tag{1.16}$$

Now there are only two arbitrary constants (Re C and Im C). By expanding the exponentials in (1.16) in terms of sines and cosines with the aid of de Moivre's theorem

$$e^{ix} = \cos x + i \sin x$$

and comparing the result with (1.14), we obtain the relation between these new constants and the previous ones,

$$\text{Re } C = \tfrac{1}{2}B_p = \tfrac{1}{2}A \cos \phi$$
$$\text{Im } C = -\tfrac{1}{2}B_q = \tfrac{1}{2}A \sin \phi \tag{1.17}$$

Form D. Since we have arranged that the expression on the right of (1.16) is real, it must be possible to write

$$\psi(t) = \text{Re } [C \exp (i\omega_0 t)] + \text{Re } [C^* \exp (-i\omega_0 t)]$$
$$= 2 \text{ Re } [C \exp (i\omega_0 t)] \tag{1.18}$$

By defining a new complex constant $D \equiv 2C$, we reach form D,

$$\psi(t) = \text{Re } [D \exp (i\omega_0 t)] \tag{1.19}$$

We shall call D the *complex amplitude* of the vibration. Like C, however, it contains information about the phase constant.

The arbitrary constants of forms A and D are linked through the relations

$$\text{Re } D = A \cos \phi$$
$$\text{Im } D = A \sin \phi \tag{1.20}$$

which follow immediately from (1.17). For a vibration started as before with $\phi = 0$, D is purely real; for a vibration in quadrature with that one ($\phi = \pm\tfrac{1}{2}\pi$) D is imaginary.

Form D provides an extremely powerful way of handling harmonic motion. Its chief merit is the ease with which it can be differentiated and integrated. To find $\dot{\psi}$, for example, we could differentiate (1.16) with respect to t and proceed as we did in (1.18) and (1.19). The result

$$\dot{\psi}(t) = \text{Re } [i\omega_0 D \exp (i\omega_0 t)]$$

is, however, just what we would get if we first differentiated the complex function $D \exp (i\omega_0 t)$ and *then* took the real part. The same is true of the acceleration

$$\ddot{\psi}(t) = \text{Re } [-\omega_0^2 D \exp (i\omega_0 t)] \tag{1.21}$$

At each differentiation we merely have to multiply the complex function by $i\omega_0$.

It is instructive to examine the behaviour of the controlling function $D \exp (i\omega_0 t)$ on an Argand diagram. First we write the complex amplitude in conventional modulus-and-argument form,

$$D = A \cos \phi + i(A \sin \phi) = A \exp(i\phi)$$

The whole function then becomes

$$D \exp(i\omega_0 t) = A \exp[i(\omega_0 t + \phi)]$$

On the Argand diagram (fig. 1.7) this is represented by a radius vector of length A rotating anticlockwise with angular speed ω_0. Its initial ($t = 0$) position makes an angle ϕ measured anticlockwise from the real axis.

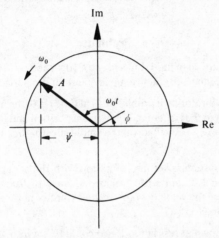

Fig. 1.7 Argand diagram representing $\psi(t)$. The real axis is equivalent to the reference axis of fig. 1.4.

This is, of course, the same rotating vector as the one in fig. 1.4. In complex number terminology we describe ψ as the projection on the real axis.

In this book form D will always be written as in (1.19); but it is common practice to write ψ *equal to* a complex function such as $D \exp(i\omega_0 t)$, leaving implicit the instruction to take the real part. It will be obvious that this is being done whenever a simple physical quantity such as a length appears to be complex.

Summary. We collect together for reference purposes the four forms of the solution,

$$\psi(t) = A \cos(\omega_0 t + \phi)$$

$$\psi(t) = B_p \cos \omega_0 t + B_q \sin \omega_0 t$$

$$\psi(t) = C \exp(i\omega_0 t) + C^* \exp(-i\omega_0 t) \qquad (1.22)$$

$$\psi(t) = \text{Re}[D \exp(i\omega_0 t)]$$

Each version contains two real constants whose values can be adjusted to fit boundary conditions.

It will sometimes be convenient to change from one form to another. The simplest relations connecting the four pairs of constants are

$$A \cos \phi = B_p = 2 \operatorname{Re} C = \operatorname{Re} D$$
$$A \sin \phi = -B_q = 2 \operatorname{Im} C = \operatorname{Im} D \tag{1.23}$$

Other relationships, such as the expressions giving A and ϕ individually in terms of the other constants, can readily be obtained from (1.23).

Problems

1.1   If the system shown in fig. 1.1 has $m = 0.010$ kg and $s = 36$ N m^{-1}, calculate (a) the angular frequency, (b) the frequency, and (c) the period.

1.2   For the same vibrator as in problem 1.1, at time $t = 0$, the mass is observed to be displaced 50 mm to the right of its equilibrium position and to be moving to the right at a speed 1.7 m s^{-1}. Calculate (a) the amplitude, (b) the phase constant, and (c) the energy.

1.3   An identical system is set into vibration with the same amplitude as the vibrator in problem 1.2, but with a phase advance of 90°. Calculate (a) the displacement, and (b) the velocity of this second vibrator at time $t = 0$. (c) At what time will it next come to rest?

1.4   The system shown at rest in fig. 1.1(a) could be set into vibration by giving the mass a sudden momentum impulse to the left: by tapping it with a hammer, for example. If the magnitude of the impulse is p_1 and it is given at time $t = 0$, find (a) the amplitude and (b) the phase constant of the ensuing motion.

1.5   Calculate (a) the amplitude, (b) the phase constant, and (c) the complex amplitude, for the vibration given by

$$\psi = (10 \text{ mm}) \cos \omega_0 t + (17 \text{ mm}) \sin \omega_0 t$$

1.6   During a vibration with a frequency of 50 Hz, the displacement is observed to be 30 mm at time $t = 0$, and -14 mm at $t = 12$ ms. Find the complex amplitude.

1.7   Calculate the maximum acceleration (in units of g) of a pickup stylus reproducing a frequency of 16 kHz, with an amplitude of 0.01 mm.

2
Free vibrations in physics

We now take a quick look at a few representative physical systems, so that we can appreciate the great variety of circumstances in which the mathematics of harmonic motion can be applied. The model system has taught us what properties we must look for if we wish to know whether vibration is possible in any given real system. The mathematics of the prototype can be used in a production-line spirit once we have isolated the ingredients (inertia and stiffness) necessary for vibration.

As well as appreciating the essential similarity of different kinds of vibration, we shall in this chapter be vitally interested in the numerical sizes of the physical quantities involved, and particularly in the values of the free vibration frequencies of the various systems. One of the most striking things we shall discover is the very wide spread of frequencies to be found.

2.1. Angular vibrations

The mass in the model system vibrates to and fro along a straight line. Some of the most familiar vibrations involve massive objects rotating back and forth under the influence of a *return torque*. We shall consider two kinds of return torque: first an elastic torque provided by a stiff suspension, and then one due to gravity.

Torsional vibrations. The system shown in fig. 2.1 is the exact rotational analogue of the model system. The coordinate ψ now measures the *angular* displacement of the mass from its equilibrium position. If the return torque exerted by the spring is T_s the angular acceleration $\ddot{\psi}$ will be

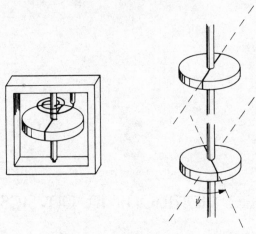

Fig. 2.1 A torsional vibrator. The mass is mounted in frictionless bearings. When it is rotated out of its equilibrium position, the spiral spring tends to move it back again. The coordinate ψ measures the angular displacement from equilibrium.

given by

$$I\ddot{\psi} = T_s \tag{2.1}$$

where I is the moment of inertia of the mass for rotations about the axis.

The angular vibration will be harmonic if the return torque is proportional to the angular displacement. In that case we have

$$T_s = -c\psi \tag{2.2}$$

where the torsional stiffness c is a constant. Then (2.1) takes the form

$$I\ddot{\psi} = -c\psi \tag{2.3}$$

These three equations are completely analogous to (1.1), (1.2) and (1.3). We can obtain the standard differential equation of motion (1.4) simply by writing†

$$\omega_0 = (c/I)^{1/2}$$

If the system is set into motion, ψ will vary harmonically with the frequency $\omega_0/2\pi$, in accordance with (1.6). It is important to realise that the amplitude A is now, like ψ itself, an *angle*.

Vibrations of this kind provide the basic time-keeping mechanism of spring-driven clocks and watches: suitable values of c and I can be obtained without difficulty. (See problem 2.1.) The possibility of such

† To simplify the notation we shall omit the modulus signs from now on; but ω_0 is always positive.

vibrations also has to be allowed for when designing instruments such as moving-coil ammeters and voltmeters, since the vibrational characteristics may interfere with the use of the instrument for static measurements. We return to this topic in chapter 4.

Pendulum vibrations. There are many familiar examples of angular vibration in which the return torque is provided not by a spring but by gravity. These include the swing of a pendulum, the rocking of a chemical balance about its pivot, and the roll of a ship. As typical of all these vibrators we shall consider the idealized system known as the simple pendulum (fig. 2.2).

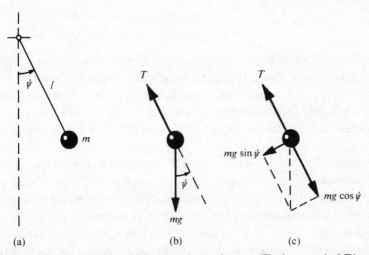

Fig. 2.2 (a) A simple pendulum. (b) Forces acting on the mass. The force marked T is equal to the tension in the string. (c) Forces acting on the mass, with its weight resolved into components along the string and perpendicular to the string. The pendulum is presumed to be swinging in the plane of the page.

The system consists of a mass m hung from a pivot by an arm or string of length l and negligible mass. The moment of inertia in this case is

$$I = ml^2$$

The return torque produced by gravity is

$$T_s = -(mg \sin \psi)l \tag{2.4}$$

when the angular displacement is ψ. Thus (2.1) becomes

$$ml^2\ddot{\psi} = -mgl \sin \psi$$
$$\ddot{\psi} + (g/l) \sin \psi = 0 \tag{2.5}$$

It is not possible to turn the last equation into (1.4) by the kind of substitution we made in (1.5). This difficulty has arisen because gravity always acts vertically, whereas the motion is rotational; the return torque is not proportional to ψ as in (2.2), but has the more complicated dependence of (2.4). We say that the system is *non-linear*.

Non-linearity is the subject of chapter 7. Here we simply note that the vibration will be *approximately* harmonic if the amplitude is small. If $\psi \ll 1$ at all times, we may put $\sin \psi \approx \psi$ in (2.5) to get

$$\ddot{\psi} + (g/l)\psi \approx 0$$

The angular frequency of these approximately harmonic vibrations is thus

$$\omega_0 = (g/l)^{1/2}$$

Since g for an earth-bound pendulum is always close to 10 m s^{-2}, and τ varies only as the square root of the length, the range of periods available with a simple pendulum is quite restricted: changing l from 100 mm to 25 m will only increase τ from about 0.6 s to about 10 s. We have no reason to expect real pendulums to be very different in this respect.

Summary. A rotating mass attached to a spring which provides a return torque proportional to the angular displacement can be treated in a parallel way to the model system: we merely have to turn all the equations into their angular counterparts. A rotational system whose 'stiffness' is gravitational, such as a pendulum, is non-linear. Its free vibrations are approximately harmonic if their amplitude is small. The period of a simple pendulum is proportional to the square root of its length, and is of the order of 1 s for any pendulum of reasonable length.

2.2. Acoustic vibrations

We consider a flask (fig. 2.3) which has a large bulb of volume v, and a narrow neck of length l and uniform cross-section a. The volume of the

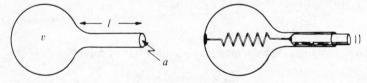

Fig. 2.3 An acoustic vibrator. An approximate calculation of the frequency assumes that the air in the neck does all the vibrating, while the air in the bulb merely contributes stiffness.

neck la is very small compared with v. The flask is exposed to the atmosphere, which has density ρ.

We can imagine a free vibration in which the air in the neck of the flask moves in and out like a solid piston, while the air in the bulb is alternately compressed and expanded: the air in the neck is behaving like the mass of the model system, and the air in the bulb like the spring. Such a description assumes that there is insignificant motion inside the bulb or outside the mouth of the flask, and that the air in the neck is incompressible. These are crude approximations, and we must not expect a very accurate answer. Later in the book (section 10.2) we shall see how the mass and the stiffness of *all* the air can be taken into account.

The mass of the air in the neck is lap, and we measure its displacement *inward* from equilibrium by ψ. If the pressure difference between the gas inside and outside the flask is p', then the net *outward* force on this air is ap'. Newton's second law says that

$$(la\rho)\ddot{\psi} = -ap'$$

To proceed further we require, as always, a relationship connecting the return force (ap' in this case) and the displacement. The relevant property of the air is its *compressibility*

$$\kappa = -\frac{1}{V}\left(\frac{\partial V}{\partial p}\right) \tag{2.6}$$

or the fractional decrease in volume per unit pressure increase. For the air in the flask, the fractional volume change is $a\psi/v$, and is brought about by the pressure difference p'. Thus, provided ψ and p' are reasonably small, we may write

$$\kappa \approx a\psi/vp'$$

which leads to

$$\omega_0 \approx (a/lv\rho\kappa)^{1/2} \tag{2.7}$$

Which compressibility? Some thought is required before we can use (2.7) to calculate ω_0 numerically. There is in fact no such thing as 'the compressibility of air'. The volume of a mass of gas depends not only on the pressure but also on the temperature. Before we can evaluate κ from (2.6) we must specify what is happening to the temperature as the pressure is changed.

One simple assumption we could make is that there are *no* temperature changes; we could then use the *isothermal* compressibility κ_T. This would be justified if the pressure changes occurred slowly enough to allow the flask to exchange heat freely with its surroundings.

It seems more likely, however, that the pressure changes in an acoustic vibration will heat and cool the air as well as compressing and expanding it. We may even have the other extreme in which the rapidity of the pressure changes is too great to allow the exchange of any heat at all; in that case we would use the *adiabatic* (or isentropic) compressibility κ_S. Clearly $\kappa_S < \kappa_T$, since it takes more pressure to compress the gas if it is also being made hotter. (See problem 2.3.)

The extreme assumption of adiabatic changes is, in fact, a valid one. The most convenient experimental test is to measure the speed of acoustic *waves* (section 10.2); under normal conditions the results of such measurements agree with a formula (10.12) which uses the adiabatic compressibility.

Numerical evaluation. We know that air at room temperature and atmospheric pressure behaves like a perfect gas. For adiabatic changes in a perfect gas pV^γ is a constant, where $\gamma = C_p/C_V$ is the ratio of the heat capacities at constant pressure and at constant volume. Thus, from (2.6), we obtain for κ_S the expression $1/\gamma p$. The product $\rho \kappa_S$ which is required in (2.7) can be found with the help of the perfect gas law

$$pV = RT$$

In terms of the molar mass M, the molar volume V and the molar gas constant R, we have

$$\rho = M/V = Mp/RT$$

$$\rho\kappa_S = M/\gamma RT \tag{2.8}$$

The angular frequency (2.7) may now be written

$$\omega_0 \approx \left[\left(\frac{a}{lv} \right) \left(\frac{\gamma RT}{M} \right) \right]^{1/2} \tag{2.9}$$

and is seen to increase with absolute temperature as $T^{1/2}$. It is, however, independent of the average pressure p. This is because both the mass $la\rho$ and the stiffness, which is inversely proportional to the compressibility $1/\gamma p$, are proportional to p. We may expect the frequencies of other acoustic systems to be given by similar formulas, but with different geometrical factors appearing in place of a/lv.

When we insert atmospheric data (table 2.1) and typical values for the dimensions of a flask ($v = 10^{-3}$ m^3, $a = 10^{-4}$ m^2, $l = 5 \times 10^{-2}$ m) we find that $\omega_0 \approx 500$ s^{-1}, or $\nu_0 \approx 80$ Hz. An audible note of about this frequency

TABLE 2.1

Data for the air in a flask

Ratio of heat capacities γ	1.4
Gas constant R	8.3 J mol^{-1} K^{-1}
Molar mass M	0.029 kg mol^{-1}
Temperature T	≈ 300 K

can be heard when an empty flask of this size is suddenly uncorked. (In musical terms, it is a note roughly two octaves below middle C.)

Summary. The free vibration of the air in a flask occurs with a frequency which is proportional to $T^{1/2}$ but independent of the average air pressure. For a flask of typical dimensions the frequency lies at the lower end of the audible range. At such frequencies the alternating compressions and decompressions of the air occur so quickly that they can be treated as adiabatic.

2.3. Plasma vibrations

A plasma is a gas which consists partly or wholly of charged particles. The numbers of positive and negative charges present are always equal, so that the plasma as a whole is electrically neutral.

To understand how a plasma can vibrate, we imagine the idealized experiment illustrated in fig. 2.4. The first stage of the experiment is to create a horizontal slab of plasma by exposing some ordinary gas to a beam of ultraviolet light which is sharply delineated in the vertical direction by means of suitable collimators. Ultraviolet light is an ionizing radiation. When it shines through the gas, energy is transferred to some of the molecules. Electrons (charge $-e$ and mass m_e) become detached and move about, leaving behind an equal number of positive ions (charge $+e$ and mass $\gg m_e$). After a while the light is switched off, leaving N electrons and N positive ions per unit volume of the plasma. There will also be some neutral molecules which have not been ionized, but they can be ignored here.

Next we switch on, for a brief moment, an external electric field pointing vertically downwards. The electrons receive an upward momentum impulse, and continue to move upwards after the field is switched off. The positive ions receive an equal downward impulse, but their velocity is much smaller because of their much greater mass: we shall in fact neglect the motion of the positive ions entirely.

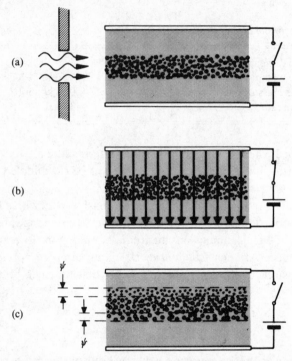

Fig. 2.4 (a) A slab of plasma is created by producing ionization in a slab of gas. (b) An electric field is switched on momentarily. (c) The plasma at some later moment during the vibration. Open circles represent positive ions, and dots represent electrons; neutral gas molecules are not shown.

At some later time t, the electrons have all moved upwards a distance ψ. They produce a sheet of unbalanced negative charge $-Ne\psi$ per unit area at the top of the slab. Similarly, the positive ions that are left behind produce a sheet of unbalanced positive charge $+Ne\psi$ per unit area at the bottom of the slab. Because of these charges there is now an electric field $\boldsymbol{E}$ in the plasma. Its direction is vertically upwards, and its magnitude is

$$E = (Ne/\varepsilon_0)\psi$$

where ε_0 is the permittivity of a vacuum. *This field produces an electric return force* $-(Ne^2/\varepsilon_0)\psi$ *acting on each electron inside the slab.* The electrons all have the same equation of motion

$$\ddot{\psi} + (Ne^2/m_e\varepsilon_0)\psi = 0$$

This is exactly of the standard form (1.4), showing that the electrons vibrate vertically, at the *plasma (angular) frequency*

$$\omega_0 = (e^2/m_e\varepsilon_0)^{1/2}N^{1/2} \qquad (2.10)$$

The vibration is one involving all the electrons, which move up and down as a body. We can imagine them to be rather like a fluid, surging up and down in a sponge composed of the nearly-stationary positive ions. Since all the quantities in (2.10) except N are fundamental constants, the plasma frequency is essentially a measure of the electron density in the particular plasma under discussion. To enable us to calculate ω_0 from N in the various examples below, we evaluate the constant

$$e^2/m_e\varepsilon_0 = 3.18 \times 10^3 \text{ m}^3 \text{ s}^{-2} \tag{2.11}$$

The ionosphere. Plasma occurs naturally in the upper atmosphere, through the ionizing action of ultraviolet light and x-rays from the sun during the day. The highest electron (and positive ion) densities occur in the region known as the ionosphere, which extends upwards from about 60 km above sea level. Although the density of the atmosphere simply falls off with increasing height, the electron density varies in a more complicated way since the intensity of the ionizing radiation reaching a given level depends on how much has been absorbed by the various ionizing processes taking place higher up. In fact the ionosphere is found to have a layered structure. In the lowest layer (the so-called D layer) electron densities build up to $N \approx 10^9 \text{ m}^{-3}$ at mid-day. In the highest (the F_2 layer) a figure of about 10^{12} m^{-3} is reached. From these numbers we can estimate the corresponding plasma frequencies with the aid of (2.10) and (2.11). The results are $\omega_0 \approx 2 \times 10^6 \text{ s}^{-1}$ ($\nu_0 \approx 300$ kHz) for the D layer, and $\omega_0 \approx 6 \times 10^7 \text{ s}^{-1}$ ($\nu_0 \approx 10$ MHz) for the F_2 layer.

These are typical radio frequencies. The presence of plasma profoundly affects the propagation of radio waves in the atmosphere. In particular the F_2 layer is responsible for reflecting Short Wave radio signals back to the ground, enabling them to be received at great distances from the transmitter in spite of the curvature of the earth. We shall see why in section 14.3.

Plasma vibrations in metals. A metal contains a number of 'conduction electrons' (roughly one per atom) which are permanently detached from their parent atoms. The remaining positive ions form a rigid framework through which the free electrons can travel, acting as carriers of electric charge when a current is passed through the metal. Some properties of metals can be understood in terms of a plasma vibration involving the conduction electrons, treated as classical particles.

The plasma frequency depends, as we have seen, on the electron density and on nothing else. We choose as a typical metal copper, and estimate N on the assumption that each copper atom contributes exactly one conduction electron to the metal. The number of copper atoms per unit volume is

$$N = 1000 N_A \rho / A$$

where N_A is the Avogadro constant $(6.02 \times 10^{23} \text{ mol}^{-1})$, A is the atomic weight and ρ is the density. Copper has an atomic weight of 63.5 and a density of 8900 kg m^{-3}; these figures give $N = 8.4 \times 10^{28} \text{ m}^{-3}$.

The conduction electron density in a metal is thus enormously greater than the electron densities found in the ionosphere. The plasma frequency is correspondingly high: the result for copper, obtained by putting the estimated value of N into (2.10) and using (2.11), is $\omega_0 \approx 2 \times 10^{16} \text{ s}^{-1}$ $(\nu_0 \approx 3 \times 10^{15} \text{ Hz})$. Such a frequency is characteristic of ultraviolet radiation; we shall use this fact later (section 14.3) to show why metals are shiny.

Summary. The frequency of a free plasma vibration is proportional to the square root of the electron density in the plasma. The electron densities found in the ionosphere lead to results in the radio frequency range. Those of the conduction electrons in metals give plasma frequencies in the ultraviolet region.

2.4. Molecular vibrations

The vibrational systems we have considered so far have involved familiar, easily calculated return forces. Not every system is so simple. The return force may have several different origins: it may be part elastic and part gravitational, for example. Sometimes the general nature of the force may be known, but not its detailed dependence on ψ. In such cases it may be possible to learn something about the force experimentally, by studying a vibration which takes place under its influence.

The force that binds atoms together to make molecules or solid materials is one such example. The attractive part of the force holding the atoms together is electric, and in principle calculable if we know where all the atomic electrons are. But the distribution of the electrons in space is determined by quantum mechanics, and the results can be quite different for different combinations of atoms.

If the atoms are pushed towards each other, a strong repulsive force rapidly comes into play. This repulsion is what prevents the molecule or the solid from collapsing. It is largely due to a quantum effect known as the exclusion principle, and is hard to calculate from first principles.

The equilibrium distance between two atoms bound together is that at which the attractive and repulsive forces exactly balance each other. If the distance is slightly increased or slightly decreased, a return force will act, and vibration becomes a possibility. By studying the vibration we can obtain information about the binding forces at inter-atomic separations close to the equilibrium value.

Motion near a position of stable equilibrium. This is a convenient point to discover how much can be said in general about a vibration when the return force does not necessarily obey (1.2).

A complex force is more conveniently discussed in terms of its *potential energy curve*. This is because potential energy is a scalar quantity and therefore simpler to handle than the force itself, which is a vector. The force can always be found by taking the negative gradient of the potential energy.

We consider a particle of mass m whose position along a straight line is measured by the coordinate r. We suppose that the force $F(r)$ acting on the particle can be described by a potential energy function $V(r)$ such that

$$F(r) = -\frac{dV}{dr} \tag{2.12}$$

This function has a minimum at the equilibrium position $r = R$ (fig. 2.5).

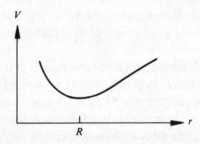

Fig. 2.5 A potential energy curve. The negative gradient of the curve at any point r gives the force acting on the particle when it is at r. At $r = R$ the particle is in stable equilibrium. If it is displaced slightly, it will vibrate harmonically with a frequency which is proportional to the square root of d^2V/dr^2, measured at $r = R$.

When the mass is at r, its displacement ψ is equal to $r - R$. The potential energy in the region of the minimum ($\psi = 0$) can be expressed by means of a Taylor series in ψ,

$$V = V_R + \frac{1}{2}\left(\frac{d^2V}{dr^2}\right)_R \psi^2 + \frac{1}{6}\left(\frac{d^3V}{dr^3}\right)_R \psi^3 + \ldots$$

This series contains no term proportional to ψ, because V is at a minimum when $\psi = 0$; for the same reason $(d^2V/dr^2)_R$ is positive. The force itself (2.12) is given by

$$-F = \left(\frac{d^2V}{dr^2}\right)_R \psi + \frac{1}{2}\left(\frac{d^3V}{dr^3}\right)_R \psi^2 + \ldots \qquad (2.13)$$

In these series we have used the fact that

$$\frac{dV}{d\psi} = \frac{dV}{dr} \cdot \frac{dr}{d\psi} = \frac{dV}{dr}$$

and similarly for higher derivatives.

If the particle were motionless at R, it would be in stable equilibrium; a move later on to a nearby position could not conserve total energy. It gains access to a finite range of ψ when the total energy exceeds V_R. If the excess is small, $|\psi|$ must also be small, and we may neglect all terms but the first in the series for F. Comparison of (1.2) and (2.13) then leads to approximate expressions for the stiffness

$$s \approx \left(\frac{d^2V}{dr^2}\right)_R \qquad (2.14)$$

and the angular frequency

$$\omega_0{}^2 \approx \frac{1}{m}\left(\frac{d^2V}{dr^2}\right)_R$$

We may deduce that *a particle vibrates harmonically when it is not quite in stable equilibrium. By measuring the frequency of the vibration we can determine* d^2V/dr^2 *at the minimum of its potential energy curve.*

Ionic molecules. We consider diatomic molecules of the form AB. In one kind of binding, known as *ionic*, an electron is transferred from atom A to atom B, forming a positive ion A^+ and a negative ion B^- which carry charges $+e$ and $-e$ respectively. Ion formation tends to happen when neutral atom A has one electron outside a closed shell, and neutral atom B requires just one extra electron to form a closed shell of its own. Both ions thus have the stable closed-shell structure and are spherical. The force that binds A to B is simply the electrostatic force of attraction between oppositely charged spheres.

A suitable formula for the mutual potential energy of two ions is

$$V(r) = -\frac{e^2}{4\pi\varepsilon_0 r} + \frac{B}{r^9} \qquad (2.15)$$

This function is plotted in fig. 2.6.

The second term in (2.15) represents a repulsive force that increases very rapidly as r decreases. Whereas we recognize the first term as the

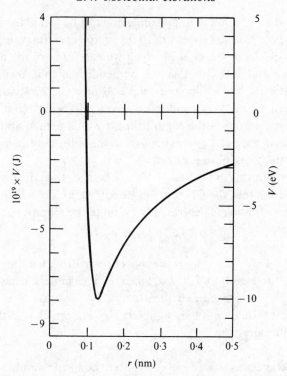

Fig. 2.6 The potential energy curve for a pair of ions: formula (2.15), with B given by (2.16) and R chosen as 0.13 nm (the measured value for HCl). The zero of potential energy is the value when the ions are infinitely far apart. (It is customary to shift the curve upwards by an amount which allows for the energy required to form the ions from neutral atoms, but this makes no difference to the vibrational behaviour discussed in this book.)

mutual potential energy of two oppositely charged spheres separated by a distance r, we make no claim to understand why the repulsive term should have the form given. We have merely chosen a convenient function which increases steeply as r decreases. It happens that the results we shall obtain are rather insensitive to the exact form chosen for the repulsive term.

The value of B in (2.15) is fixed if we know the equilibrium separation (the *bond length*) R, because dV/dr must vanish at $r = R$ and so

$$B = e^2 R^8 / 36\pi\varepsilon_0 \qquad (2.16)$$

The stiffness (2.14) can be found by differentiating (2.15) twice and setting $r = R$. The result

$$s \approx 2e^2 / \pi\varepsilon_0 R^3$$
$$= (1.84 \times 10^{-27}\,\mathrm{N\,m^2})/R^3 \qquad (2.17)$$

depends only on the bond length of the molecule.

The bond length of a diatomic molecule can be found from measurements on its 'rotational spectrum'. The energy excitations involved in a rotational spectrum are ones in which we can think of the molecule as turning over and over about an axis perpendicular to its own axis.

If we take HCl as a simple example of an ionic molecule, we have $R = 0.13$ nm, which gives a stiffness of about 840 N m^{-1}. (It is interesting to note that a spring of the same stiffness would extend approximately 12 mm when a mass of 1 kg was hung on it: the stiffness of an ionic binding force is nothing out of the ordinary.)

Since the chlorine ion is about 35 times heavier than the hydrogen ion, we can assume that the Cl$^-$ remains nearly stationary while the H$^+$ ion does all the vibrating. Then we may estimate the vibration frequency as

$$\omega_0 \approx (s/m_H)^{1/2} = 7.1 \times 10^{14}\,s^{-1} \tag{2.18}$$

where $m_H = 1.67 \times 10^{-27}$ kg is the mass of a hydrogen atom. This result corresponds to $\nu_0 \approx 1.1 \times 10^{14}$ Hz. Since the HCl molecule has an electric dipole moment which varies with its length r, we should expect that, when it vibrates, it should emit light at about 10^{14} Hz, which is in the infrared region of the spectrum.

Vibrational spectra. Something as small as a molecule should not, strictly, be discussed in terms of classical mechanics at all. Nevertheless, the result we have found is strikingly similar to what quantum mechanics predicts.

The quantum treatment of the harmonic oscillator (our model system) leads to the result that its energy must have one of the values

$$W = (n + \tfrac{1}{2})\hbar\omega_0$$

where the 'quantum number' n can be any positive integer or zero, *and ω_0 is given by* (1.5). The system can, moreover, absorb or emit radiation only when n changes† by 1. Energy must thus be absorbed or emitted in quanta $\hbar\omega_0$, which means that the light has the classical frequency $\omega_0/2\pi$.

The value of ν_0 deduced from the vibrational spectrum of HCl is 8.9×10^{13} Hz. This is remarkably close to our estimate of 1.1×10^{14} Hz, and tells us that the 'ionic' potential used (2.15) must describe the real binding force fairly well in the case of HCl, at least in the region near $r = R$.

---

† A slight complication arises from another quantum rule which says that a vibrational transition must be accompanied by a unit change in a second quantum number describing the rotational energy of the molecule. The vibrational spectrum does not comprise a single line but a whole collection of lines associated with the various rotational possibilities.

Summary. A particle which is not quite in stable equilibrium will vibrate harmonically. The frequency is fixed by the second derivative of the potential energy with respect to the position coordinate, at the equilibrium position.

This result can be applied to deduce the curvature of the potential well for the HCl molecule (or any other diatomic molecule) from its vibration frequency, which is in the infrared region.

2.5. Circuit oscillations

Figure 2.7 shows a simple electrical circuit made by connecting a capacitor (capacitance C) across a coil (self inductance L, resistance negligible). We consider this circuit at a moment when there is a charge ψ on the live plate of the capacitor, and a charge $-\psi$ on the earthed plate.

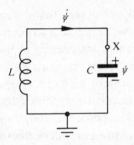

Fig. 2.7 An LC circuit in oscillation. The charge on the capacitor is ψ, and a current $\dot{\psi}$ flows clockwise round the circuit.

The potential of the live plate is then ψ/C and the current flowing clockwise round the circuit is $\dot{\psi}$. We wish to know how ψ varies with time.

If we consider the voltage changes encountered in going round the circuit, the second of Kirchhoff's rules states that

$$L\left(\frac{d\dot{\psi}}{dt}\right) + \left(\frac{1}{C}\right)\psi = 0 \qquad (2.19)$$

This equation,† which must be satisfied by ψ at all times, can be readily put into the now familiar form

$$\ddot{\psi} + (1/LC)\psi = 0$$

† Electrical engineers would write (2.19) in the form

$$L\left(\frac{dI}{dt}\right) + \frac{1}{C}\int_0^t I\, dt = 0$$

where I is the current $\dot{\psi}$.

Comparison with (1.4) leads us to write

$$\omega_0 = (1/LC)^{1/2} \tag{2.20}$$

Here we have a vibration without moving parts. The harmonically varying quantity is the charge on the capacitor. We could observe the variation by connecting an oscilloscope between point X (fig. 2.7) and earth, so monitoring the voltage ψ/C; we can describe it as an *oscillation* to distinguish it from vibrations in which masses are physically moved.†

A very wide range of frequencies can be obtained by adjusting the values of L and C. A 100 mH coil and a 100 μF capacitor, for example, will produce vibrations at approximately 50 Hz, well down in the Audio Frequency (AF) range. With $L = 1$ μH and $C = 10$ pF, on the other hand, the frequency is approximately 50 MHz and lies in the Very High Frequency (VHF) region used for stereo radio transmissions.

A scheme of analogies. We have seen that the charge ψ on the capacitor can be regarded as the basic variable describing an electrical oscillation; in this it corresponds to the displacement of the mass in the prototype mechanical vibration. To exploit the work we have already done on the prototype, it is helpful to draw up a list of correspondences between other pairs of mechanical and electrical quantities in the two systems. Such a list is provided in table 2.2. The entries relating to a circuit with resistance, which we have neglected at present, will be discussed later; but the rest of the table can be understood now.

The first three entries in the table follow directly from our preceding discussion. The remaining ones depend on an arbitrary decision which we make now. We note an obvious similarity between the mass of the prototype and the inductance of the coil: the inductance keeps the current flowing in the coil when the charge on the capacitor has reached zero, producing the inertial overshoot essential to any vibration. It is natural, therefore, to choose L as the electrical counterpart of m. This choice obliges us to match C to $1/s$, making (2.20) consistent with (1.5).

If we make these substitutions in the equation of motion (1.3) we obtain equation (2.19) whose terms are voltages. The correspondences between these voltages and the forces in (1.3) are easy to memorize: there is, for example, a clear analogy between the spring force $s\psi$ tending to reduce the displacement and the voltage $C^{-1}\psi$ tending to reduce the charge.

---

† In other books the words 'vibration' and 'oscillation' may be found in use without this distinction.

TABLE 2.2

Analogous quantities in the prototype vibrator and an oscillating circuit

Prototype vibrator		Oscillating circuit	
Displacement of mass	ψ	Charge on capacitor	ψ
Velocity of mass	$\dot\psi$	Current	$\dot\psi$
Acceleration of mass	$\ddot\psi$	Rate of increase of current	$\ddot\psi$
Mass	m	Inductance	L
Stiffness	s	Inverse capacitance	C^{-1}
Resistance	b	Resistance	R
Force applied by spring	$s\psi$	Voltage across C	$C^{-1}\psi$
Drag force	$b\dot\psi$	Voltage across R	$R\dot\psi$
Force accelerating mass	$m\ddot\psi$	Voltage across L	$L\ddot\psi$
Potential energy	$\frac{1}{2}s\psi^2$	Electric energy	$\frac{1}{2}C^{-1}\psi^2$
Kinetic energy	$\frac{1}{2}m\dot\psi^2$	Magnetic energy	$\frac{1}{2}L\dot\psi^2$

The entries involving b and R are explained in section 4.1.

We can also see that the kinetic energy $\frac{1}{2}m\dot\psi^2$ of the mass becomes $\frac{1}{2}L\dot\psi^2$, the energy stored in the inductance when the current is $\dot\psi$; and the potential energy $\frac{1}{2}s\psi^2$ becomes $\frac{1}{2}C^{-1}\psi^2$, the energy stored in the capacitor. Again these are plausible analogies, since we may associate the energy of the inductance with charges *in motion* (as current through the coil), and ascribe the capacitance energy to the *positions* of charges (on the capacitor plates).

These correspondences are not to be taken literally, however. The variable ψ represents something which is purely electrical (a quantity of charge); it does not measure the displacement in space of a mass, a molecule, or even an electron. The table is merely a scheme of analogies (one of several possible schemes), useful mainly for memory purposes. Such analogies are to be found in other areas of physics.

There is a practical significance, however. Faced with a mechanical vibrator of unknown properties, we may sometimes be able to understand it and predict its behaviour with the aid of an analogue computer in the form of the analogous electrical circuit.† When using such a technique we are free to scale the component values up or down to any convenient sizes, since the table does not lay down any particular systems of units.

---

† The rules for translating between complex mechanical and electrical systems are not totally straightforward. We shall not go into them in this book.

Summary. In a simple LC circuit the charge on the capacitor satisfies the same differential equation as the displacement of the mass in the prototype vibrator, and therefore varies with time in exactly the same way. The mathematics of harmonic motion can be taken over to describe circuit oscillations.

Prototype properties are most easily translated into electrical language by using a scheme of analogies. Table 2.2 lists the correspondences within one such scheme.

Problems

2.1   A watch ticks 5 times per second. Its balance wheel has a moment of inertia 2×10^{-6} kg m^2. Calculate the torsional stiffness of the balance spring. (Assume that the period is 2 ticks.)

2.2   A grandfather clock ticks once per second. Show that it must be at least 1 m high.

2.3   Show that the isothermal compressibility κ_T is equal to $1/p$ for a perfect gas. Estimate the percentage difference which the use of κ_T instead of κ_S would make to the calculated value of ω_0 for a flask containing air.

2.4   Estimate the approximate change in the pitch of the organ if the temperature in church falls by 20°C. (The pitch of a vibrator rises by one semitone every time its frequency rises by 5.9 per cent.)

2.5   A miniature loudspeaker unit has a cone of diameter 80 mm mounted in a hole of the same diameter in a sealed cabinet of internal dimensions 150 mm × 150 mm × 300 mm. The mass of the cone is 5.0 g, and the mounting is such that the stiffness of the suspension may be neglected. Estimate the free vibration frequency of the cone.

2.6   Two masses m_1 and m_2 are joined by a spring of stiffness s. They can vibrate along the line of their centres, moving alternately towards and away from each other. For this vibration, show that $\omega_0^2 = s/\mu$ where

$$\mu = m_1 m_2/(m_1 + m_2)$$
$$\approx m_1 \quad \text{if } m_1 \ll m_2$$

(We used the approximation in (2.18).)

2.7   Figure 2.8 shows an arrangement which could be used to set an LC circuit into oscillation. The capacitor is first charged to a voltage V_1 by means of the battery. At time $t = 0$ the switch is thrown to connect the charged capacitor across the coil. Derive (a) the amplitude, and (b) the phase constant of the resulting oscillation.

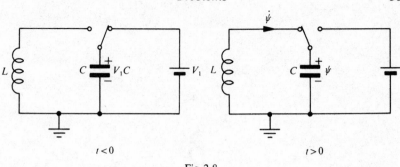

Fig. 2.8.

2.8 Show that vertical vibrations of a mass m *suspended* on a spring of stiffness s whose other end is fixed have angular frequency $(s/m)^{1/2}$. (Hint: measure displacements from the equilibrium position of the mass, where its weight is balanced by the spring force.)

2.9 An astronaut on the surface of the moon weighs rock samples using a light spring balance. The balance, which was calibrated on earth, has a scale 100 mm long which reads from 0 to 1.0 kg. He observes that a certain rock gives a steady reading of 0.40 kg and, when disturbed, vibrates with a period of 1.0 s. What is the acceleration due to gravity on the moon?

2.10 A certain pickup cartridge has a mass of 12 g, and the vertical stiffness of the stylus mounting is 80 N m^{-1}. It is attached to a light arm at a point 220 mm from the pivot, with a balance weight on the other side of the pivot to reduce the vertical tracking force on the record.

Such an assembly can undergo free angular vibration in a vertical plane. For minimum effect on the sound reproduction, the frequency of this vibration should be below the limit of audibility. Calculate the required mass of the balance weight, and its distance from the pivot, to give a frequency of 12 Hz and a tracking force of 0.020 N. (Neglect the mass of the arm.)

2.11 A mass moves under a potential $V(x) = V_0 \cosh (x/x_0)$ where V_0 and x_0 are constants. (a) Find the position of stable equilibrium. (b) Show that the frequency of small vibrations about this point is the same as it would be if the same mass was vibrating on a spring of stiffness V_0/x_0^2.

3

Damping

The preceding chapters have highlighted the parts played by stiffness and inertia in any vibrating system. Vibration can occur only if these two quantities (or two analogous quantities) are present.

In a real system, forces other than those due to stiffness will also be present, and these will not generally behave like return forces. An obvious example is friction, which always tends to oppose motion, so that its direction will sometimes be opposite to that of the spring force. To discover how such forces can affect the vibration, we now add to our model system a standard type of *damping force* which acts whenever the mass is moving.

In the damped system (fig. 3.1) the mass moves inside a lubricated cylinder. We assume that it experiences a drag force F_d which always

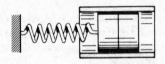

Fig. 3.1 A model vibrator with damping.

opposes the motion, and whose size is always proportional to the instantaneous speed $|\dot{\psi}|$. Thus we have

$$F_d = -b\dot{\psi} \tag{3.1}$$

where b is a positive constant which we shall call the *resistance*. Its value, which will be measured in $N(m\,s^{-1})^{-1} = kg\,s^{-1}$, can in principle be adjusted to any size we wish by changing the grade of oil used to lubricate the cylinder.

When we write down Newton's second law we must now include the damping force as well as the spring force (1.2) so that

$$m\ddot{\psi} = F_s + F_d = -s\psi - b\dot{\psi} \tag{3.2}$$

As before, we divide through by m to obtain

$$\ddot{\psi} + \gamma\dot{\psi} + \omega_0^2\psi = 0 \tag{3.3}$$

where we have defined

$$\gamma \equiv b/m \tag{3.4}$$

The units of γ will be s^{-1}, like those of ω_0. For reasons which will become clear later (section 5.1) we shall call γ the *width*.

Equation (3.3) is, like (1.4), linear and homogeneous, and has constant coefficients. We may therefore substitute a trial solution of the form

$$\psi = C e^{pt} \tag{3.5}$$

This leads to an algebraic equation

$$p^2 + \gamma p + \omega_0^2 = 0$$

with two roots

$$p = -\tfrac{1}{2}\gamma \pm (\tfrac{1}{4}\gamma^2 - \omega_0^2)^{1/2} \tag{3.6}$$

Depending on the relative sizes of γ and ω_0, the expression in brackets on the right of (3.6) can be positive, negative or zero. There will be major differences in the nature of the motion in the three cases. From now on, therefore, we distinguish *light damping* ($\gamma < 2\omega_0$), *heavy damping* ($\gamma > 2\omega_0$) and *critical damping* ($\gamma = 2\omega_0$).

3.1. Light damping

Here the square root of a negative quantity is called for in (3.6), leading to a complex exponent in (3.5). We write

$$p = -\tfrac{1}{2}\gamma \pm i\omega_f$$

where

$$\omega_f \equiv (\omega_0^2 - \tfrac{1}{4}\gamma^2)^{1/2} = \omega_0[1 - (\gamma/2\omega_0)^2]^{1/2} \tag{3.7}$$

and we take the positive square roots.

We now follow the derivation of form C in section 1.2, introducing two complex coefficients C and C' to multiply exponentials involving the two possible values of p. As before, we make $C' = C^*$ to avoid a complex ψ, obtaining finally

$$\psi(t) = \exp\left(-\tfrac{1}{2}\gamma t\right)[C \exp\left(i\omega_f t\right) + C^* \exp\left(-i\omega_f t\right)]$$

The expression in brackets varies harmonically at the angular frequency ω_f. The optional forms may be written down immediately as

$$\psi(t) = A \exp\left(-\tfrac{1}{2}\gamma t\right) \cos\left(\omega_f t + \phi\right)$$

$$\psi(t) = \exp\left(-\tfrac{1}{2}\gamma t\right)(B_p \cos \omega_f t + B_q \sin \omega_f t) \tag{3.8}$$

$$\psi(t) = \mathrm{Re}\left[D \exp\left(-\tfrac{1}{2}\gamma t + i\omega_f t\right)\right]$$

What we are dealing with here is a *damped vibration*, whose amplitude† dies away exponentially with time. It falls by a factor $1/e = 0.368$ whenever the time increases by $2/\gamma$.

These free vibrations have an angular frequency ω_f which is, not surprisingly, smaller than the value ω_0 that we had with $\gamma = 0$. The difference is usually slight, however. When $\gamma = \omega_0/10$, for example, ω_f is only 0.1 per cent smaller than ω_0. In such cases we may calculate ω_f using an approximate form of (3.7),

$$\omega_f \approx \omega_0[1 - \tfrac{1}{2}(\gamma/2\omega_0)^2] \tag{3.9}$$

When $\gamma \ll \omega_0$, a condition we shall refer to as *very light damping*, it may be sufficiently accurate simply to replace ω_f by ω_0.

Initial conditions. As before, we have two arbitrary constants whose values can be adjusted to fit initial conditions. We illustrate with the starting procedure discussed previously, in which we set up a vibration by drawing the mass out a distance A_1 to the right and then letting it go. We have to replace (1.7) by

$$\psi = A \cos \phi = A_1$$

$$\dot{\psi} = -\tfrac{1}{2}\gamma A \cos \phi - \omega_f A \sin \phi = 0$$

On this occasion the phase constant ϕ is not zero; we have instead

$$\tan \phi = -\gamma/2\omega_f$$

This leads to an initial amplitude

$$A = A_1 \sec \phi = A_1(1 + \tan^2 \phi)^{1/2}$$

$$= A_1[1 + (\gamma/2\omega_f)^2]^{1/2} = (\omega_0/\omega_f)A_1$$

† It is convenient now to refer to the time dependent quantity $A \exp\left(-\tfrac{1}{2}\gamma t\right)$ as the amplitude, and to the constant A as the *initial amplitude*.

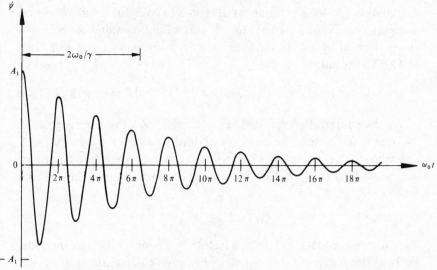

Fig. 3.2 A lightly damped vibration. The mass is pulled a distance A_1 to the right, and released from rest at $t = 0$. In this example $\gamma = \omega_0/10$.

This type of vibration is illustrated in fig. 3.2. Although the motion does not repeat itself as an undamped vibration does, it is still useful to describe as 'a cycle' the events that take place while the phase angle $\omega_f t + \phi$ increases by 2π. Thus successive zeros of ψ, for example, are separated in time by one period

$$\tau \equiv 2\pi/\omega_f$$

Decay of the energy. Since the damping force (3.1) always opposes the motion, it continually removes energy from the system and never gives any back. We are not, therefore, entitled to our previous assumption (1.13) that the total (kinetic plus potential) energy is constant. Using (1.12), we have instead

$$\frac{dW}{dt} = \frac{dW}{d\dot{\psi}} \cdot \frac{d\dot{\psi}}{dt} + \frac{dW}{d\psi} \cdot \frac{d\psi}{dt}$$

$$= (m\ddot{\psi} + s\psi)\dot{\psi}$$

Now we find, from Newton's second law (3.2), that

$$\frac{dW}{dt} = -b\dot{\psi}^2 \leq 0 \tag{3.10}$$

Thus the energy decreases as time passes, which is why the vibrations die away. The rate at which the energy decays is large when the resistance b is large, as we might expect.

Equation (3.10) is valid under all conditions: light, heavy or critical damping. An explicit formula for W under light damping conditions is easily derived by substituting the form A expression for ψ (3.8) into (1.12). The result is

$$W = \tfrac{1}{2}mA^2 e^{-\gamma t}\{\omega_0^2 + \tfrac{1}{2}\gamma\omega_f \sin\left[2(\omega_f t + \phi)\right] + \tfrac{1}{4}\gamma^2 \cos\left[2(\omega_f t + \phi)\right]\} \quad (3.11)$$

We shall usually be interested in the decay of $\langle W\rangle$, the average value of W over many cycles. The second and third terms on the right of (3.11) describe small fluctuations of W within each cycle, but they contribute almost nothing to an average over many complete cycles. Thus we may write

$$\langle W\rangle \approx \tfrac{1}{2}m\omega_0^2 A^2 e^{-\gamma t} \quad (3.12)$$

We can now see that $\langle W\rangle$ falls by a factor $1/e$ whenever the time increases by $1/\gamma$. The *fractional* rate of energy loss is given at all times by

$$-\frac{1}{\langle W\rangle}\cdot\frac{d\langle W\rangle}{dt} \approx \gamma \quad (3.13)$$

This is a very useful formula for the width of a system whose energy decay characteristics can be found.

Q-values. We have been using the smallness of $\gamma/2\omega_0$, a pure number, to indicate the lightness of the damping. A related number, which is large when the damping is light, is the quality factor

$$Q \equiv \omega_0/\gamma \quad (3.14)$$

usually known as the 'Q-value'. The system whose vibrations are shown in fig. 3.2 has a Q-value of 10. A very lightly damped system has

$$Q \approx \omega_f/\gamma \gg 1$$

Properties of lightly damped vibrations are commonly expressed in terms of Q. For example, the fractional decrease per cycle in the mean energy is, from (3.13),

$$\gamma\tau = 2\pi\gamma/\omega_f \approx 2\pi/Q$$

We can say that *the system loses approximately $2\pi/Q$ of its energy during each cycle.* Alternatively, since the energy decreases by a factor $1/e$ in a time $1/\gamma$, and in that time the vibration goes through $\omega_f/2\pi\gamma \approx Q/2\pi$ cycles, we can say that *the energy falls by a factor $1/e$ in the course of about $Q/2\pi$ cycles of vibration.*

Summary. Light damping ($\gamma < 2\omega_0$) slightly reduces the frequency of a vibrator and introduces a decay factor $\exp(-\frac{1}{2}\gamma t)$ in the amplitude. The average energy also decays exponentially, losing a fraction γ of its value per unit time.

The controlling quantity γ is the damping force per unit speed per unit mass. A convenient alternative parameter for describing the lightness of damping is the Q-value, defined as ω_0/γ.

3.2. Heavy damping

We return to (3.6) and consider what happens if $\gamma > 2\omega_0$. The two values of p are both real and negative, and we label them

$$p_1 = -[\tfrac{1}{2}\gamma + (\tfrac{1}{4}\gamma^2 - \omega_0^2)^{1/2}] \equiv -\mu_1$$
$$p_2 = -[\tfrac{1}{2}\gamma - (\tfrac{1}{4}\gamma^2 - \omega_0^2)^{1/2}] \equiv -\mu_2 \tag{3.15}$$

We observe that $\mu_1 > \tfrac{1}{2}\gamma > \omega_0$, that $\mu_1\mu_2 = \omega_0^2$ and therefore that $\mu_2 < \omega_0$.

The general form of ψ is now

$$\psi(t) = C_1 \exp(-\mu_1 t) + C_2 \exp(-\mu_2 t)$$

Here the arbitrary constants C_1 and C_2 must be real, to avoid a complex ψ. They can be adjusted as usual to fit initial conditions. We consider again our usual example; putting

$$\psi(0) = C_1 + C_2 = A_1$$
$$\dot\psi(0) = -(\mu_1 C_1 + \mu_2 C_2) = 0$$

we obtain the result

$$\psi(t) = \frac{A_1[\mu_1 \exp(-\mu_2 t) - \mu_2 \exp(-\mu_1 t)]}{\mu_1 - \mu_2} \tag{3.16}$$

The second term, with the larger decay constant in the exponent, will decay the faster. Thus, in the long run, the motion becomes a simple exponential decay,

$$\psi(t) \approx \left(\frac{A_1\mu_1}{\mu_1 - \mu_2}\right) \exp(-\mu_2 t)$$

The solution (3.16) is plotted in fig. 3.3 for a system with $\gamma = 4\omega_0$. It is not really a vibration at all: we call such motion *aperiodic*.†

---

† A system whose free motion is aperiodic can be made to undergo *forced* vibration (chapter 5).

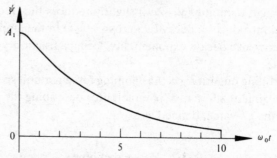

Fig. 3.3 Heavily damped motion. The system is set into motion in the same way as in fig. 3.2. In this example $\gamma = 4\omega_0$, giving $\mu_1 = 3.73\omega_0$ and $\mu_2 = 0.268\omega_0$. The horizontal scale is finer than in fig. 3.2.

In this example the displacement never changes sign during the motion. With different starting arrangements, ψ can presumably be made to cross the axis at least once; we can show, however, that it will not do so more than once. We suppose that, at some moment, the mass is at $\psi = 0$ and is moving to the right at some speed v_1. We may label this moment $t = 0$ and write down the initial conditions

$$\psi(0) = C_1 + C_2 = 0$$

$$\dot{\psi}(0) = -(\mu_1 C_1 + \mu_2 C_2) = v_1$$

These lead to the solution

$$\psi(t) = \frac{v_1[\exp(-\mu_2 t) - \exp(-\mu_1 t)]}{\mu_1 - \mu_2}$$

which is plotted in fig. 3.4. It shows that the mass never passes its equilibrium point again. After ψ reaches its maximum value the mass returns to $\psi = 0$ in the same way as before.

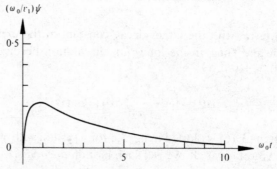

Fig. 3.4 The mass of a heavily damped vibrator never passes its equilibrium point more than once. The same system as in fig. 3.3, but now the mass is assumed to be passing $\psi = 0$ from left to right with speed v_1 at $t = 0$.

Very heavy damping. A very heavily damped system will have $\gamma \gg \omega_0$. For such a system the decay factors (3.15) become

$$\mu_1 \approx \gamma$$

$$\mu_2 = \tfrac{1}{2}\gamma - \tfrac{1}{2}\gamma(1 - 4\omega_0^2/\gamma^2)^{1/2}$$

$$\approx \tfrac{1}{2}\gamma - \tfrac{1}{2}\gamma(1 - 2\omega_0^2/\gamma^2) = \omega_0^2/\gamma$$

Putting these values in the expression (3.16) for $\psi(t)$ under our usual starting conditions, we find that the motion

$$\psi(t) = \frac{A_1[\gamma^2 \exp(-\omega_0^2 t/\gamma) - \omega_0^2 \exp(-\gamma t)]}{\gamma^2 - \omega_0^2}$$

$$\approx A_1 \exp(-\omega_0^2 t/\gamma)$$

(3.17)

is, from the outset, a nearly exponential decay.

Exponential motion may be characterized by a *relaxation time* τ_r: the time it takes for the displacement to be reduced by a factor $1/e$. In the present case we have

$$\tau_r = \gamma/\omega_0^2 = b/s \qquad (3.18)$$

The value of τ_r does not depend on m. We can see why if we calculate the drag force (3.1) in terms of ψ. We obtain, for the motion described by (3.17),

$$F_d \approx s\psi = -F_s$$

Thus F_d nearly balances F_s: in this case the spring has to work so hard against the drag force that the mass hardly influences the motion at all.

Summary. Heavy damping ($\gamma > 2\omega_0$) leads to aperiodic motion. The mass passes its equilibrium point at most once, before returning asymptotically to rest. In its final stages the return is essentially exponential. If the damping is very heavy ($\gamma \gg \omega_0$) the relaxation time is independent of the mass.

3.3. Critical damping

What limiting form does the heavily damped solution (3.16) to our standard example take when γ is reduced to the critical value $2\omega_0$? In (3.15) we then have

$$\mu_1 = \mu_2 = \tfrac{1}{2}\gamma = \omega_0$$

but if these conditions are applied directly to (3.16) both the numerator and the denominator become zero, making $\psi(t)$ indeterminate.

We can find the limit by defining

$$\omega_1 \equiv (\tfrac{1}{4}\gamma^2 - \omega_0^2)^{1/2}$$

and writing (3.16) in the form

$$\psi(t) = \tfrac{1}{2}A_1\{\exp(\omega_1 t) + \exp(-\omega_1 t)$$
$$+ (\omega_0/\omega_1)[\exp(\omega_1 t) - \exp(-\omega_1 t)]\}\exp(-\omega_0 t)$$

When ω_1 approaches zero (as $\gamma \to 2\omega_0$) the quantity in square brackets tends to the limit $2\omega_1 t$. (The easiest way to see this is to write the exponentials in series form.) Thus the critically damped motion is given by

$$\psi(t) = A_1(1 + \omega_0 t)\exp(-\omega_0 t)$$

This is plotted in fig. 3.5.

It is instructive to review figs. 3.2, 3.3 and 3.5 as a group. The same initial conditions were assumed in each case, the systems differing only in their degrees of damping.

By comparing fig. 3.5 with fig. 3.3 we see that, in the long run, ψ decays faster when the damping is critical than when it is heavy. The reason is that ω_0, the amplitude decay constant for critically damped motion, is always larger than μ_2. By comparing fig. 3.5 with fig. 3.2 we see that, in the long run, ψ decays faster when the damping is critical than the maxima of ψ do when the damping is light. In this case the reason is that, with light damping, ω_0 is larger than $\tfrac{1}{2}\gamma$, the amplitude decay constant.

We conclude that *a system with a given value of ω_0 settles down most quickly after a disturbance if it is critically damped.* This is a point of great

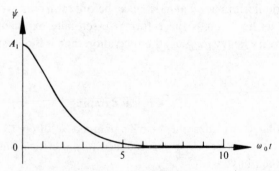

Fig. 3.5 Critically damped motion ($\gamma = 2\omega_0$). The system is set into motion as in figs. 3.2 and 3.3.

importance in the design of any equipment containing parts that are capable of vibrating.

Summary. The aperiodic motion of a critically damped system (one with $\gamma = 2\omega_0$) has an amplitude decay constant equal to ω_0. The final asymptotic return to equilibrium is faster than in heavily damped motion, and faster than the decay of the amplitude in a lightly damped vibration.

Problems

3.1 For a vibrator with $m = 0.010$ kg and $s = 36$ N m^{-1}, (a) what value of b would make the amplitude decrease from A to A/e in 1.0 s? (b) What is the Q-value of such a system? (c) What value of b would produce critical damping?

3.2 Show that the amplitude of a damped vibration is halved in a time $1.39/\gamma$.

3.3 Show that successive maxima of ψ during a damped vibration are separated in time by $2\pi/\omega_f$.

3.4 A convenient way of measuring γ for a lightly damped vibrator is to record the period τ and the ratio r of any maximum to the next. Show that γ is given by $(2 \ln r)/\tau$. (The quantity $\ln r$ is called the *logarithmic decrement*.)

3.5 For a lightly damped vibrator, show that $\omega_f \approx \omega_0(1 - 1/8Q^2)$.

3.6 A lightly damped system is set into vibration with initial conditions $\psi(0) = 0$, $\dot{\psi}(0) = v_1$. Show that the subsequent motion is given by

$$\psi(t) = (v_1/\omega_f) \exp\left(-\tfrac{1}{2}\gamma t\right) \sin \omega_f t$$

3.7 A certain system has stiffness 10 N m^{-1} and is very heavily damped. The mass is observed to be moving to the left at 10 mm s^{-1} when its position is 30 mm to the right of its equilibrium position. Calculate the resistance.

3.8 A critically damped system is set into motion with initial conditions $\psi(0) = 0$, $\dot{\psi}(0) = v_1$. (a) Show that the subsequent motion is given by

$$\psi(t) = v_1 t \exp\left(-\omega_0 t\right)$$

(b) Show that the maximum displacement is $v_1/e\omega_0$.

3.9 Verify by direct substitution that the general solution of (3.3) under critical damping conditions may be written

$$\psi(t) = (C_1 + C_2\omega_0 t) \exp\left(-\omega_0 t\right)$$

where C_1 and C_2 are arbitrary constants.

4

Damping in physics

In this chapter we consider four kinds of damping to be met with in real systems. In three of these examples the damping produces results very similar to the 'standard' behaviour of the damped prototype. The fourth (friction) gives quite different behaviour. We shall come across other kinds of damping elsewhere in the book.

4.1. Resistance damping

We return first to the system with which we closed chapter 2: the oscillating circuit. The reason for starting here is that the damping effect of resistance in the circuit turns out to be *exactly* analogous to that of our standard damping force of chapter 3. We shall see that the results derived for the model system can be applied without modification, once we have made the necessary addition to our list of analogies connecting a vibrator and an oscillating circuit.

Figure 4.1 is the same as fig. 2.7, except that it includes the resistance R of the coil (and of the various connecting wires). Kirchhoff's rule for the voltages in fig. 4.1 states

$$L\left(\frac{d\dot{\psi}}{dt}\right) + R\dot{\psi} + \left(\frac{1}{C}\right)\psi = 0 \qquad (4.1)$$

and may be rewritten

$$\ddot{\psi} + (R/L)\dot{\psi} + (1/LC)\psi = 0$$

In this form it may be compared directly with (3.3) to give

$$\omega_0 = (1/LC)^{1/2}$$

$$\gamma = R/L \qquad (4.2)$$

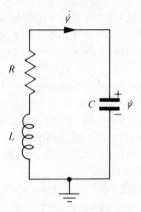

Fig. 4.1 An *LCR* circuit. As in fig. 2.7, the instantaneous charge on the capacitor is ψ and a current $\dot\psi$ flows clockwise round the circuit.

All the results of chapter 3 are now available to us. It is immediately apparent that, for lightly damped circuit oscillations of nearly the same frequency (i.e. for given values of L and C), the amplitude will decay at a rate which depends on the resistance. The Q-value (3.14) is $\omega_0 L/R$.

If the resistance is increased sufficiently, the variation of ψ will become aperiodic. It is sometimes useful to know the value of R which will just make this happen; by equating γ and $2\omega_0$ we find that the *critical damping resistance is equal to* $2\omega_0 L$.

Analogies. Fitting the resistance into our existing scheme of analogies is equally straightforward. The width for circuit oscillations (4.2) is R/L; that for mechanical vibrations (3.4) is b/m. We have already paired the inductance L with the mass m, and so the two resistances R and b obviously go together (table 2.2). We thus see that the new voltage $R\dot\psi$ in (4.1) is the exact counterpart of the damping force $b\dot\psi$ in (3.2).

Summary. The mathematics describing the behaviour of the damped prototype vibrator can be taken over without any modification at all to describe the variation of ψ, the charge on the capacitor in a series LCR circuit.

4.2. Electromagnetic damping

Most galvanometers consist of a current-carrying coil mounted on an axis in a magnetic field. The field is produced by a permanent magnet, shaped so that the coil experiences the same size of field at all angles. Thus a steady current in the coil gives rise to a torque which is proportional to the current. The coil comes to rest in a position where this electromagnetic torque is balanced by a return torque due to the stiffness of the suspension.

Mechanically, the suspended coil is an example of an angular vibrator (section 2.1). It will be subject to various mechanical damping processes. For example, the viscosity of the atmosphere will produce a damping torque which we may expect to be 'standard', i.e. proportional to the angular velocity of the coil. Usually such effects will be small enough to allow lightly damped vibrations.

There is also a specifically electromagnetic source of damping. As the coil rotates towards its new equilibrium position, the magnetic field will induce voltages proportional to the instantaneous angular speed $|\dot{\psi}|$. According to Lenz's law these voltages will reduce the current flowing in the coil, so reducing the electromagnetic torque by an amount $b_e\dot{\psi}$ where b_e is a constant (the electromagnetic rotational resistance). Using ψ to denote the displacement from the *new* equilibrium position, we may write

$$I\ddot{\psi} = -c\psi - b_e\dot{\psi}$$

If we now define

$$\gamma_e \equiv b_e/I$$

we obtain

$$\ddot{\psi} + \gamma_e\dot{\psi} + \omega_0^2\psi = 0$$

which is of the standard form (3.3).

The total width γ for the system will include the mechanical contribution γ_m as well as γ_e. Because γ_e is intrinsically more controllable than γ_m, it will usually pay to keep γ_m as small as possible.

Ballistic galvanometer. The use of electromagnetic damping to control the motion of the suspension is an important feature of the ballistic galvanometer. This device exploits the free vibration of the suspension following an angular impulse, as a means of measuring that impulse. By calibrating the instrument the cause of the impulse (usually an unknown quantity of charge pulsed through the coil as a short-lived current) can be measured. For these purposes the damping must be light.

If the impulse is applied at $t = 0$ and its size is J, the initial conditions are

$$\psi(0) = A \cos \phi = 0$$

$$\dot{\psi}(0) = -\tfrac{1}{2}\gamma A \cos \phi - \omega_f A \sin \phi = J/I$$

Here we have used $\psi(t)$ in form A, taken from (3.8). We see immediately that $\phi = -\frac{1}{2}\pi$. (The value $+\frac{1}{2}\pi$ is ruled out because $\sin \phi$ must be negative.) Thus

$$\psi(t) = (J/I\omega_f) \exp\left(-\tfrac{1}{2}\gamma t\right) \cos\left(\omega_f t - \tfrac{1}{2}\pi\right)$$

This vibration is shown in fig. 4.2, where we have taken $\gamma = \omega_0/10$.

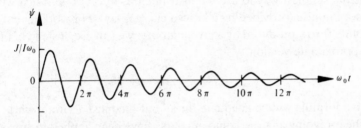

Fig. 4.2 The vibration of a ballistic galvanometer with $Q = 10$. The instrument is given an angular momentum impulse J at $t = 0$.

The impulse J is most easily found by measuring the value of ψ at the first maximum. If this maximum (the 'throw') occurs at time t' and has the value ψ_{max}, we know that

$$\psi(t') = (J/I\omega_f) \exp\left(-\tfrac{1}{2}\gamma t'\right) \cos\left(\omega_f t' - \tfrac{1}{2}\pi\right) = \psi_{max}$$

$$\dot{\psi}(t') = (J/I\omega_f) \exp\left(-\tfrac{1}{2}\gamma t'\right)\left[-\tfrac{1}{2}\gamma \cos\left(\omega_f t' - \tfrac{1}{2}\pi\right) - \omega_f \sin\left(\omega_f t' - \tfrac{1}{2}\pi\right)\right] = 0 \quad (4.3)$$

From the second of these conditions we find

$$\cot \omega_f t' = -\tan\left(\omega_f t' - \tfrac{1}{2}\pi\right) = \gamma/2\omega_f$$

$$\cos\left(\omega_f t' - \tfrac{1}{2}\pi\right) = \left[1 + (\gamma/2\omega_f)^2\right]^{-1/2} \quad (4.4)$$

$$\approx 1 - \tfrac{1}{2}(\gamma/2\omega_f)^2$$

The last approximation relies on the very light damping condition $\gamma \ll \omega_0$.

It is obvious that t' is slightly smaller than $\frac{1}{4}\tau_f$. Thus the phase angle $\omega_f t' - \tfrac{1}{2}\pi$ is small and we may write

$$\omega_f t' - \tfrac{1}{2}\pi \approx \tan\left(\omega_f t' - \tfrac{1}{2}\pi\right) = -\gamma/2\omega_f \quad (4.5)$$

The exponential decay factor may now be expressed

$$\exp\left(-\tfrac{1}{2}\gamma t'\right) = \exp\left[(\gamma/2\omega_f)^2 - \pi\gamma/4\omega_f\right]$$

$$\approx \left[1 + (\gamma/2\omega_f)^2\right] \exp\left(-\pi\gamma/4\omega_f\right) \quad (4.6)$$

Combining (4.3), (4.4) and (4.6), we obtain finally

$$\psi_{max} \approx (J/I\omega_f)[1 - \tfrac{1}{2}(\gamma/2\omega_f)^2][1 + (\gamma/2\omega_f)^2] \exp(-\pi\gamma/4\omega_f)$$
$$\approx (J/I\omega_f)[1 + \tfrac{1}{2}(\gamma/2\omega_f)^2] \exp(-\pi\gamma/4\omega_f) \qquad (4.7)$$

If the vibrational characteristics of the galvanometer (namely the values of I, ω_f and γ) do not change, ψ_{max} will always be proportional to J. Thus the simplest way to use the instrument is to calibrate it first with a known impulse (provided by a known charge, say). For working out the absolute throw produced by a given impulse we can use, instead of (4.7), an approximate version

$$\psi_{max} \approx (J/I\omega_f) \exp(-\pi\gamma/4\omega_f) \qquad (4.8)$$

Such a formula will be adequate for all but the most accurate work.

The foregoing analysis could of course have been applied to any other lightly damped system. The special feature of the ballistic galvanometer is the flexible control over the damping which is possible.

If the coil is open circuited as soon as the impulse has been delivered, γ_e will be zero. Then $\gamma = \gamma_m$ and the damping will be light, making (4.8) valid. With light damping, however, the vibration may go on for an inconveniently long time after the measurement. To get the system ready for a new measurement as quickly as possible we wish critical damping. This can be contrived by connecting a suitable external resistance across the galvanometer terminals. The smaller the total resistance in the circuit, the larger will be the induced current, and therefore b_e and γ_e. The total resistance can never be made smaller than the resistance of the coil; if that is not too large, it will be possible to choose an external resistance which will make the damping critical.

Other applications of electromagnetic damping. Now we can see how we should arrange the damping of ordinary direct-current galvanometers. In these instruments the coil must be left in the circuit while the reading is taken. Clearly the needle will take up its steady deflection most speedily if the combined resistance of the coil and the external circuit are such as to give critical damping.

Electromagnetic damping can be applied to any mechanical system. It is not necessary to have a loop of wire: *eddy currents* can be induced in the vibrating mass itself, if it is electrically conducting. These currents will always be proportional to $\dot{\psi}$, and so will the damping force. Eddy-current damping, being both 'standard' and controllable, is frequently the most convenient kind of damping to build into a vibrating system.

Summary. Electromagnetic damping arises when a closed circuit moves in a magnetic field, and is of the standard type. We have examined its application to the ballistic galvanometer, a device which relies on the fact that the 'throw' of any lightly damped vibrator is proportional to the initial impulse it receives. Light damping is achieved by open circuiting the galvanometer; after the throw has been measured, the instrument may be critically damped by connecting the appropriate resistance across its terminals.

4.3. Collision damping

To illustrate the concept of collision damping we return to plasma vibrations, previously discussed in section 2.3.

The essential feature of a plasma vibration is the cooperative behaviour of the participating electrons: they all move together, with the same $\psi(t)$. This collective motion is actually superimposed on the much more vigorous but random motion of thermal agitation, and every electron runs the risk of colliding with another randomly moving particle. If an electron does have a collision it will be scattered in a random direction, losing the special value of its speed $\dot{\psi}$ in the vibration direction. It thus joins the background of particles which have random motion only, and plays no further part in the vibration.

The rate at which electrons leak away from the vibration in this way must be proportional to the number taking part at any given moment. In fact the fraction colliding per unit time will be $1/\tau_c$ where τ_c is the *collision lifetime*: the average length of time for which an electron survives before having a collision. Since every colliding electron is 'lost' as far as the plasma vibration is concerned, we can say that the fractional rate of decrease of the number of participating electrons is

$$-\frac{1}{N} \cdot \frac{dN}{dt} = \frac{1}{\tau_c}$$

Unlike other kinds of damping we have met, there is in this case no decay of the amplitude; the damping takes the form of a steady draining away of electrons from the vibration. Each electron lost takes with it its share of the organized vibrational energy, just as a viscous damping force causes dissipation of the energy stored in a simple vibrator. The fractional rate of energy loss will, on average, be the same as that of electron loss. We now see, by comparison with (3.13), that the *collision width* must be

$$\gamma_c = 1/\tau_c$$

Estimating the collision width for any particular system is thus a matter of estimating τ_c.

Collision damping in the ionosphere. In the ionosphere the electrons are most likely to collide with neutral molecules. The number of collisions per unit time should be proportional to the number of molecules per unit volume N_m, and to the speed of the electrons. The latter is essentially the speed of their thermal motion v_{th} which, as we have seen, always outweighs $\dot{\psi}$. We therefore write

$$\gamma_c = \sigma_c(N_m v_{th}) \tag{4.9}$$

The constant of proportionality σ_c is known as the *collision cross-section*. It has the dimensions of area, and may be thought of as the effective target area offered by the molecule to the electrons: each hit on the target scores one collision.

The molecular density N_m will be high if the pressure p is high and the absolute temperature T is low. A perfect gas would have

$$N_m = p/kT \tag{4.10}$$

where k is the Boltzmann constant.

The average thermal kinetic energy of an electron at temperature T is $\frac{3}{2}kT$. Thus a reasonable estimate of v_{th} can be obtained by writing

$$\tfrac{1}{2}m_e v_{th}^2 \approx \tfrac{3}{2}kT$$

which gives

$$v_{th} \approx (3kT/m_e)^{1/2} \tag{4.11}$$

The collision cross-section σ_c can be measured in a laboratory experiment; its value is of the order of 10^{-19} m^2. Putting this value in (4.9) along with N_m and v_{th} from (4.10) and (4.11) gives

$$\gamma_c \approx (3\sigma_c^2/m_e k)^{1/2} pT^{-1/2} = (4.9 \times 10^7 \text{ kg}^{-1} \text{ m s K}^{1/2})pT^{-1/2}$$

The pressure p in the D layer of the ionosphere is about 0.5 N m^{-2}, compared with 10^5 N m^{-2} at sea level. The temperature T is approximately 200 K, somewhat lower than on the ground. These figures lead to the value $\gamma_c \approx 2 \times 10^6$ s^{-1}. This is the same as our previous estimate for ω_0, so that plasma vibrations in the D layer have $Q \sim 1$.

The F_2 layer, on the other hand, is not only much more rarified ($p \sim 10^{-10}$ N m^{-2} or less) but its temperature is also about 10 times greater than that of the D layer. We therefore expect γ_c to be much smaller. Since we also found previously that the plasma frequency of the

F_2 layer is about a million times larger than that of the D layer, we may be sure that damping in the F_2 layer is very light indeed ($Q \gg 1$). This is a fact of great importance for the propagation of radio waves (section 14.3).

Collision damping in metals. Most of the collisions involving the conduction electrons in a metal also involve the positive ion lattice. The frequency of these same collisions determines how well the metal conducts electricity: frequent collisions lead to low conductivity. We can therefore estimate the collision width γ_c for a particular metal from its measured conductivity.

When an electric field of magnitude E is set up in the metal, the conduction electrons acquire a small 'drift velocity' v_{dr}. Like $\dot{\psi}$ in a plasma, v_{dr} is superimposed on the much greater but randomly directed velocity of thermal agitation. The conductivity of the metal is

$$\sigma = Nev_{dr}/E \qquad (4.12)$$

where N is the conduction electron density.

A collision does not necessarily bring an electron to a standstill, but it does randomize its velocity component in the drift direction. Between collisions the electron receives an acceleration Ee/m_e due to the field. A reasonable estimate of the drift velocity is thus

$$v_{dr} \approx (Ee/m_e)\tau_c \qquad (4.13)$$

where τ_c is the collision lifetime, whose reciprocal is γ_c. Combining (4.12) and (4.13) gives

$$\gamma_c \approx Ne^2/m_e\sigma = (2.8 \times 10^{-8}\ \mathrm{C^2\,kg^{-1}})N/\sigma \qquad (4.14)$$

The quantities we therefore need to know, to be able to estimate γ_c for a particular metal, are its conduction electron density N and its conductivity σ. We have already found $N \approx 8.4 \times 10^{28}\ \mathrm{m^{-3}}$ for the example of copper. The conductivity of copper is $5.8 \times 10^7\ \Omega^{-1}\,\mathrm{m^{-1}}$, and these two values give $\gamma_c \approx 4 \times 10^{13}\ \mathrm{s^{-1}}$. When we remember our value ($\omega_0 \approx 2 \times 10^{16}\ \mathrm{s^{-1}}$) for the plasma frequency in copper, we see that $Q \approx 500$. We therefore expect plasma vibrations in metals to be very lightly damped.

Summary. Plasma vibration depends on a collective motion of the electrons taking part, and any electron which has a collision ceases to make its due contribution to the vibration. The width for collision damping is $1/\tau_c$ where τ_c is the collision lifetime of an electron in the plasma under consideration.

TABLE 4.1

Estimated plasma properties

Plasma	Electron density	Plasma frequency	Q
Ionosphere (D layer)	$\leqslant 10^9 \text{ m}^{-3}$	$\approx 300 \text{ kHz}$	~ 1
Ionosphere (F_2 layer)	$\leqslant 10^{12} \text{ m}^{-3}$	$\approx 10 \text{ MHz}$	$\gg 1$
Metal (copper)	$8.4 \times 10^{28} \text{ m}^{-3}$	$\approx 3 \times 10^{15} \text{ Hz}$	$\gg 1$

Properties of the plasmas we have discussed are summarized in table 4.1. We shall use these properties in section 14.3, where we discuss the propagation of electromagnetic waves in plasmas.

4.4. Friction damping

If there is no lubricating oil in the cylinder (fig. 3.1) the drag force F_d will be one due to the friction between the dry metal parts, rather than an essentially viscous force obeying (3.1).

The simplest assumption we can make about a frictional drag force is that its magnitude $|F_d|$ is *independent of the speed*. Its direction, however, always opposes the motion: we have, whenever the mass is moving towards the right ($\dot{\psi} > 0$),

$$m\ddot{\psi} = -s\psi - F_{sl}$$

but when the mass is moving to the left ($\dot{\psi} < 0$) the relation is

$$m\ddot{\psi} = -s\psi + F_{sl}$$

In these equations we have replaced $|F_d|$ by F_{sl}, the magnitude of the force of sliding friction, assumed to be a constant.

These equations are more closely related to the equation of motion for an *undamped* free vibration (1.3) than to the corresponding equation with standard damping (3.2). In fact, a simple pair of coordinate shifts

$$\psi_r \equiv \psi + F_{sl}/s$$

$$\psi_l \equiv \psi - F_{sl}/s$$

allows us to write

$$\ddot{\psi}_r + \omega_0^2 \psi_r = 0$$

$$\ddot{\psi}_l + \omega_0^2 \psi_l = 0$$

for motion to the right and to the left respectively.†

† Compare the motion of a mass on a vertical spring (problem 2.8).

Both ψ_r and ψ_l must vary harmonically with angular frequency ω_0. The motion of the mass, given by ψ, also appears harmonic during a given half cycle (to the right or to the left), but is centred on an apparent equilibrium position which is displaced a distance F_{sl}/s to the left (for motion to the right) or to the right (for motion to the left). Each half cycle takes a time π/ω_0 exactly.

Example. We examine our usual example of a vibration started by pulling the mass a distance A_1 to the right and releasing it from rest. The course of the motion can be constructed half-cycle by half-cycle. It is easy to see that the first swing (to the left) stops when $|\psi| = A_1 - 2F_{sl}/s$, the next swing (to the right) stops when $|\psi| = A_1 - 4F_{sl}/s$, and so on. The effective amplitude decreases *linearly* with the number of completed half cycles, and therefore with time. The motion is illustrated in fig. 4.3.

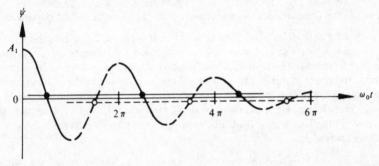

Fig. 4.3 Friction damping. The vibration is started by pulling the mass a distance A_1 to the right and releasing it from rest. The effective amplitude decays linearly with time, not exponentially. In this example $A_1 = 40F_{sl}/3s$ and $F_{st} = 3F_{sl}/2$. Swings to the left (solid curve) are harmonic but are centred on the apparent equilibrium position indicated by a solid line. Swings to the right (broken curve) are centred on the apparent equilibrium position indicated by a broken line.

Sticking friction. When the mass comes to rest at the end of a swing to the right or the left, it will set off again in the other direction only if the spring is able to overcome sticking friction, which we have so far ignored. For a given vibrator there is a fixed size F_{st} for the force of sticking friction at the threshold of motion, and $F_{st} > F_{sl}$. There will ultimately come a time when the size of the spring force at a maximum of $|\psi|$ is too small for F_{st}, and motion will then stop completely.

In fig. 4.3 we have taken $F_{st} = \frac{3}{2}F_{sl}$, and the motion ceases at the end of the sixth half cycle. There is a striking contrast with the asymptotic behaviour shown in fig. 3.2.

Summary. Sliding friction causes the amplitude of a free vibration to fall linearly with time rather than exponentially. Sticking friction ultimately brings the motion to a stop, always with the mass at a maximum of $|\psi|$.

Problems

4.1   Show that the relaxation time for a very heavily damped LCR circuit is RC.

4.2   A capacitor is charged to a voltage V_1 and is then connected across a coil. If the damping is critical, show that the current rises to a maximum value $2V_1/eR$, where R is the total resistance of the circuit made by the capacitor and the coil.

4.3   A steady torque of 2.0×10^{-6} N m, applied to the suspension of a ballistic galvanometer (by passing a suitable steady current), produces a steady deflection of $50°$. Free vibrations of the suspension have a period of 2.5 s. Calculate the throw produced by an angular impulse of 5.5×10^{-7} N m s (applied by discharging a capacitor through the coil). Neglect damping.

4.4   A ballistic galvanometer has a period of 4.0 s and a Q-value of 5.0. Calculate the time taken, after the impulse, for $|\psi|$ to reach its first maximum.

4.5   A mass of 0.10 kg is attached to a spring. It is pulled 200 mm to the right of its position when the spring is neither stretched nor compressed, and then released from rest. The resulting free vibrations, which are damped by friction, have a frequency of 2.0 Hz. It is observed that each swing to the right takes the mass to a point 30 mm to the left of its previous limit. The mass finally comes to rest 235 mm to the left of the point from which it was released. (a) Calculate the force of sliding friction F_{sl}. (b) Calculate upper and lower limits for the force of sticking friction F_{st}.

4.6   In fig. 4.4 values of the energy W at successive maxima of $|\psi|$ are plotted against the time t, for the friction-damped vibration of fig. 4.3. Show that the smooth curve through the points is a parabola (not the exponential curve obtained with standard damping).

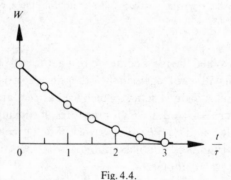

Fig. 4.4.

5

Forced vibrations

The discussion so far has concerned free vibrations (damped or undamped). No forces other than a return force and a drag force were acting while the vibrations were under way, though external forces presumably played a part in setting the vibrations going in the first place.

We now turn to the more general problem of a vibrator acted upon by a time-varying *driving force* $F(t)$. Initially we consider a single harmonic driving force

$$F(t) = F_0 \cos \omega t \tag{5.1}$$

Later we shall investigate the effect of two harmonic driving forces acting simultaneously.

We consider forced vibrations of the damped model system (fig. 3.1). One way of applying a force like (5.1) is to move the anchor point of the spring back and forth harmonically, with amplitude F_0/s (fig. 5.1). We

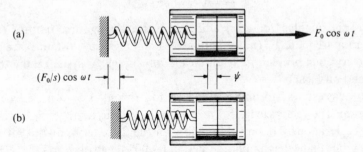

Fig. 5.1 A driving force $F_0 \cos \omega t$ can be applied to a mass on a spring either (a) directly, or (b) by moving the anchor point of the spring according to $(F_0/s) \cos \omega t$. In each case the driving force is additional to the return force $-s\psi$ due to the displacement of the mass from its equilibrium position.

can therefore expect our analysis to predict the kind of behaviour expected when any massive object is attached by means of something springy to something which is vibrating: a very common practical situation.

The new equation of motion

$$m\ddot{\psi} = -s\psi - b\dot{\psi} + F_0 \cos \omega t \tag{5.2}$$

is simply (3.2) with the driving force added. When we divide through by m, we obtain

$$\ddot{\psi} + \gamma\dot{\psi} + \omega_0^2\psi = (F_0/m) \cos \omega t \tag{5.3}$$

which would revert to (3.3) if the driving force were removed.

5.1. Steady states

If the driving force has been acting for a time which is very long compared with $1/\gamma$, we can expect the system to have settled into a regime in which the mass is vibrating harmonically, at the same frequency as the driving force. (This statement is justified in section 5.3, where we consider events immediately after the driving force is switched on.) The displacement will not necessarily be in phase with the driving force, but the phase difference should at least have achieved a constant value. This state of affairs will be permanent only if the mass and the force have the same frequency.

We call this precisely repetitive motion a *steady state*. The free vibration of an undamped system is an example of a steady state. The free vibration of a damped system is not a steady state, since it does not consist of identical cycles.

In the steady state, the displacement of the forced vibrator is

$$\psi = A \cos (\omega t + \phi) \tag{5.4}$$

where the angular frequency ω is that of the driving force (*not* the free vibration frequency). The amplitude A and the phase constant ϕ are not arbitrary: our problem is to discover, for any given system, how they depend on F_0 and ω.

The easiest way to find expressions for A and ϕ is to use a vector diagram. The vectors in fig. 5.2 represent the various terms in (5.3), which are accelerations. To draw the vectors for the three terms on the left, we remember that $\dot{\psi}$ has amplitude ωA and is 90° ahead of ψ, and that $\ddot{\psi}$ has amplitude $\omega^2 A$ and is 90° ahead of $\dot{\psi}$. The values of A and ϕ are chosen so that a closed figure is formed when we add the fourth vector, representing the term on the right of (5.3). This all works because the motion is a

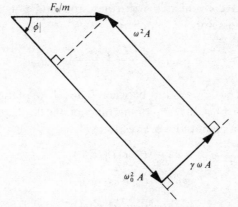

Fig. 5.2 A vector diagram for calculating A and ϕ in the steady state. Each vector represents an acceleration at $t = 0$. Three of these accelerations are the terms on the left of the equation of motion (5.3). The fourth vector (at the top) is drawn so as to close the figure.

steady state: as time passes the whole figure will rotate without changing its shape, and we can therefore match up the four vectors at any time we choose. (As usual we have picked $t = 0$ for convenience.)

The first thing we notice from the diagram is that ϕ must lie in the range

$$-\pi < \phi \leq 0$$

Since we chose the driving force as our standard of phase by writing it (5.1) with a zero phase constant, ϕ is the phase advance of ψ relative to the force. Since we have found that ϕ is always negative (clockwise in the vector diagram) *the displacement always lags the driving force* – as we might indeed expect. The value of ϕ can be found by examining the right-angled triangle marked in the figure. We see that

$$\tan \phi = -\gamma\omega/(\omega_0^2 - \omega^2) \tag{5.5}$$

This relation does not contain F_0; thus *the phase lag depends on the frequency but not on the size of the driving force.*

We can find an expression for A by applying Pythagoras's theorem

$$(\omega_0^2 - \omega^2)^2 A^2 + \gamma^2\omega^2 A^2 = (F_0/m)^2$$

to the same right-angled triangle. We obtain the result

$$A = \frac{F_0}{m}\left[\frac{1}{(\omega_0^2 - \omega^2)^2 + \gamma^2\omega^2}\right]^{1/2} \tag{5.6}$$

Corresponding expressions for the velocity and acceleration amplitudes are easily found by multiplying (5.6) by ω and ω^2 respectively.

We shall find it convenient to write these and similar quantities in terms of a dimensionless *response function*

$$R(\omega) \equiv \frac{\gamma^2\omega^2}{(\omega_0^2 - \omega^2)^2 + \gamma^2\omega^2} \qquad (5.7)$$

The value of $R(\omega)$ always lies between 0 and 1, the latter figure being reached when $\omega = \omega_0$. In terms of this function the displacement, velocity and acceleration amplitudes are respectively

$$A = (F_0/b\omega)[R(\omega)]^{1/2}$$
$$\omega A = (F_0/b)[R(\omega)]^{1/2} \qquad (5.8)$$
$$\omega^2 A = (F_0\omega/b)[R(\omega)]^{1/2}$$

Light damping: resonance. The three amplitudes are plotted against ω in fig. 5.3 for a system with $Q = 5$. Graphs such as these are known as *response curves*. The way in which the phase constant ϕ changes with frequency is shown for the same system in fig. 5.4.

Figure 5.3 illustrates the very important phenomenon of *resonance*, which appears when a lightly damped system is made to undergo forced vibration. The response curves all show large maxima at frequencies near ν_0, to which we can now give the name of *resonance frequency*.

Since the velocity amplitude ωA is proportional to the square root of $R(\omega)$, it has its maximum value exactly at the resonance frequency. The displacement amplitude A has ω in the denominator; this pulls down the curve at high frequencies more than at low frequencies, and so A peaks at a frequency slightly below the resonance frequency. The acceleration amplitude $\omega^2 A$, on the other hand, contains ω in the numerator, and therefore peaks slightly above the resonance frequency. These features can just be seen in fig. 5.3. Finding the precise frequencies at which A and $\omega^2 A$ have their maxima is left as an exercise: see problems 5.3 and 5.4.

At the resonance frequency the phase constant (fig. 5.4) is $-\frac{1}{2}\pi$; then the displacement lags the driving force by exactly 90°. We know that the displacement of any harmonic vibration lags the velocity by 90°. Thus *the velocity is in phase with the driving force at resonance*.

At very low driving frequencies ($\omega \ll \omega_0$) we have

$$\phi \approx 0$$
$$A \approx F_0/m\omega_0^2 = F_0/s \qquad (5.9)$$
$$\psi \approx (F_0/s)\cos\omega t$$

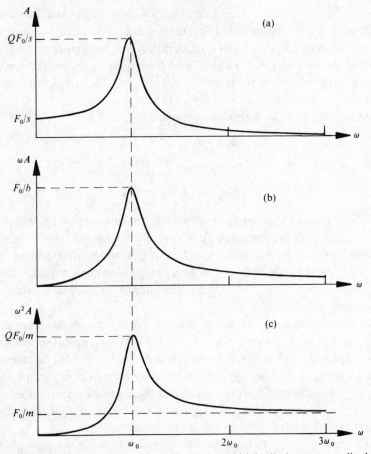

Fig. 5.3 Response curves for a system with $Q = 5$: plots of (a) the displacement amplitude A, (b) the velocity amplitude ωA, and (c) the acceleration amplitude $\omega^2 A$, against the driving angular frequency ω.

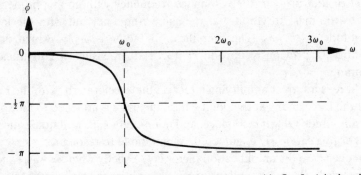

Fig. 5.4 Variation of ϕ with driving frequency for a system with $Q = 5$; ϕ is the phase advance of ψ relative to the driving force, and is always negative.

Neither m nor γ is involved here, and we say that the low frequency motion is *stiffness controlled*. The situation is easy to understand physically. The mass has a very small acceleration which requires only a small part of the driving force, most of which therefore goes to balance the spring force $-s\psi$. Since the latter is a return force, the driving force must be nearly in phase with ψ, as (5.5) confirms.

At high frequencies ($\omega \gg \omega_0$) we have

$$\phi \approx -\pi$$

$$A \approx F_0/m\omega^2 \qquad\qquad (5.10)$$

$$\psi \approx -(F_0/m\omega^2) \cos \omega t$$

Now the motion is *mass controlled*. The spring force is negligible in comparison with the force required to give the mass the large accelerations involved at these frequencies: the spring would hardly be missed if it were removed completely. Since the driving force must provide almost the entire return force, it is naturally almost in antiphase with the displacement.

At low frequencies the acceleration is small, but the displacement amplitude A is never smaller than F_0/s. At high frequencies the acceleration amplitude $\omega^2 A$ is never smaller than F_0/m, but the displacement is small. (See fig. 5.3.) The high frequency behaviour can be used to provide *vibration insulation*. To protect an object from vibration of frequency ω taking place at the other end of a supporting spring, we should, if possible, choose a spring whose stiffness makes $\omega_0 \ll \omega$.

We can see from (5.6) that A would become infinite at $\omega = \omega_0$ if γ were zero. We can therefore describe the motion at resonance as *resistance limited*.

Resonance provides *amplification*. For example, vibrating the anchor point of the spring at the resonance frequency with an amplitude F_0/s causes the mass to vibrate, at the same frequency but with the larger amplitude $F_0/m\gamma\omega_0$. The amplification factor is ω_0/γ, which is the quantity Q previously introduced (3.14) to measure the lightness of damping.

There is a similar amplifying effect on the acceleration, with the result that the force acting at the point where the mass is attached to the spring has amplitude QF_0 at resonance, and not merely F_0: a potential source of danger in a carelessly engineered system liable to resonance.

In each case the amplification factor is Q exactly when $\omega = \omega_0$ exactly. Since the displacement and the acceleration have their maxima at slightly different frequencies, they are amplified by slightly larger factors at these frequencies. (See problems 5.6 and 5.7.)

Power absorption. The appearance of Q as the amplification factor for a system in resonance highlights the basic connection between free and forced vibrations of the same system: the lighter the damping, and therefore the more long-lived the free vibrations, the more pronounced the resonance. The most striking expression of this relationship emerges when we consider the power that has to be put into the system to keep it vibrating.

The maintenance of the steady-state vibration requires a sustained supply of energy by the driving force, to replace the energy dissipated by the drag force. When the mass moves from ψ to $\psi + \Delta\psi$, the work done against the drag force is $-F_d\,\Delta\psi$. If the movement takes time Δt, the rate at which energy is dissipated is $-F_d(\Delta\psi/\Delta t)$. In the limit $\Delta t \to 0$ this becomes the instantaneous power absorption

$$P = -F_d\dot{\psi} = b\dot{\psi}^2 \tag{5.11}$$

We are interested in $\langle P \rangle$, the average of P over many cycles. Since the velocity $\dot{\psi}$ varies harmonically and has amplitude ωA, the average value of $\dot{\psi}^2$ over a complete cycle is

$$\langle \dot{\psi}^2 \rangle = \tfrac{1}{2}(\omega A)^2 = \tfrac{1}{2}(F_0/b)^2 R(\omega)$$

We thus find

$$\langle P \rangle = (F_0^2/2b)R(\omega)$$

$$= \frac{F_0^2}{2m\gamma}\left[\frac{\gamma^2\omega^2}{(\omega_0^2 - \omega^2)^2 + \gamma^2\omega^2}\right]$$

This is the average power absorbed from the driving force when the system is driven with an angular frequency ω; it is plotted against the frequency in fig. 5.5 for the same system as in the previous diagrams.

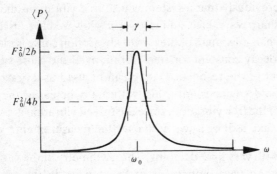

Fig. 5.5 The average power absorption as a function of driving frequency, for a system with $Q = 5$. The maximum is at the resonance frequency, and the width of the curve at half maximum height is γ.

The maximum value of $\langle P \rangle$ is $F_0^2/2b$, reached at the resonance frequency exactly. The fact that it is inversely proportional to the resistance is of some importance. The absorbed power produces heat in the system, and this may present a problem. It is worth noting that any such problem will become increasingly serious as the damping is made *lighter*.

The shape of the power absorption curve (fig. 5.5) is essentially the shape of $R(\omega)$. As ω decreases from the resonance value, $R(\omega)$ falls from its maximum value of 1. At some angular frequency ω_1 we shall have

$$R(\omega_1) = \tfrac{1}{2}$$
$$\gamma^2 \omega_1^2 = (\omega_0^2 - \omega_1^2)^2 \tag{5.12}$$
$$\gamma\omega_1 = \omega_0^2 - \omega_1^2$$

In the last equation we have taken the positive square root on the right because $\omega_1 < \omega_0$.

As ω increases from resonance, $R(\omega)$ again falls. There will be another angular frequency ω_2 at which

$$R(\omega_2) = \tfrac{1}{2}$$
$$\gamma^2 \omega_2^2 = (\omega_0^2 - \omega_2^2)^2 \tag{5.13}$$
$$\gamma\omega_2 = -(\omega_0^2 - \omega_2^2)$$

This time we have taken the negative square root because $\omega_2 > \omega_0$.

We now add the last equations in (5.12) and (5.13), finding

$$\gamma(\omega_1 + \omega_2) = \omega_2^2 - \omega_1^2 = (\omega_2 - \omega_1)(\omega_1 + \omega_2)$$
$$\gamma = \omega_2 - \omega_1$$

Thus γ *is the size of the angular frequency range within which* $\langle P \rangle$ *is greater than half its maximum value.* This is, of course, the origin of the term 'width'.

We saw previously that a system whose free vibrations die out slowly will have a sharp resonance, and *vice versa*. Now we have discovered that the width at half maximum of the power absorption curve is identical with the energy decay constant for free vibrations of the same system. Thus measurement of the resonance curve can be used as a practical way of obtaining the decay constant. This will usually be easier than finding it directly from the free vibration, since the forced vibration is a steady state and we can take as long as we wish over the measurement.

Resonance with very light damping. The asymmetry of the various curves in fig. 5.3 is not very noticeable at frequencies in the region of the resonance itself, and it becomes even less noticeable as γ is decreased. If the damping is very light ($Q \gg 1$) we can use simpler formulas involving a symmetric approximation to $R(\omega)$.

When ω is close to ω_0 and $Q \gg 1$ we have

$$\omega/\omega_0 \approx 1$$

$$\omega_0^2 - \omega^2 = (\omega_0 + \omega)(\omega_0 - \omega) \approx 2\omega_0(\omega_0 - \omega)$$

(5.14)

These approximations allow us to write

$$R(\omega) \approx \frac{\frac{1}{4}\gamma^2}{(\omega_0 - \omega)^2 + \frac{1}{4}\gamma^2} \equiv L(\omega)$$

The new response function $L(\omega)$, known as a Lorentzian, is symmetric about $\omega = \omega_0$. Its width is the same as that of $R(\omega)$, since

$$L(\omega_0 \pm \tfrac{1}{2}\gamma) = \tfrac{1}{2}$$

(5.15)

Replacing $R(\omega)$ by $L(\omega)$, and substituting ω_0 for ω whenever it occurs elsewhere, we obtain for the three amplitudes (5.8) the convenient expressions

$$A \approx \frac{QF_0}{s}\left[\frac{\frac{1}{4}\gamma^2}{(\omega_0 - \omega)^2 + \frac{1}{4}\gamma^2}\right]^{1/2}$$

$$\omega A \approx \frac{F_0}{b}\left[\frac{\frac{1}{4}\gamma^2}{(\omega_0 - \omega)^2 + \frac{1}{4}\gamma^2}\right]^{1/2}$$

(5.16)

$$\omega^2 A \approx \frac{QF_0}{m}\left[\frac{\frac{1}{4}\gamma^2}{(\omega_0 - \omega)^2 + \frac{1}{4}\gamma^2}\right]^{1/2}$$

These approximations can be used if the damping is very light and if we are concerned only with events in the resonance region. A corresponding expression involving the phase constant,

$$\tan\phi \approx -\tfrac{1}{2}\gamma/(\omega_0 - \omega)$$

(5.17)

is obtained by applying the approximations of (5.14) to (5.5).

A different set of approximations is available for use when $Q \gg 1$ and the interesting frequencies are far from resonance ($\omega \ll \omega_0$ or $\omega \gg \omega_0$). In that case

$$R(\omega) \approx \left(\frac{\gamma\omega}{\omega_0^2 - \omega^2}\right)^2$$

(5.18)

The most interesting consequence of using this form of $R(\omega)$ in (5.8) is that none of the amplitudes then depends on γ, whose precise value is therefore unimportant away from resonance.

We can easily verify that (5.18) leads to the correct limiting forms F_0/s and $F_0/m\omega^2$ for A at low and high frequencies respectively.

Heavy damping. We know that free vibrations cannot occur when the damping is heavy ($\gamma > 2\omega_0$). The system, when disturbed, merely returns asymptotically to $\psi = 0$. Forced vibrations are possible, however, and it is worth thinking about how a heavily damped system will respond to a harmonic driving force. The behaviour is most easily illustrated by considering the case of very heavy damping ($\gamma \gg \omega_0$).

We expect appreciable movement of the mass to take place only at the lowest frequencies. If we assume that $\omega \ll \omega_0$ the exact expression (5.7) for $R(\omega)$ may be approximated by

$$R(\omega) \approx \frac{\gamma^2\omega^2/\omega_0^4}{1+\gamma^2\omega^2/\omega_0^4} = \frac{\tau_r^2\omega^2}{1+\tau_r^2\omega^2}$$

Here we have reintroduced the relaxation time τ_r obtained in (3.18). As we saw previously, the relaxation time is the characteristic property of a very heavily damped system.

Now $R(\omega)$ has nothing like a resonance shape (fig. 5.6). All the interest is in the frequency region near

$$\omega = 1/\tau_r = Q\omega_0 \ll \omega_0$$

There the absorbed power, for example, will rise from a very small value to its maximum possible value of $F_0^2/2b$. Clearly the velocity must be swinging into phase (and therefore the displacement must be swinging into quadrature) with the driving force at these frequencies. We confirm that this is so by noting that (5.5) may now be written

$$\tan\phi \approx -\tau_r\omega$$

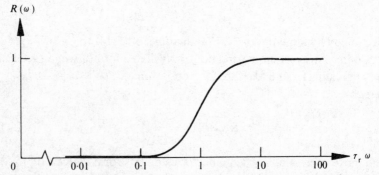

Fig. 5.6 The response function $R(\omega)$ for a very heavily damped system with relaxation time τ_r, at low driving frequencies. Note that the horizontal scale is logarithmic.

under our very heavy damping and low frequency assumptions ($\omega \ll \omega_0 \ll \gamma$). Whereas, with light damping, ϕ passed from 0 to $-\pi$ *through* the power absorbing value of $-\frac{1}{2}\pi$ at $\omega = \omega_0$, now it remains near $-\frac{1}{2}\pi$ for all frequencies $\omega \gtrsim 1/\tau_r$.

Although power is absorbed efficiently, the amplitude (5.6) is small at these higher frequencies, since ω appears in the denominator of the factor multiplying $R(\omega)$. The very heavy damping, low frequency approximation

$$A \approx \frac{F_0}{s}\left(\frac{1}{1+\tau_r^2\omega^2}\right)^{1/2}$$

is plotted in fig. 5.7. At the lowest frequencies the amplitude is, as always, limited to the value F_0/s fixed by the spring. But this is now the largest value the amplitude ever has; as soon as the period of the driving force becomes comparable with the relaxation time, A falls rapidly.

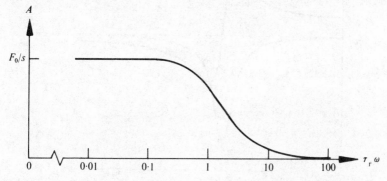

Fig. 5.7 The amplitude response curve for a very heavily damped system with relaxation time τ_r, at low driving frequencies. Note that the horizontal scale is logarithmic.

Complex response functions. Forced-vibration calculations can often be simplified by the use of certain complex functions describing the steady-state response of the system. These functions all involve in some way the complex amplitude of the forced vibration which, in form D, is

$$\psi = \text{Re}\,(D\,e^{i\omega t})$$

Of the various functions that have been invented we consider only two: the compliance and the impedance.

The *compliance* $K(\omega)$ is simply the complex amplitude D divided by the (real) driving force amplitude F_0. The real and imaginary parts of D are given by the rules (1.20) connecting form D with form A. With the help of fig. 5.2 we find

$$\cos \phi = m(\omega_0^2 - \omega^2)A/F_0$$
$$\sin \phi = m\gamma\omega A/F_0$$

and so

$$\operatorname{Re} K(\omega) = \frac{A \cos \phi}{F_0} = \frac{\omega_0^2 - \omega^2}{m[(\omega_0^2 - \omega^2)^2 + \gamma^2\omega^2]}$$

$$-\operatorname{Im} K(\omega) = + \frac{A \sin \phi}{F_0} = \frac{\gamma\omega}{m[(\omega_0^2 - \omega^2)^2 + \gamma^2\omega^2]} \tag{5.19}$$

The shapes of the $\operatorname{Re} K(\omega)$ and $\operatorname{Im} K(\omega)$ curves are shown in fig. 5.8 for the same system as in previous examples.

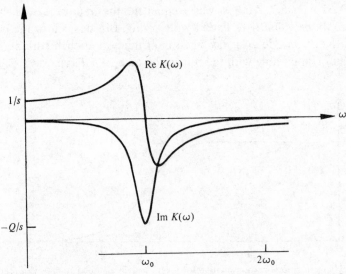

Fig. 5.8 The real and imaginary parts of the compliance $K(\omega)$, for a system with $Q = 5$.

The expressions in (5.19) are rather clumsy. The complete function $K(\omega)$ can be written more neatly as

$$K(\omega) = \frac{1}{(s - m\omega^2) + ib\omega} \tag{5.20}$$

where we have made the denominator complex. The usefulness of the compliance lies in the relative simplicity of this function. Rationalization into real and imaginary parts can usually be delayed until the end of the calculation, and sometimes it is not necessary to rationalize at all. (See for example problem 5.13.)

Approximate forms of $K(\omega)$ are available for calculations in very lightly damped systems. In the resonance region we may use

$$K(\omega) \approx -i/\gamma\omega_0$$

Here the factor $-i$ reminds us that the displacement lags the driving force by $\frac{1}{2}\pi$.

Off resonance, we have

$$K(\omega) \approx 1/(s - m\omega^2)$$

In this approximation the compliance is real, reflecting the fact that the phase lag is either zero (at low frequencies) or π (at high frequencies).

The *impedance* $Z(\omega)$ is the driving force amplitude F_0 divided by the complex amplitude for the velocity.[†] Since the latter is $i\omega D$ we have

$$Z(\omega) = 1/i\omega K(\omega)$$
$$= b + i(m\omega - s/\omega) \tag{5.21}$$

For frequencies in the resonance region with $Q \gg 1$ the impedance is almost the same as the resistance b, and $Z(\omega) = b$ exactly when $\omega = \omega_0$. A real impedance is a reflection of the fact that the velocity is in phase with the driving force; finding the frequency that makes $Z(\omega)$ real may sometimes be the easiest way of obtaining ω_0 for a complicated system whose impedance can be calculated.

Summary. This section has introduced some of the most important subject matter in the book. We summarize here the essential physical properties of steady-state forced vibrations.

In the steady state the motion is harmonic, and has the same frequency as the driving force.

When $\omega \ll \omega_0$ the displacement is almost in phase with the driving force, and the motion is stiffness controlled. When $\omega \gg \omega_0$ the displacement is almost in antiphase with the driving force, and the motion is mass controlled. When $\omega = \omega_0$ the velocity is exactly in phase with the driving force, and the motion is resistance limited.

When the damping is light, all amplitudes show resonance near $\omega = \omega_0$. The velocity resonance occurs at $\omega = \omega_0$ exactly. The most convenient formulas to use in the resonance region are those in (5.16), valid when the damping is very light.

Power absorption is significant only when the velocity is nearly in phase with the driving force. This is the case in the resonance region for a lightly damped system, and the width of the power absorption resonance at half maximum is exactly γ. In a very heavily damped system the condition is fulfilled at angular frequencies higher than the inverse relaxation time.

---

† Not the other way up; $1/Z(\omega)$ is known as the admittance.

5.2. Superposition

We continue the study of steady-state forced vibrations by thinking about what happens when two harmonic driving forces act simultaneously on the same system. Obviously we have to analyze an extended version of (5.3) with two terms on the right.

First we assume that the driving forces have the same angular frequency ω, and write

$$\ddot{\psi} + \gamma\dot{\psi} + \omega_0^2\psi = (F_1/m) \cos (\omega t + \alpha_1) + (F_2/m) \cos (\omega t + \alpha_2) \qquad (5.22)$$

The forces are not, in general, in phase with each other. It turns out that their phase difference $|\alpha_2 - \alpha_1|$ affects the behaviour crucially.

The principle of superposition. If F_2 were zero we could immediately write down the steady-state solution

$$\psi_1 = A_1 \cos (\omega t + \alpha_1 + \phi) \qquad (5.23)$$

where A_1 comes from (5.6) with F_0 replaced by F_1, and ϕ comes from (5.5). Similarly if F_1 were zero we would have

$$\psi_2 = A_2 \cos (\omega t + \alpha_2 + \phi) \qquad (5.24)$$

We have put the same ϕ in both equations because the two driving forces have the same frequency.

Now it is a simple matter to verify that $\psi = \psi_1 + \psi_2$ is the steady-state solution when both of the driving force amplitudes F_1 and F_2 are finite. This is true because (5.22) is a linear equation: the variable ψ and its derivatives appear only in first power and in separate terms.

This is our first encounter with a very important property of 'linear systems', the *principle of superposition. If one driving force alone produces the displacement $\psi_1(t)$, and a second driving force alone produces the displacement $\psi_2(t)$, then both forces acting together will produce the displacement $\psi_1(t) + \psi_2(t)$.* Clearly the argument can be extended to any number of driving forces.

The resultant amplitude. Because ψ_1 and ψ_2 have the same frequency, ψ is harmonic. Its amplitude can be found with the help of the vector diagram shown in fig. 5.9. The vector representing ψ_1 is added to the vector representing ψ_2, so producing the vector representing ψ which has the amplitude A. As usual the diagram shows the state of affairs at $t = 0$. It

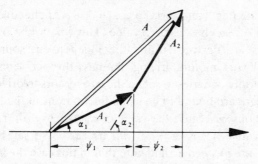

Fig. 5.9 A vector diagram for calculating the resultant amplitude A when two driving forces of the same frequency act simultaneously on the system.

rotates anticlockwise with angular velocity ω and at all times the projection of the resultant on any axis is equal to the sum of the projections of the vectors representing ψ_1 and ψ_2. The value of A is given by the standard trigonometric formula

$$A^2 = A_1^2 + A_2^2 + 2A_1A_2 \cos(\alpha_2 - \alpha_1) \qquad (5.25)$$

We shall not go so far as to substitute for A_1 and A_2 in terms of F_1, F_2 and ω. The resultant amplitude A will go up and down with frequency in exactly the same way as A_1 and A_2. If, for example, the system is lightly damped, there will be a resonance near $\omega = \omega_0$. What we are more interested in here is the *size* of A, relative to A_1 and A_2.

The influence of the phase difference $|\alpha_2 - \alpha_1|$ is clear from (5.25). Depending on the value of $\cos(\alpha_2 - \alpha_1)$, which can be anything from -1 to $+1$, A can have any value between $|A_1 - A_2|$ and $(A_1 + A_2)$. If A comes out smaller than the larger of A_1 and A_2 we say the superposition is *destructive*; if A comes out larger than both we say the superposition is *constructive*. Driving forces in antiphase $(\alpha_2 - \alpha_1 = \pm\pi)$ provide *full destructive superposition* and the minimum possible amplitude; driving forces in phase $(\alpha_1 = \alpha_2)$ provide *full constructive superposition* and the maximum possible amplitude.

The most striking effects arise when $F_1 = F_2$ and hence $A_1 = A_2$. Then we have

$$A^2 = 2A_1^2[1 + \cos(\alpha_2 - \alpha_1)]$$

Since the energy stored in the vibrator is proportional to the square of the amplitude, we see that it is quadrupled when a single driving force is replaced by two similar forces acting in phase with each other. On the other hand, there is no motion at all, and no stored energy, when equal forces act in antiphase with each other.

Coherence. The important term in $\cos(\alpha_2 - \alpha_1)$ is effective over a long period only if $\alpha_2 - \alpha_1$ has a stable value. This will not be true if the two driving forces are derived from quite independent sources, such as separate clockwork motors driving the mass through separate springs. However carefully one tries to set up the two motors to drive at the same frequency, there are bound to be small fluctuations in their speeds, and these fluctuations will cause the value of $\alpha_2 - \alpha_1$ to drift. Consequently the average value of $\cos(\alpha_2 - \alpha_1)$ will be zero. We say then that the driving forces are *incoherent*, and note that in this case the average value of A^2 is simply

$$\langle A^2 \rangle = A_1^2 + A_2^2$$

When two incoherent driving forces are active the energies stored in the vibrator by the separate motions are, on average, simply added together. The result may be either smaller or larger than that for coherent driving forces (5.25).

The obvious way to stabilize $\alpha_2 - \alpha_1$, and so to make the driving forces *coherent*, is to derive both forces from the same ultimate source. For example, we could have a single clockwork motor connected to the system through separate gear trains. Alternatively, we could use two synchronous electric motors driven from the same AC supply. The important thing is that the driving frequencies are locked to a common reference frequency. Only under these conditions can we observe superposition effects.

Driving forces of different frequency. When the driving forces have different angular frequencies ω_1 and ω_2, equation (5.22) takes the modified form

$$\ddot{\psi} + \gamma\dot{\psi} + \omega_0^2\psi = (F_1/m)\cos(\omega_1 t + \alpha_1) + (F_2/m)\cos(\omega_2 t + \alpha_2)$$

To be definite we shall assume that $\omega_2 > \omega_1$.

The principle of superposition makes it easy to write down the steady-state solution. If F_2 were zero the motion would be

$$\psi_1 = A_1 \cos(\omega_1 t + \alpha_1 + \phi_1)$$

with A_1 again coming from (5.6) and ϕ_1 from (5.5). Similarly, if F_1 were zero the motion would be

$$\psi_2 = A_2 \cos(\omega_2 t + \alpha_2 + \phi_2)$$

This time ϕ_1 and ϕ_2 are unequal.

When both forces are active we superpose these two motions and obtain

$$\psi = A_1 \cos(\omega_1 t + \alpha_1 + \phi_1) + A_2 \cos(\omega_2 t + \alpha_2 + \phi_2) \qquad (5.26)$$

Beats. What kind of motion does (5.26) describe? It will certainly not be harmonic in general; it may not even be periodic.

We can investigate the question of periodicity by looking for a time $t + \tau$ at which ψ_1, ψ_2, $\dot{\psi}_1$ and $\dot{\psi}_2$ are all exactly the same as they were at time t. This requires that ψ_1 and ψ_2 should each execute an integral number of cycles in the time interval τ. This in turn requires that

$$\omega_1 = n_1(2\pi/\tau)$$
$$\omega_2 = n_2(2\pi/\tau)$$

where n_1 and n_2 are integers. The motion will therefore be periodic if we can write

$$\omega_1/\omega_2 = n_1/n_2$$

The number of cycles of ψ_1 or ψ_2 that must elapse before the motion begins to repeat itself will be large if the integers n_1 and n_2 are large. During a period of such motion, there will be two occasions, half a period apart, when ψ_1 and ψ_2 are almost in step with each other and produce constructive superposition; at one of these times ψ_1 and ψ_2 will both be positive, and at the other they will both be negative. There will be two other occasions, intermediate between the first pair, when ψ_1 and ψ_2 are almost out of step and produce destructive superposition.

The most spectacular effects occur when ψ_1 and ψ_2 have equal amplitudes. The destructive superposition is then complete, and the motion will cease altogether twice in each period. To examine this case in more detail we put $A_2 = A_1$ in (5.26), which gives

$$\psi = A_1[\cos(\omega_1 t + \alpha_1 + \phi_1) + \cos(\omega_2 t + \alpha_2 + \phi_2)] \qquad (5.27)$$

By defining

$$\Omega_+ \equiv \tfrac{1}{2}(\omega_1 + \omega_2)$$
$$\Omega_- \equiv \tfrac{1}{2}(\omega_2 - \omega_1)$$
$$\Phi_+ \equiv \tfrac{1}{2}(\alpha_1 + \alpha_2 + \phi_1 + \phi_2) \qquad (5.28)$$
$$\Phi_- \equiv \tfrac{1}{2}(\alpha_2 - \alpha_1 + \phi_2 - \phi_1)$$

and using the identities

$$\cos(x \pm y) = \cos x \cos y \mp \sin x \sin y$$

we can rewrite (5.27) in the form

$$\psi = A_1[\cos(\Omega_+ t + \Phi_1 - \Omega_- t - \Phi_-) + \cos(\Omega_+ t + \Phi_+ + \Omega_- t + \Phi_-)]$$

$$= 2A_1 \cos(\Omega_- t + \Phi_-) \cos(\Omega_+ t + \Phi_+)$$

If $\omega_1 \approx \omega_2$ (so that $\Omega_+ \gg \Omega_-$) the motion appears to be *nearly harmonic* at the angular frequency Ω_+, but with an amplitude which itself varies harmonically between the values $2A_1$ and zero, at the much slower angular frequency Ω_-. Figure 5.10 shows an example with $\omega_2/\omega_1 = 11/9$.

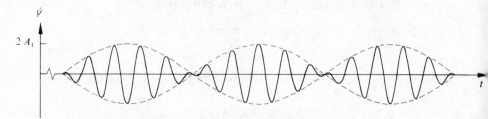

Fig. 5.10 Beats: the driving forces, acting separately, would produce equal amplitudes A_1. In this example $\omega_2/\omega_1 = 11/9$, giving $\Omega_+ = 10\Omega_-$, $\Phi_+ = \frac{1}{4}\pi$ and $\Phi_- = 0$.

The solid line represents the value of ψ, and the two broken lines show the constraints on the maximum excursions of ψ imposed by the *modulating factor* $2A_1 \cos(\Omega_- t + \Phi_-)$. Peaks in activity, known as *beats*, occur at the frequency Ω_-/π, that is, at the *difference frequency* $(\omega_2 - \omega_1)/2\pi$. The phenomenon provides a valuable means of measuring, in a sensitive manner, the difference between two frequencies.

Summary. The principle of superposition allows us to obtain the steady-state motion when two driving forces act simultaneously on a linear system, by adding together the separate motions produced by the forces acting separately.

When the two driving forces have the same frequency and are coherent, their superposition may be destructive or constructive, depending on their phase difference. If they are incoherent their phase difference has, on average, no effect; in that case the average stored energies of the separate motions are additive.

Driving forces whose frequencies are not the same produce a more complicated motion. When the frequencies are only slightly different, beats occur: the motion is nearly harmonic, but the amplitude waxes and wanes at the difference frequency.

5.3. Transients

In the first section we explored the steady-state behaviour of the system when a single harmonically varying force was driving it. We end the chapter by considering very briefly the condition of a lightly damped vibrator *before* the steady state is established. We imagine the system to be at rest with no displacement until the driving force is switched on, at time $t = 0$ say. We wish to know the nature of the *transient* motion immediately subsequent to $t = 0$.

Mathematically, the problem is one of finding the general solution of (5.3). It is clear that the steady-state solution

$$\psi_s = A_s \cos{(\omega t + \phi_s)}$$

is not the general solution: it contains no arbitrary constants, whereas the general solution should have two.

We found in chapter 3 the general solution of the *homogeneous* equation (3.3). One form of that solution (3.8) is

$$\psi_f = A_f \exp{(-\tfrac{1}{2}\gamma t)} \cos{(\omega_f t + \phi_f)}$$

where ω_f is the angular frequency of free vibrations, and both A_f and ϕ_f are arbitrary. We can now use the theorem which states that, because the inhomogeneous equation (5.3) is linear, its general solution is the superposition of ψ_s (a particular solution) and ψ_f (the general solution of the homogeneous equation). The fact that $\psi_s + \psi_f$ *satisfies* (5.3) is easily verified; it must be the *general* solution because it has two arbitrary constants (A_f and ϕ_f).

At $t = 0$ we have

$$\psi(0) = A_s \cos{\phi_s} + A_f \cos{\phi_f} = 0$$

$$\dot{\psi}(0) = -\omega A_s \sin{\phi_s} - A_f(\tfrac{1}{2}\gamma \cos{\phi_f} + \omega_f \sin{\phi_f}) = 0$$

We shall investigate the particular case in which the damping is very light and the driving frequency is close to the resonance frequency. Then we have

$$\omega \approx \omega_f$$

$$\gamma \ll \omega_f$$

and the initial conditions become

$$A_s \cos{\phi_s} + A_f \cos{\phi_f} = 0$$

$$A_s \sin{\phi_s} + A_f \sin{\phi_f} \approx 0$$

They imply that

$$A_f \approx A_s$$

$$\phi_f \approx \phi_s - \pi$$

and lead to

$$\psi(t) \approx A_s[\cos(\omega t + \phi_s) - \exp(-\tfrac{1}{2}\gamma t)\cos(\omega_f t + \phi_s)] \qquad (5.29)$$

At $t = 0$ the motion is a superposition of two harmonic terms with nearly equal frequencies ($\omega \approx \omega_f$). Thus the initial motion consists of beats between the free and forced vibrations, with a tendency of ψ_f and ψ_s to cancel whenever they are out of step with each other. This explains the apparently chaotic behaviour of a lightly damped vibrator just after a harmonic driving force is switched on.

If there were no damping these beats would continue for ever, and it would not be possible to establish a steady state. Damping causes the

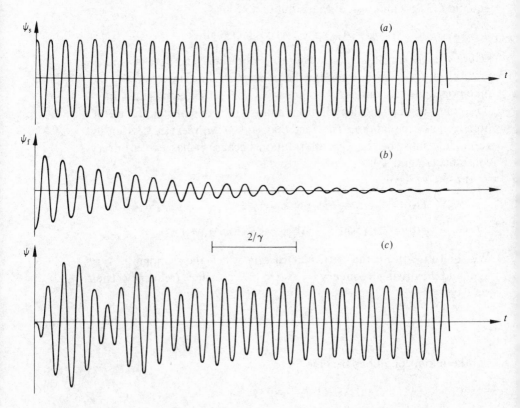

Fig. 5.11 Transient response of a system with $Q = 20$. The system is driven at an angular frequency $\omega_f + 5\gamma$. (a) Steady-state term. (b) Transient term. (c) Complete motion.

second term in (5.29) to diminish with the passage of time, so making the beats less violent. Thus the free vibration ψ_f becomes small when $\gamma t \gg 1$, and the steady-state solution becomes the dominant one after a time which is long compared with $1/\gamma$. (We made this assumption at the beginning of section 5.1.)

The transient response with $\omega = \omega_f + 5\gamma$ is shown in fig. 5.11. Whenever we have

$$(\omega - \omega_f)t \approx \pi, 3\pi, 5\pi, \ldots$$

the components ψ_s and ψ_f reinforce each other, causing the final amplitude A_s to be momentarily exceeded. This is another point that must be watched when engineering a system of high Q. (See problem 5.18.)

If the driving frequency is exactly matched to the free vibration frequency ($\omega = \omega_f$) the vibration is

$$\psi(t) = A_s[1 - \exp(-\tfrac{1}{2}\gamma t)] \cos(\omega t + \phi_s)$$

and there are now no beats: the amplitude simply builds up exponentially and safely to its final value, as shown in fig. 5.12. A slight frequency discrepancy can be tolerated; all that is necessary for a smooth build-up is that the damping time should be short compared with the beat period.

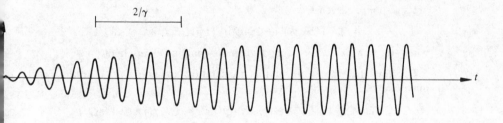

Fig. 5.12 Transient response of a system with $Q = 20$, driven at its free vibration frequency.

Summary. The transient motion of a lightly damped system acted upon by a single harmonic driving force is a superposition of:

(1) the free vibration, which is damped harmonic and disappears when $\gamma t \gg 1$;

(2) the steady-state forced vibration, which is harmonic at the driving frequency.

The most characteristic feature of a transient with $\omega \approx \omega_f$ is the beating between these two components.

Problems

5.1   For a very lightly damped system, make sketches like fig. 5.2 for (a) $\omega \ll \omega_0$, (b) $\omega = \omega_0 - \frac{1}{2}\gamma$, (c) $\omega = \omega_0$, (d) $\omega = \omega_0 + \frac{1}{2}\gamma$, and (e) $\omega \gg \omega_0$. Use these sketches to verify directly the values of A and ϕ derived elsewhere.

5.2   A system with $m = 0.010$ kg, $s = 36$ N m^{-1} and $b = 0.50$ kg s^{-1} is driven by a harmonically varying force of amplitude 3.6 N. Find the amplitude A and the phase constant ϕ of the steady-state motion when the angular frequency is (a) 8.0 s^{-1}, (b) 80 s^{-1}, and (c) 800 s^{-1}.

5.3   Show that the displacement amplitude A is a maximum at the driving frequency given by

$$\omega^2 = \omega_0^2 - \tfrac{1}{2}\gamma^2$$

5.4   Show that the acceleration amplitude $\omega^2 A$ is a maximum at the driving frequency given by

$$\omega^2 = \omega_0^2/(1 - \gamma^2/2\omega_0^2) \approx \omega_0^2 + \tfrac{1}{2}\gamma^2$$

where the approximation is good when the damping is very light.

5.5   A certain system has $\nu_0 = 50$ Hz exactly and $Q = 10$ exactly. Compare the numerical values of (a) the resonance angular frequency ω_0, (b) the angular frequency of free vibrations ω_f, (c) the value of ω at which the displacement amplitude A is a maximum, and (d) the value of ω at which the acceleration amplitude $\omega^2 A$ is a maximum.

5.6   Show that the value of the displacement amplitude A at the exact maximum of its response curve is

$$A_{max} = (QF_0/s)(1 - 1/4Q^2)^{-1/2} \approx (QF_0/s)(1 + 1/8Q^2)$$

where the approximation is good when the damping is very light.

5.7   Show that the value of the acceleration amplitude $\omega^2 A$ at the exact maximum of its response curve is

$$(\omega^2 A)_{max} = (QF_0/m)(1 - 1/4Q^2)^{-1/2} \approx (QF_0/m)(1 + 1/8Q^2)$$

where the approximation is good when the damping is very light.

5.8   Figure 5.13 shows the mean power absorption $\langle P \rangle$ in watts, as a function of driving frequency in the resonance region. Find the numerical values of (a) ω_0,

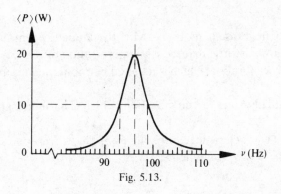

Fig. 5.13.

(b) γ, and (c) Q. (d) If the driving force is removed, after how many subsequent cycles will the energy of the system be $1/e$ of its initial value?

5.9   If the system in problem 5.8 has $m = 0.010\,\text{kg}$, calculate the force amplitude F_0.

5.10   In form B the steady-state forced vibration (5.5) is written

$$\psi = B_\text{p} \cos \omega t + B_\text{q} \sin \omega t$$

Show that $\langle P \rangle = \frac{1}{2}\omega F_0 B_\text{q}$.

5.11   Consider the average energy $\langle W \rangle$ stored in a system undergoing steady-state forced vibration. Show (a) that $\langle W \rangle$ is mostly potential energy when $\omega \ll \omega_0$, (b) that potential energy and kinetic energy are equal when $\omega = \omega_0$, and (c) that $\langle W \rangle$ is mostly kinetic energy when $\omega \gg \omega_0$. (d) Show also that $\langle P \rangle = \gamma \langle W \rangle$ at resonance, where $\langle P \rangle$ is the mean power supplied by the driving force.

5.12   A very heavily damped system has a relaxation time of $0.10\,\text{s}$. In steady-state forced vibration at a frequency $0.10\,\text{Hz}$ it has an amplitude of $20\,\text{mm}$. (a) If the driving force frequency is increased to $7.0\,\text{Hz}$ but the maximum force is unchanged, what is the new steady-state amplitude? (b) What is the new phase constant?

5.13   By finding the value of ω which makes the compliance $|K(\omega)|$ a maximum, confirm the result of problem 5.3.

5.14   By finding the value of ω which makes the impedance $|Z(\omega)|$ a minimum, confirm that the velocity amplitude is a maximum when $\omega = \omega_0$.

5.15   For the motion

$$\psi = (7.5\,\text{mm}) \cos \left[(6.28\,\text{s}^{-1})t + 27° \right] - (2.3\,\text{mm}) \sin \left[(6.20\,\text{s}^{-1})t - 121° \right]$$

find (a) the frequency, and (b) the time interval separating successive beats.

5.16   A single driving force produces steady-state forced vibrations with amplitude $10\,\text{mm}$. A second driving force of the same frequency as the first, acting on the same system, produces a steady-state amplitude of $20\,\text{mm}$. When both forces are acting simultaneously, the steady-state amplitude is $15\,\text{mm}$. What is the phase difference between the two forces?

5.17   A simple seismometer consists of a mass hung on a spring attached to a rigid framework, which is fixed to the ground, with critical damping. The vertical displacement of the mass relative to the framework is recorded.
  (a) Show that the measured amplitude A of the steady-state vibration resulting from a vertical displacement $H \cos \omega t$ of the earth's surface is given by

$$A/H = (\omega/2\omega_0)[R(\omega)]^{1/2}$$

  (b) Show that, if the frequencies of the detected disturbances lie in the region of mass controlled motion, the mass remains almost stationary when the ground moves.

5.18   A system initially at rest is set into vibration by a harmonic driving force whose frequency is 1 per cent greater than the free vibration frequency of the system. Estimate the Q-value the system must have if its amplitude during the build-up is not to exceed the steady-state amplitude by more than 10 per cent.

6

Forced vibrations in physics

In this chapter we see how our careful analysis of steady-state forced vibrations can help us to understand phenomena as diverse as the response of a resonant circuit, the scattering of light, and the dielectric properties of gases and liquids.

6.1. Resonant circuits

We begin with the forced oscillations of an LCR circuit. Reinterpretation of previous results is particularly straightforward in this case, and most of what follows will take the form of categorical statements which you can check at each point against the basic theory of section 5.1.

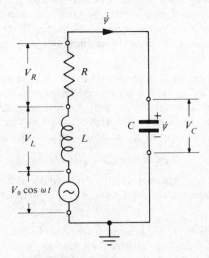

Fig. 6.1 A resonant circuit: the circuit of fig. 4.1 with a voltage generator added.

The system under discussion is the one shown in fig. 6.1. It is simply our previous circuit of fig. 4.1 with an alternating voltage generator connected in series. We take the voltage output of the generator to be $V_0 \cos \omega t$; the frequency is under our control, and we assume that the voltage amplitude V_0 remains the same for all values of the current drawn from the generator.†

Since we are dealing with a series circuit, the sum of the potential differences across the three passive components must always be equal to the generator voltage. In an obvious notation we write

$$V_C(t) + V_R(t) + V_L(t) = V_0 \cos \omega t \qquad (6.1)$$

The three voltages on the left are (table 2.2)

$$V_C(t) = C^{-1}\psi$$
$$V_R(t) = R\dot{\psi} \qquad (6.2)$$
$$V_L(t) = L(\mathrm{d}\dot{\psi}/\mathrm{d}t) = L\ddot{\psi}$$

where ψ is, as usual, the charge on the capacitor and $\dot{\psi}$ is the clockwise current. Thus V_C varies with time like the displacement of the mass, V_R varies like the velocity, and V_L like the acceleration. Recognizing these analogies allows us to make a number of statements about the behaviour of the circuit.

Voltage response curves. In a lightly damped circuit the response curves for the amplitudes of V_C, V_R and V_L are like those for the displacement, velocity and acceleration amplitudes respectively (fig. 5.3). Thus V_C has its maximum amplitude slightly below the resonance frequency; V_R has its maximum amplitude exactly at the resonance frequency; and V_L has its maximum amplitude slightly above the resonance frequency.

Amplification factors. By writing (5.8) in terms of electrical parameters (table 2.2) and putting $\omega = \omega_0$, we can see that the amplitudes of ψ, $\dot{\psi}$ and $\ddot{\psi}$ at resonance are respectively $V_0/\omega_0 R$, V_0/R and $V_0\omega_0/R$. From these we can find the amplitudes of the various voltages (6.2) at the resonance frequency. For both V_C and V_L the result is QV_0; thus we can obtain voltage amplification by a factor Q with the aid of a resonant circuit.

Phase differences. The phase differences involved are merely those between ψ, $\dot{\psi}$ and $\ddot{\psi}$. We can therefore say immediately that V_R leads V_C

† We say that the generator 'has zero internal impedance'.

by 90°, and V_L leads V_R by a further 90°. (These relationships are equivalent to the statement in electricity textbooks that the voltage across L leads the *current* by 90°, and the voltage across C lags the current by 90°, for a series *LCR* circuit.)

Returning to our original voltage equation (6.1), we can now see that V_C and V_L, which are always in antiphase, must cancel each other at resonance where they have the same amplitude QV_0. The whole generator voltage V_0 must then appear across R.

Tuning the circuit. A useful feature of a resonant circuit is the ease with which its resonance frequency can be altered. A common way of doing this is to make the capacitor C variable.

Let us suppose that the *LCR* circuit is part of a radio receiver, and that the driving voltage (represented in fig. 6.1 by the generator) appears because the radio is responding to electromagnetic radiation produced by a number of transmitters within its range. We wish to select the signal from the particular transmitter which broadcasts at the angular frequency ω, and to reject all the others.

We take the voltage $V_C(t)$ as our output, which will be passed to another circuit for decoding and further amplification. The voltage amplification of our circuit is given by (5.6), suitably adapted, as

$$\frac{V_{C0}}{V_0} = \frac{1}{LC}\left[\frac{1}{(\omega_0^2 - \omega^2)^2 + \gamma^2\omega^2}\right]^{1/2} = \omega_0^2\left[\frac{1}{(\omega_0^2 - \omega^2)^2 + \gamma^2\omega^2}\right]^{1/2}$$

where V_{C0} is the amplitude of the output voltage V_C.

As we adjust the knob that controls the value of C, we alter ω_0 but not γ (4.2). The effect on the amplification can be seen by plotting V_{C0}/V_0 against ω_0 while keeping ω and γ fixed. The resulting *tuning curve* (fig.

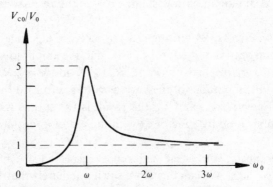

Fig. 6.2 The tuning curve for the circuit in fig. 6.1. The 'generator' frequency is fixed, and the resonance frequency is varied by varying C. In this example $\omega = 5\gamma$.

6.2) is similar in shape to a resonance curve. When the circuit is tuned to a resonance frequency ω_0 equal to the transmitter frequency ω, the amplification is ω/γ. Increasing or decreasing ω_0 rapidly reduces V_{C0}/V_0, which falls to zero at low resonance frequencies, and to 1 at high resonance frequencies. We are thus able to select the transmission of frequency ω in preference to those of other frequencies: to tune to one station at a time.

Clearly the circuit will discriminate most effectively in favour of the selected frequency when ω/γ is high. The ratio ω/γ is just Q for the circuit, calculated at $\omega_0 = \omega$. Here, then, is yet another way of looking at the Q-value: as a measure of the sharpness of tuning, or *selectivity*, of a circuit.

Summary. The resonance curves for V_C, V_R and V_L in a series LCR circuit in the steady state have the respective shapes of those for the displacement, velocity and acceleration of the model system. The phase relationships are also the same. At resonance, V_C and V_L are a factor Q larger than the generator voltage, and V_R is equal to the generator voltage.

A circuit can be tuned, by varying its resonance frequency, to respond selectively to signals of a given frequency.

6.2. Scattering of light

When electromagnetic radiation, such as light, passes through a region of space, it provides at each point an electric field whose value oscillates with time. The space will usually contain some electrons, attached to any atoms or molecules present, and these electrons will be set into forced vibration by the field. The shaken electrons will in turn radiate waves of their own.

The net result of this activity will depend on whether the electrons are driven coherently, or whether they can be considered to act independently of one another. We saw in section 5.2 that only with coherently driven vibrators can we observe superposition effects, in which relative phases play a vital part.

Before we can treat the case of coherently driven electrons, we must know something about electromagnetic waves (chapter 14). We can, on the other hand, make good headway with the simpler, incoherent process, known as *scattering*, on the basis of much less knowledge.

The details of how an electron is bound to an atom or a molecule can only be described correctly by quantum theory. Classical mechanics cannot provide reliable values for ω_0 or γ in the case of a vibrating electron. To understand light scattering, however, it turns out that we only need to know these values very roughly indeed, and therefore crude classical estimates have their uses.

An estimate of ω_0. We first make a reasoned guess at the order of magnitude to be expected for the value of ω_0 for an electron belonging to a hydrogen atom. The values for other atoms, or for molecules, should not be grossly different.

The modern picture of a hydrogen atom shows the electron smeared out around the nucleus (a single proton) in a way which depends on the wave function describing the 'state' it happens to be in. To simplify matters we shall replace this picture by a 'cartoon' atom (fig. 6.3) in which the electron cloud is rigid and spherical, the electron's charge $-e$ being spread uniformly through it.

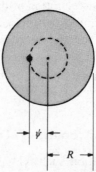

Fig. 6.3 A model hydrogen atom. The electron is smeared out uniformly through a sphere of radius R. In free vibration the centre of the sphere is displaced from the position of the proton. (The displacement is greatly exaggerated in the diagram.)

When the electron cloud is displaced a distance ψ, how big is the return force $|F|$? It is actually easier to calculate $|F|$ as the force exerted by the electron cloud on the proton; that force is proportional to the charge q_ψ inside the sphere of radius ψ drawn round the centre of the displaced electron cloud, and inversely proportional to ψ^2. If R is the radius of the electron cloud, we have

$$q_\psi = (|\psi|/R)^3 e$$

$$|F| = eq_\psi/4\pi\varepsilon_0\psi^2 = (e^2/4\pi\varepsilon_0 R^3)|\psi|$$

from which we can obtain

$$s = e^2/4\pi\varepsilon_0 R^3$$

$$\omega_0 = e/(4\pi m_e\varepsilon_0 R^3)^{1/2}$$

where m_e is the electron mass (9.1×10^{-31} kg).

If for R we use the approximate radius of the hydrogen atom (0.05 nm), we find $\omega_0 \approx 4.5 \times 10^{16} \, \text{s}^{-1}$, or $\nu_0 \sim 10^{16}$ Hz. Electromagnetic radiation having a frequency of this order would lie in the ultraviolet region of the spectrum. The frequencies in visible light are more than 10 times smaller than this, ranging from 4.2×10^{14} Hz (red) to 7.5×10^{14} Hz (blue). We conclude that visible light probably drives the electrons at a frequency well below their resonance frequency.

An estimate of γ. The free vibration of our model atom would be damped, because energy would be continuously radiated away by the vibrating electron. Here we introduce a standard expression for the average power radiated in all directions by an electric dipole whose dipole moment is varying harmonically at the angular frequency ω,

$$\langle P \rangle = p_0^2 \omega^4 / 12 \pi \varepsilon_0 c^3 \tag{6.3}$$

In this formula p_0 is the amplitude of the dipole moment (equal to eA for an electron vibrating with amplitude A), and c is the velocity of light. From our present point of view the most important factor on the right is ω^4.

Equation (6.3) tells us how fast the vibrating electron loses energy; to find γ we require the *fractional* rate of energy loss (3.13). When the amplitude is A, the mean stored energy (3.12) is

$$\langle W \rangle = \tfrac{1}{2} m_e \omega_0^2 A^2$$

Now we can find the width

$$\gamma = \frac{\langle P \rangle}{\langle W \rangle} \approx \frac{\omega_0^2 e^2}{6 \pi \varepsilon_0 m_e c^3} \tag{6.4}$$

We have put $\omega \approx \omega_0$ in (6.3) because at the moment we are considering free vibrations.

Substituting in (6.4) our previous estimate for ω_0, together with the various constants, we obtain for the estimated damping width $\gamma \approx 1.3 \times 10^{10} \, \text{s}^{-1}$ (or $Q \sim 10^6$, indicating very light damping).

Our estimated ω_0 was at least 10 times higher than the driving frequency with visible light; now we have found that that frequency difference is equivalent to something like 10^6 widths. We can clearly afford to be quite badly wrong in our estimates and still be entitled to assume that the damping is light and that *visible light drives the electrons at a frequency well below resonance.*

Rayleigh scattering. Off-resonance scattering is known as Rayleigh scattering. Since $\omega \ll \omega_0$ for visible light driving electrons, we can say that the forced vibration will be stiffness controlled, and that the amplitude must therefore be independent of the driving frequency (5.9). Thus *the amplitude with which the electron is shaken does not depend on the colour of the light.*

The electron's efficiency as a radiator does, however, depend very strongly indeed on the colour: for a given amplitude, the radiated power (6.3) increases as ω^4. The intensity of scattered blue light will thus be greater than that of scattered red light by a factor $(7.5/4.2)^4 \approx 10$.

We assumed at the outset that the scattering electrons act independently, and so we can simply add their contributions to the radiated power without having to worry about phase differences. To achieve these conditions we must have a scattering medium in which the atoms or molecules (and therefore the electrons attached to them) are randomly spaced, with an average spacing which is large in comparison with the wavelength of light. If they are regularly spaced, or if they are so close together that there are several within the space of a single wavelength, the electrons will be driven coherently, with quite different results which we shall investigate later.

The wavelength of visible light is about 500 nm. In the upper part of the atmosphere (100 km or so above sea level) the molecules in the air are sufficiently far apart to produce incoherent scattering, giving rise to the blue appearance of the overhead daytime sky in clear weather. The mean molecular spacing at sea level is only about 3 nm, but incoherent scattering can be produced by more widely spaced particles in suspension. Scattering from dust and mist particles is responsible for the red or orange colour of the western sunset sky, from which we tend to receive only the residual unscattered light.

Summary. Crude classical estimates suggest that an electron attached to an atom or a molecule behaves as a very lightly damped vibrator with a resonance frequency in the ultraviolet region. Forced vibrations driven by visible light are therefore stiffness controlled, with an amplitude which is independent of the driving frequency. The amount of light scattered thus depends only on the efficiency with which a vibrating electron radiates, and this increases as the fourth power of the frequency.

6.3. Dielectric susceptibility

In the previous section we estimated the stiffness with which an electron might be bound to an atom or a molecule. The stiffness can also be measured, and there are ways of doing this which involve large-scale effects produced by an electric field on an assembly of identical atoms or molecules.

In this section we set up the formal connection between the 'microscopic' properties of the atom or molecule, and the 'macroscopic' properties of the material they compose. For simplicity we deal only with a *non-polar gas*: we assume that the molecules are symmetric (possibly monatomic) and thus have no intrinsic dipole moment, and that they are far enough apart not to be affected by the fields produced by their polarized neighbours.

Steady fields. We first remind ourselves of the basic electrical relations which describe what happens when an insulating material is placed in a steady electric field of magnitude E. The field gives rise to a polarization field, whose magnitude $\mathscr{P}$ at any point in the material is the induced dipole moment per unit volume. We define a dimensionless quantity

$$\chi_e \equiv \mathscr{P}/\varepsilon_0 E \tag{6.5}$$

called the *electric susceptibility* of the material. There are several effects which allow χ_e to be measured. The best known is the fact that, when the material completely fills the space between the plates of a capacitor, the capacitance is increased from its vacuum value by a factor $1 + \chi_e$ (the 'dielectric constant').

If there are N molecules per unit volume, and each has one 'polarizable' electron,† bound to it with a stiffness s, the displacement produced by a field E will be eE/s, giving a polarization field

$$\mathscr{P} = Ne^2E/s$$

The susceptibility is then

$$\chi_e = Ne^2/\varepsilon_0 s = (\omega_p/\omega_0)^2 \tag{6.6}$$

To obtain the last expression we have defined a new quantity

$$\omega_p^2 \equiv (e^2/m_e\varepsilon_0)N = (3.18 \times 10^3 \text{ m}^3 \text{ s}^{-2})N \tag{6.7}$$

which is essentially another way of quoting the number density N. There is a superficial similarity between ω_p and the angular frequency of plasma vibrations (2.10). In both cases N is the number of 'active' electrons per unit volume; but in the plasma the electrons are free, whereas here they are still attached to their parent molecules.

---

† Or N/z molecules per unit volume, each with z polarizable electrons; N is essentially the density of polarizable electrons.

Using (6.6) we can deduce the resonance frequency from χ_e. As an example we consider the simplest monatomic gas, helium. The susceptibility of helium, measured at atmospheric pressure, is 6.8×10^{-4}. If we assume that both electrons on a helium atom are polarizable, we have $N = 5.4 \times 10^{25}$ m^{-3} at atmospheric pressure and a temperature of 273 K. Combining (6.6) and (6.7) and putting in these values, we obtain a resonance frequency of 2.5×10^{15} Hz for an electron belonging to a helium atom. Like our estimated value ($\sim 10^{16}$ Hz) for a hydrogen atom, this is safely in the ultraviolet region, and justifies our off-resonance treatment of visible-light scattering in section 4.2.

Alternating fields. The susceptibility in an alternating field is presumably a function of the frequency $\chi_e(\omega)$. We can adapt (6.6) for alternating field purposes merely by replacing $1/s$ by the molecular compliance (5.20) and writing

$$\chi_e(\omega) = m_e \omega_p^2 K(\omega)$$

$$= \frac{\omega_p^2}{(\omega_0^2 - \omega^2) + i\gamma\omega}$$

(6.8)

This makes the susceptibility not only frequency dependent, but also complex. Its real and imaginary parts vary with frequency like those of $K(\omega)$ (fig. 5.8). A complex $\chi_e(\omega)$ is of course an indication that the polarization will lag behind the driving field. As usual, a phase lag between displacement and driving force means power absorption, and the dielectric will tend to heat up.

These effects are appreciable only in the resonance region, however. Clearly we cannot produce driving frequencies anywhere near 10^{15} Hz with conventional electrical apparatus. The measurement of $\chi_e(\omega)$ at these frequencies depends on its effect on electromagnetic waves (section 14.2).

Summary. The stiffness of the binding between an electron and its parent molecule can be found by measuring the electric susceptibility χ_e for a gas of these molecules, in a steady field; χ_e is proportional to the number density of polarizable electrons, and inversely proportional to s. In an alternating field the complex susceptibility $\chi_e(\omega)$ is proportional to the compliance of the electron.

6.4. Absorption of microwaves by water

Water is in many ways one of the most remarkable substances known. One of its most striking properties is its electric susceptibility: Re χ_e has

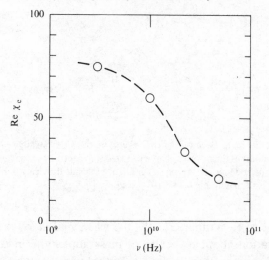

Fig. 6.4 Frequency variation of the real part of the susceptibility for water at 20°C. Note that the frequency scale is logarithmic. (Based on data of C. H. Collie, J. B. Hasted and D. M. Ritson, *Proc. Phys. Soc.* **60** (1948) 145; and T. A. Lane and J. A. Saxton, *Proc. Roy. Soc.* A **213** (1952) 400.)

very high values, around 80, at low frequencies, but falls dramatically as the frequency is raised (fig. 6.4). The decrease occurs entirely in the microwave frequency range ($\sim 10^9$ to 10^{11} Hz) and is accompanied by an equally marked rise in the efficiency with which the water absorbs power from the alternating field. At frequencies around 10^{11} Hz water and water-containing materials (such as food) become 'black', a characteristic which is exploited in microwave cookery.

We shall see that both phenomena can be understood if we think of a water molecule in an alternating field as undergoing very heavily damped forced angular vibrations. We know that a very heavily damped system is completely characterized by its relaxation time τ_r. The amplitude of forced vibration is small, and power absorption is efficient, when we drive the system at a frequency for which $\tau_r \omega \gtrsim 1$. (See figs. 5.6 and 5.7.)

We have already met, in HCl, an example of a *polar* molecule, having an intrinsic dipole moment due to its lack of symmetry. The water molecule H_2O is another example, since the two hydrogen atoms do not lie at 180° to each other, but at 105° (fig. 6.5). In a material composed of polar molecules there are three effects contributing to the polarization in an alternating field. The first of these is the varying dipole moment induced in the molecule when the *electrons* vibrate relative to the nuclei: the effect discussed in the preceding sections. The second is the fluctuating moment associated with molecular vibrations, in which the *atoms*

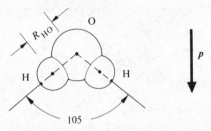

Fig. 6.5 A water molecule. Because H_2O has a bent structure, the negative charge centroid is not at the same point as the positive charge centroid (the centre of gravity of the molecule) but is slightly displaced towards the oxygen end. The molecule thus has a permanent electric dipole moment p, of magnitude 6.14×10^{-30} C m.

move relative to each other. Finally, the *whole molecule* takes part in an angular vibration about its centre of mass, under the influence of the fluctuating torque exerted by the electric field on the dipole moment.

We have seen previously that molecular vibrations are characterized by frequencies in the infrared region, and resonances due to electronic vibrations are expected to lie even higher. Neither of these processes can be responsible for power absorption in the much lower microwave region, since we are very far off resonance in both cases. For present purposes we therefore restrict our attention to rotations.

The plan of this section is as follows:

(1) We estimate the apparent torsional stiffness c which opposes the tendency of an electric field to line up a water molecule in a certain direction; this will allow us to estimate ω_0.

(2) We estimate the angular resistance d due to drag forces opposing rotation (the rotational analogue of the linear resistance b); this will allow us to estimate γ, and it will appear that the damping is very heavy ($\gamma \gg \omega_0$).

(3) We combine our estimates of c and d to obtain the relaxation time τ_r; we shall find that $1/2\pi\tau_r \sim 10^{10}$ Hz.

Thermal stiffness. In the absence of any electric field, the polar water molecules point in completely random directions. When a steady field is applied, there is a tendency for the dipoles to line up so that they all point in the same direction as the field. This effect is only partial, because thermal agitation of the molecules tends to restore the randomness. At ordinary temperatures and field strengths, the equilibrium situation is still a mixture of dipoles pointing in all directions, but now those angles which give rise to a component of dipole moment in the field direction are slightly favoured.

If all the molecules have dipole moment p, a molecule whose moment points at some angle θ with respect to the field E will contribute an amount

$$p_E = p \cos \theta$$

to the total dipole moment of the assembly. A standard result in electricity textbooks states that the average value of p_E at absolute temperature T is

$$\langle p_E \rangle = p^2 E / 3kT \tag{6.9}$$

For a dipole moment p of 6.1×10^{-30} C m (the measured value for H_2O) at room temperature in a field of 10^5 V m^{-1}, this is only a fraction

$$pE/3kT \sim 10^{-4}$$

of the maximum possible average moment of p.

The same thermal effect influences the forced angular vibrations of the molecules in an alternating field. In order to bring our existing knowledge of forced vibrations to bear on this problem, we invent a model in which the water molecules do not interact with each other, the randomizing effect of the collisions being simulated by attaching each molecule to an imaginary wire suspension of torsional stiffness c. We have to decide what value of c will give the same results as the real thermal effect.

We consider a suspended molecule initially making an angle θ with the direction of the field. Before the field is switched on the values of θ are completely random. When the field is switched on the molecule will rotate to a smaller angle $\theta - \Delta\theta$ where

$$\Delta\theta = (pE \sin \theta)/c$$

This molecule thus increases its contribution to the polarization by an amount

$$p'_E \equiv \Delta p_E = -\frac{\mathrm{d}}{\mathrm{d}\theta}(p \cos \theta)\Delta\theta$$

$$= p \sin \theta \, \Delta\theta$$

$$= (p^2 E \sin^2 \theta)/c$$

We must find the average value of this quantity for all the molecules.

Since we began with a random distribution of angles, the θ-distribution is just proportional to the available solid angle $2\pi \sin \theta \, \mathrm{d}\theta$. Thus

$$\langle p'_E \rangle = \frac{(p^2 E/c) \int_0^\pi \sin^3 \theta \, \mathrm{d}\theta}{\int_0^\pi \sin \theta \, \mathrm{d}\theta}$$

$$= \frac{(p^2 E/c)[\frac{1}{3} \cos^3 \theta - \cos \theta]_0^\pi}{[\cos \theta]_0^\pi} = \frac{2p^2 E}{3c}$$

If our torsion-wire model is to give the correct results, $\langle p'_E \rangle$ must be equal to $\langle p_E \rangle$ of (6.9). This will be so if the wires have stiffness

$$c = 2kT \tag{6.10}$$

At room temperature, the required stiffness is approximately 8×10^{-21} N m.

We can now compute the resonance frequency if we know the moment of inertia I. We are interested in the order of magnitude only; for water we have

$$I \approx 2m_p R_{HO}^2$$

where m_p is the mass of a proton $(1.67 \times 10^{-27}$ kg) and R_{HO} is the hydrogen–oxygen bond length (0.103 nm). Thus $I \approx 3.5 \times 10^{-47}$ kg m^2, which gives $\omega_0 \sim 10^{13}$ s^{-1} when we combine it with our room temperature value for c.

An estimate of γ. The motion of the water molecule is damped because it drags round its neighbours, thereby diminishing its own energy supply which has to be continually topped up by the driving field. The transfer of momentum between neighbouring molecules is what makes water viscous. We may therefore get a rough idea of the molecular damping width from the observed viscosity of the liquid.

To do this we regard the molecule as a sphere surrounded, not by other molecules, but by a continuous fluid impeding rotation of the sphere by a viscous drag on its surface. This is obviously a crude model, and therefore we perform a crude analysis rather than an exact one.

The quantity we wish to estimate is the angular resistance d, defined by

$$T_d \equiv -d\dot{\psi} \tag{6.11}$$

which is the rotational analogue of (3.1). The drag torque T_d acting on the sphere must be proportional to its radius a, and we can write

$$T_d \equiv aF_d \tag{6.12}$$

which defines an 'average' drag force F_d acting over the surface of the sphere.

This force will be given by the product of the surface area $4\pi a^2$, the viscosity η, and an 'average' velocity gradient in the liquid at the surface. If the sphere drags the fluid round without slipping, the fluid velocity round the diameter is $a\dot{\psi}$. Since we expect only those molecules which are immediate neighbours of 'our' molecule to be seriously disturbed, we shall not be far wrong if we assume that the velocity falls off linearly from

$a\dot{\psi}$ to zero in a distance a, and treat the resulting velocity gradient $\dot{\psi}$ as the 'average' value required to work out F_d. The result is

$$F_d \approx - (4\pi a^2)\eta\dot{\psi}$$

and leads to

$$d \approx 4\pi\eta a^3 \tag{6.13}$$

when we combine (6.11) and (6.12).

Water at room temperature has a viscosity coefficient η of $10^{-3}\,\mathrm{kg\,m^{-1}\,s^{-1}}$. Replacing $4\pi a^3$ by 3 times the volume per molecule $(3\times 10^{-29}\,\mathrm{m^3})$ estimated from the density $(1000\,\mathrm{kg\,m^{-3}})$, we find that d is approximately $9\times 10^{-32}\,\mathrm{N\,m\,s}$.

With the aid of our previous estimate of the moment of inertia I, we may now calculate the width

$$\gamma \equiv d/I \sim 10^{15}\,\mathrm{s^{-1}}$$

This is two orders of magnitude larger than our estimate for ω_0, showing that the damping is very heavy.

The relaxation time. Because $\gamma \gg \omega_0$ the response of the molecules to the driving field at low frequencies ($\omega \ll \omega_0$) is determined entirely by the relaxation time τ_r. The rotational analogue of (3.18) is

$$\tau_r = d/c$$

Substitution of (6.10) and (6.13) for c and d gives

$$\tau_r \approx 2\pi\eta a^3/kT$$

When we insert our previous values for η and πa^3 we find that water at room temperature should have a relaxation time in the region of 10 ps. Thus the angular amplitude of the average molecule should become small, and the power absorption should become large, when the driving frequency is above $1/2\pi\tau_r$, or about 10^{10} Hz.

A small angular amplitude implies that the average molecule makes a small contribution to the polarization. Thus we can also understand why the susceptibility falls as ω rises above $1/\tau_r$.

Summary. The best way to assimilate this section is to re-read the program of work outlined on p. 88, and then to re-examine figs. 5.6 and 5.7, bearing in mind our estimate that $\tau_r\omega \sim 1$ when the driving frequency is $\sim 10^{10}$ Hz.

Problems

6.1 For forced oscillations in an *LCR* circuit, show that the voltage across the capacitor at low frequencies ($\omega \ll \omega_0$) and the voltage across the inductance at high frequencies ($\omega \gg \omega_0$) are both equal to the generator voltage.

6.2 Derive the impedance (5.21) of the *LCR* circuit (a) at very low frequencies ($\omega \ll \omega_0$ and $\omega \ll 1/RC$), (b) at the resonance frequency ($\omega = \omega_0$), and (c) at very high frequencies ($\omega \gg \omega_0$ and $\omega \gg R/L$). (The answers should suggest the idea, familiar in AC circuit theory, of the impedance of a single component.)

6.3 For forced oscillations in an *LCR* circuit, show that the mean power absorbed is $-\frac{1}{2}V_0 I_0 \sin \phi$ where I_0 is the amplitude of the current, and the other symbols have the same meanings as in the text. (In electricity textbooks the *power factor* $-\sin \phi$ is usually written in the form $\cos \phi'$, where $\phi' = \phi + \frac{1}{2}\pi$ is the angle by which the current lags the generator voltage.)

6.4 Show that, for x-rays, the scattered power is independent of frequency ('Thomson scattering').

6.5 The complex susceptibility $\chi_e(\omega)$ for water, relative to its steady-field value $\chi_e = \chi_e(0)$, may be written

$$\chi_e(\omega)/\chi_e = sK(\omega)$$

Since this $\chi_e(\omega)$ comes from rotations only, we may also write

$$\chi_e(\omega) = \varepsilon_r(\omega) - \varepsilon_\infty$$

where $\varepsilon_r(\omega)$ is the complex dielectric function and ε_∞ is a (real) 'high-frequency dielectric constant' which includes all the effects we left out of consideration.

Derive the *Debye equations*

$$\mathrm{Re}\ \varepsilon_r \approx \varepsilon_\infty + \frac{\varepsilon_r(0) - \varepsilon_\infty}{1 + \tau_r^2 \omega^2}$$

$$-\mathrm{Im}\ \varepsilon_r \approx \left[\frac{\varepsilon_r(0) - \varepsilon_\infty}{1 + \tau_r^2 \omega^2} \right] \tau_r \omega$$

7

Anharmonic vibrations

From the beginning we have recognized that a vibration is harmonic only if the return force controlling it is directly proportional to the displacement (1.2). A system in which the return force depends on ψ in some other way is called a *non-linear system*. Vibrations of non-linear systems are *anharmonic*.

Non-linear systems are described by non-linear differential equations, and these are usually difficult to solve. We have seen that it is sometimes possible to use a linear equation as an approximation, and we know, in fact, that any *small* vibration is approximately harmonic (section 2.4).

In this chapter we think about systems which are only slightly non-linear, and vibrations which are only slightly anharmonic. As well as discovering the ways in which non-linearity typically affects a vibration, we shall gain some experience of the kind of approximation methods used to tackle non-linear problems.

We discuss one simple example of a symmetric return force (one whose magnitude varies with $|\psi|$ in the same way under both stretching and compression) and also a simple asymmetric example. In the final section we examine the effects of non-linearity on forced vibrations. For simplicity we neglect damping throughout.

7.1. A symmetric return force

We first discuss the free vibration of a mass attached to a spring which exerts a return force with a *cubic* dependence on the displacement,

$$-F_s = (1 + \alpha \psi^2)s\psi \tag{7.1}$$

where s and α are constants. This force is symmetric, as we can see by plotting its magnitude $|F_s|$ against positive and negative displacements. Curves are shown in fig. 7.1 for both positive and negative values of α. When α is positive the spring appears to become stiffer as $|\psi|$ increases; such a spring is sometimes called 'hard'. A negative α denotes a 'soft' spring, which appears to become less stiff as $|\psi|$ increases. In practice hard springs are more common than soft springs, but there are other kinds of system which have 'soft' return forces.

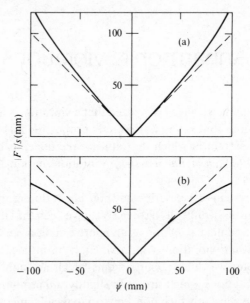

Fig. 7.1 Examples of symmetric non-linear return forces. Each of the forces shown has the cubic form given by equation (7.1), and the symmetry is indicated by plotting the magnitude of the force against ψ. Linear forces ($\alpha = 0$) are shown for comparison. (a) A 'hard' force ($\alpha = +25$ m^{-2}). (b) A 'soft' force ($\alpha = -25$ m^{-2}); in this case the force would vanish again at $\psi = \pm(-\alpha)^{-1/2}$ (± 200 mm in this example), but we consider only vibrations for which $|\alpha|\psi^2 \ll 1$.

We shall restrict our attention to situations in which the non-linearity is only slight. This means we must keep $|\alpha\psi^2| \ll 1$. Whatever the size of α, we can always ensure that $\alpha\psi^2$ is small enough by limiting the maximum value that $|\psi|$ reaches during the vibration.

The equation of motion under a cubic force may be written

$$\ddot{\psi} + (1 + \alpha\psi^2)\omega_0^2\psi = 0 \tag{7.2}$$

Although ω_0 is defined in the same way as before, we must not now presume that it represents the angular frequency of the vibration.

Properties of the vibration. To guide us in thinking about possible solutions of equation (7.2), we first note three quite general properties that the free vibration must possess.

(1) The motion must be strictly periodic, since there is no damping. We shall denote the period by τ and write

$$\psi(t) = \psi(t + \tau)$$

(2) The 'inward' part of any cycle (the part during which $|\psi|$ is decreasing) will look just like the 'outward' part run backwards. We can express this by writing

$$\psi(t_0 - t) = \psi(t_0 + t) \tag{7.3}$$

where t_0 is any value of t for which

$$\psi(t_0) = 0 \tag{7.4}$$

(3) Because $|F_s|$ is symmetric about $\psi = 0$, the motion during an excursion to the left ($\psi < 0$) will be a mirror image of the motion during an excursion to the right ($\psi > 0$). Both excursions take the same time, exactly half a period, and so

$$\psi(t + \tfrac{1}{2}\tau) = -\psi(t) \tag{7.5}$$

Solution. In looking for a solution which repeats itself with a period τ we naturally think of harmonic functions. We consider a ψ which is the sum of terms in $\cos(\omega_f t + \phi_1)$, $\cos(2\omega_f t + \phi_2)$, $\cos(3\omega_f t + \phi_3)$, ..., where ω_f is given by $2\pi/\tau$. With a symmetric force, however, we can exclude all the terms involving even multiples of $\omega_f t$, such as $\cos(2\omega_f t + \phi_2)$, since these do not satisfy (7.5).

It will simplify matters further if we choose $\dot\psi(0) = 0$ as an initial condition, thereby making all the phase constants zero. That leaves us with a series which we may write in the form

$$\psi = A(\cos \omega_f t + \varepsilon \cos 3\omega_f t + \ldots) \tag{7.6}$$

The $\cos \omega_f t$ term is known as the *fundamental*, and the other terms are called *harmonics*. We have argued that a vibration controlled by a symmetric return force cannot contain any 'even harmonics'.

The effect of adding a 'third harmonic' to a fundamental is shown in fig. 7.2. You should verify, by examining the figure, that such a vibration fulfils the three basic conditions listed above.

We can expect that vibration under a force which is only slightly non-linear will be only slightly anharmonic. If the vibration represented by (7.6) is only slightly anharmonic, the coefficient ε will be small;

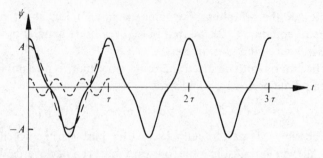

Fig. 7.2 The effect of the third harmonic. The curve plotted is $\psi = A(\cos \omega_f t + 0.2 \cos 3\omega_f t)$. This has rather a large third-harmonic content, and the approximations derived in the text would not be valid. The individual components are plotted in the first period.

contributions from higher harmonics than the third can be neglected. The maximum value of $|\psi|$ will then be approximately equal to A, the amplitude of the fundamental, and our condition for slight non-linearity will become

$$|\alpha A^2| \ll 1 \qquad (7.7)$$

It will turn out (7.10) that ε is at least as small as αA^2.

We now proceed to substitute (7.6) into (7.2). At each stage we shall allow terms containing factors like ε^2, $(\alpha A^2)^2$ or $\varepsilon \alpha A^2$, or higher powers of ε or αA^2, to be dropped, but we shall keep terms which are linear in ε or in αA^2.

First we write down expressions for ψ^3 and $\ddot{\psi}$, which are required in (7.2). We have

$$\psi^3 = A^3(\cos^3 \omega_f t + 3\varepsilon \cos^2 \omega_f t \cos 3\omega_f t + \ldots)$$

$$\ddot{\psi} = -\omega_f^2 A(\cos \omega_f t + 9\varepsilon \cos 3\omega_f t + \ldots)$$

Here it is worth noticing that $\ddot{\psi}$ is *not* proportional to ψ as it would be in a linear system. We should also be aware of the related difficulty that the differentiation rules for form D cannot be applied to non-linear systems: using a formula like (1.21) to find $\ddot{\psi}$ would give the wrong answer in this case.

Substitution for ψ, ψ^3 and $\ddot{\psi}$ in (7.2) now leads to

$$-\omega_f^2 A(\cos \omega_f t + 9\varepsilon \cos 3\omega_f t + \ldots) + \omega_0^2 A(\cos \omega_f t + \varepsilon \cos 3\omega_f t + \ldots)$$

$$+ \alpha\omega_0^2 A^3(\tfrac{1}{4}\cos 3\omega_f t + \tfrac{3}{4}\cos \omega_f t + \ldots) = 0 \qquad (7.8)$$

Here we have used the relation

$$\cos^3 \omega_f t = \tfrac{1}{4}\cos 3\omega_f t + \tfrac{3}{4}\cos \omega_f t$$

obtained from elementary trigonometry.

Equation (7.8) is not an equation to be solved for t, but one which must remain valid for *all* t, and so the terms in cos $\omega_f t$, and the terms in cos $3\omega_f t$, may be equated to zero separately. Gathering together coefficients of cos $\omega_f t$, we get

$$-\omega_f^2 A + \omega_0^2 A + \tfrac{3}{4}\alpha\omega_0^2 A^3 = 0$$

$$\omega_f^2 = \omega_0^2(1 + \tfrac{3}{4}\alpha A^2) \tag{7.9}$$

$$\omega_f \approx \omega_0(1 + \tfrac{3}{8}\alpha A^2)$$

This shows that the non-linear term in the return force (7.1) can either increase or decrease the fundamental frequency from its value with $\alpha = 0$, depending on the sign of α. We also see that the frequency shift, being proportional to A^2, becomes larger as the maxima of $|\psi|$ become larger. *The period of vibration is not fixed by the mass and the spring alone*, as it would be in the case of a linear system.

We similarly collect together the coefficients of cos $3\omega_f t$, getting

$$-9\varepsilon\omega_f^2 A + \varepsilon\omega_0^2 A + \tfrac{1}{4}\alpha\omega_0^2 A^3 = 0$$

We may now substitute for ω_f^2 from (7.9) to obtain

$$\varepsilon \approx \tfrac{1}{32}\alpha A^2 \tag{7.10}$$

As we anticipated, making αA^2 small (7.7) has ensured that ε is also small. Not surprisingly, the anharmonicity of the vibration (measured by ε) increases with the non-linearity of the return force (measured by α). But it also increases with A^2, so that *the motion is more anharmonic at large amplitudes than it is at small amplitudes.*

Validity of the approximations. We have found that (7.6) makes an acceptable solution of (7.2), the quantities ω_f and ε being given by (7.9) and (7.10) respectively. These equations are approximate: how good are they?

The question whether an approximation is 'good' or 'bad' can never be settled in isolation; it makes more sense to ask whether it is adequate for some stated purpose. In deriving (7.9) and (7.10) we have thrown away terms whose sizes are at least an order of magnitude smaller than αA^2. Any value of ω_f or ε that we calculate using these formulas will thus have an error of the order of $(\alpha A^2)^2$.

If, for example, we have a vibration with $\alpha A^2 = 0.1$, then $(\alpha A^2)^2$ is 0.01, and the errors will be of the order of 1 per cent. If we wish to know ω_f or ε more precisely than this, it will be necessary to go through a more complicated exercise in which we *retain* terms 'of second order' in αA^2. For smaller amplitude vibrations, on the other hand, with $\alpha A^2 = 0.01$ say, our 'first-order' approximations will be good to about 0.01 per cent. In that case even 'zeroth-order' approximations which ignore the non-linearity entirely will yield errors of only 1 per cent.

Example: pendulum vibrations. A simple pendulum is a 'soft' system, in which the return torque (2.4) is proportional to $\sin \psi$ (fig. 7.3). The standard series expansion for $\sin \psi$ is

$$\sin \psi = \psi - \psi^3/3! + \psi^5/5! - \ldots$$

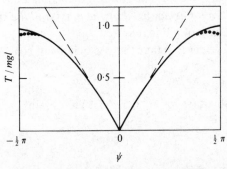

Fig. 7.3 The simple pendulum is a 'soft'. non-linear system. The magnitude of the return torque is plotted, in units mgl, against the angular displacement ψ. The diagram shows: the exact curve $|\sin \psi|$ (solid line); the linear approximation made in section 2.1 (broken line); and the cubic approximation $|(1 - \frac{1}{6}\psi^2)\psi|$ made in this section (dotted line).

Under most conditions it will be sufficient to take only the first two terms. Comparison with (7.1) then leads us to choose $\alpha = -\frac{1}{6}$, and the condition (7.7) for slight non-linearity says that A should be small compared with $\sqrt{6}$ rad (about 140°).

The first-order approximations for ω_f and ε are

$$\omega_f \approx \omega_0(1 - \tfrac{1}{16}A^2)$$
$$\varepsilon \approx \tfrac{1}{192}A^2 \tag{7.11}$$

For an amplitude of 10°, αA^2 is -0.005 and so the errors will be of the order of 0.001 per cent. (For the same amplitude, the linear approximation we made in section 2.1 would give ω_f to within about 0.2 per cent.)

Summary. A symmetric non-linear return force introduces odd harmonics into the vibration. A 'hard' force shifts the frequency upwards and a 'soft' force shifts it downwards. Approximate expressions for the frequency shift and the third-harmonic admixture due to a cubic force are given in table 7.1 on p. 102. These formulas neglect terms of second order in αA^2.

7.2. An asymmetric return force

As an example of an asymmetric return force we consider one of *quadratic* form

$$-F_s = (1 + \beta\psi)s\psi \tag{7.12}$$

and discuss vibrations for which $|\beta\psi| \ll 1$. If β is positive, as in fig. 7.4, the spring becomes stiffer the more it is stretched ($\psi > 0$), but becomes less stiff the more it is compressed ($\psi < 0$); if β is negative the reverse is true.

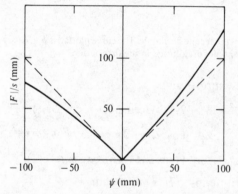

Fig. 7.4 An asymmetric non-linear return force. The force shown has the quadratic form given by equation (7.12), with $\beta = 2.5$ m^{-1}. This force would vanish again when $\psi = -1/\beta$ (−400 mm in this example), but we consider only vibrations for which $|\beta\psi| \ll 1$. The graph for $\beta = -2.5$ m^{-1} is the mirror image of the one shown. A linear force ($\beta = 0$) is shown for comparison.

The equation of motion under this force may be written

$$\ddot{\psi} + (1 + \beta\psi)\omega_0^2\psi = 0 \tag{7.13}$$

Again we look for a solution in the form of a series of harmonics. As before, we arrange zero phase constants by choosing $\dot{\psi}(0) = 0$ as an initial condition. This time we cannot exclude the even harmonics, however, since we do not have a symmetric force. The most important harmonic will now be the second, and we write

$$\psi = A_0 + A (\cos \omega_f t + \eta \cos 2\omega_f t + \ldots) \qquad (7.14)$$

We have included a constant term A_0, which can be thought of as the 'zeroth' harmonic. (A constant term would not have been consistent with a symmetric force.)

Figure 7.5 shows what the second harmonic does to the vibration. The loops on the positive ψ side are sharpened, and those on the negative ψ side are blunted, showing how the 'hard' force turns the mass round more rapidly than the 'soft' force.

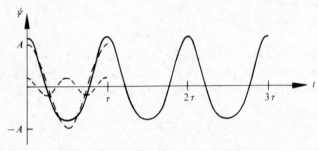

Fig. 7.5 The effect of the second harmonic. The curve plotted is $\psi = A (\cos \omega_f t + 0.2 \cos 2\omega_f t)$. The individual harmonics are plotted in the first period.

We form the expressions

$$\psi^2 = A_0^2 + 2A_0 A (\cos \omega_f t + \eta \cos 2\omega_f t)$$

$$+ A^2 (\cos^2 \omega_f t + 2\eta \cos \omega_f t \cos 2\omega_f t + \ldots)$$

$$\ddot{\psi} = -\omega_f^2 A (\cos \omega_f t + 4\eta \cos 2\omega_f t + \ldots)$$

which we substitute into (7.13) to obtain

$$-\omega_f^2 A (\cos \omega_f t + 4\eta \cos 2\omega_f t \ldots) + \omega_0^2 [A_0 + A (\cos \omega_f t + \eta \cos 2\omega_f t + \ldots)]$$

$$+ \beta \omega_0^2 [A_0^2 + 2A_0 A \cos \omega_f t + A^2 (\tfrac{1}{2} + \tfrac{1}{2} \cos 2\omega_f t + \ldots)] = 0$$

Here we have used the identity

$$\cos^2 \omega_f t = \tfrac{1}{2} (1 + \cos 2\omega_f t)$$

We have also dropped all terms in η^2, $(\beta A)^2$ or $\eta \beta A$, or in higher powers of η or βA.

Collecting coefficients of $\cos \omega_f t$ and equating them to zero as before, we find

$$-\omega_f^2 A + \omega_0^2 A + 2\beta \omega_0^2 A_0 A = 0$$

$$\omega_f^2 = \omega_0^2 (1 + 2\beta A_0) \qquad (7.15)$$

Doing the same for the coefficients of $\cos 2\omega_f t$ gives

$$-4\eta\omega_f^2 A + \eta\omega_0^2 A + \tfrac{1}{2}\beta\omega_0^2 A^2 = 0$$

into which can be substituted our expression (7.15) for ω_f^2; we find

$$-4\eta(1 + 2\beta A_0) + \eta + \tfrac{1}{2}\beta A = 0$$

$$\eta \approx \tfrac{1}{6}\beta A \tag{7.16}$$

after neglecting a term in $(\beta A)(\beta A_0)$. Since η is now seen to be smaller than βA, our neglect of terms in η^2 is justified.

On this occasion we also have constant terms, and these must also add up to zero. Thus

$$\omega_0^2 A_0 + \beta\omega_0^2 A_0^2 + \tfrac{1}{2}\beta\omega_0^2 A^2 = 0$$

$$A_0(1 + \beta A_0) = -\tfrac{1}{2}\beta A^2 \tag{7.17}$$

$$A_0 \approx -\tfrac{1}{2}\beta A^2$$

where we have dropped a term in $(\beta A)^2$. We now see that A_0 is much smaller than A, the amplitude of the fundamental, since

$$|A_0|/A \approx \tfrac{1}{2}|\beta A| \ll 1$$

The approximate expression for ω_f^2 (7.15) now becomes

$$\omega_f^2 \approx \omega_0^2(1 - \beta^2 A^2) \approx \omega_0^2$$

There would be no point in keeping the term in $(\beta A)^2$ here, because we have already thrown away other terms of the same order. We therefore write

$$\omega_f \approx \omega_0 \tag{7.18}$$

There are two significant differences between these results and the ones obtained with the cubic force. In the first place there is, within our approximation, no frequency shift.

The second difference is the presence in the solution of the small constant term A_0. The average position of the mass during the vibration is

$$\langle\psi\rangle = A_0 \approx -\tfrac{1}{2}\beta A^2 \tag{7.19}$$

since the average of each cosine term in (7.14) is zero. Equation (7.17) says that A_0 has the opposite sign to β, and so, as we might expect, the mass spends more time on the 'soft' side of $\psi = 0$ than on the 'hard' side. (Under a symmetric force the average position of the mass is of course the same as its equilibrium position.)

The results obtained with the cubic and quadratic forces are compared in table 7.1.

TABLE 7.1

Free vibrations under cubic (symmetric) and quadratic (asymmetric) return forces.
In each case the non-linear term in the force is assumed to be small

Cubic force	Quadratic force
$F_s = -(1 + \alpha\psi^2)s\psi \quad (\|\alpha\psi^2\| \ll 1)$	$F_s = -(1 + \beta\psi)s\psi \quad (\|\beta\psi\| \ll 1)$
$\psi = A(\cos\omega_f t + \varepsilon\cos 3\omega_f t + \ldots)$	$\psi = A_0 + A(\cos\omega_f t + \eta\cos 2\omega_f t + \ldots)$
$\omega_f \approx \omega_0(1 + \tfrac{3}{8}\alpha A^2)$	$\omega_f \approx \omega_0$
$\varepsilon \approx \tfrac{1}{32}\alpha A^2$	$\eta \approx \tfrac{1}{6}\beta A$
$\langle\psi\rangle = 0$	$\langle\psi\rangle = A_0 \approx -\tfrac{1}{2}\beta A^2$

Example: thermal expansion of a crystal. The force $F(r)$ between a pair of atoms or ions is asymmetric: 'soft' under stretching and 'hard' under compression. This leads to lop-sided potential energy curves such as that shown in fig. 2.6 for a pair of ions. A graph of $|F|$ against r (fig. 7.6) shows a corresponding asymmetry.

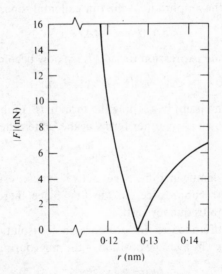

Fig. 7.6 The magnitude of the return force plotted against the separation distance r, for a pair of oppositely charged ions. The curve was obtained by differentiating the potential energy curve for HCl (fig. 2.6) in the region near the equilibrium value of r.

Because of this asymmetry, the average distance between the two atoms

$$\langle r \rangle = R + \langle \psi \rangle$$

should increase slightly if the amplitude of vibration is increased. By taking the quadratic force (7.12) as an approximation to the actual inter-atomic force, we can write an explicit formula

$$\langle r \rangle \approx R - \tfrac{1}{2}\beta A^2 \qquad (7.20)$$

which uses (7.19). The value of β will depend on the particular force involved, but it will always be negative, making $\langle r \rangle \geqslant R$.

The form of (7.20) suggests that $\langle r \rangle$ is closely related to the energy W stored in the vibration. For an undamped *harmonic* vibration of amplitude A, the stored energy is exactly equal to $\tfrac{1}{2}sA^2$. Since the largest component of our slightly anharmonic vibration is the fundamental, whose amplitude is A, we may write

$$W \approx \tfrac{1}{2}sA^2$$
$$\langle r \rangle \approx R - \beta W/s \qquad (7.21)$$

If the atoms belong to a system which is in equilibrium at absolute temperature T, the average vibrational energy per atom† will be kT. Thus (7.21) becomes

$$\langle r \rangle \approx R - \beta kT/s \qquad (7.22)$$

which shows that $\langle r \rangle$ should increase, approximately linearly, with temperature.

Finding β for a given inter-atomic force is straightforward once we know the potential $V(r)$. Comparing (2.13) and (7.12), we see immediately that

$$\beta = \frac{1}{2}\left(\frac{\mathrm{d}^3 V}{\mathrm{d}r^3}\right)_R \Big/ \left(\frac{\mathrm{d}^2 V}{\mathrm{d}r^2}\right)_R \qquad (7.23)$$

The particular example of our inter-ionic potential

$$V(r) = \frac{e^2}{4\pi\varepsilon_0 r}\left[\frac{1}{9}\left(\frac{R}{r}\right)^8 - 1\right]$$

taken from (2.15) and (2.16), gives

$$\beta = -13/2R \qquad (7.24)$$

---

† By the theorem of the equipartition of energy, each degree of freedom makes a contribution $\tfrac{1}{2}kT$ to the average energy per particle in the system. A particle vibrating in one dimension has two degrees of freedom: one kinetic and one potential. Equipartition of energy may be assumed if $kT \gg h\nu_0$, which is usually the case at room temperature and above.

From (7.22) we then find, for the rate of expansion with temperature,

$$\frac{\mathrm{d}\langle r\rangle}{\mathrm{d}T} \approx -\frac{\beta k}{s} = \frac{13k}{2Rs}$$

Into this we can substitute our previous expression (2.17) for the stiffness of the ion pair, to obtain

$$\frac{\mathrm{d}\langle r\rangle}{\mathrm{d}T} \approx \frac{13\pi\varepsilon_0 R^2 k}{4e^2}$$

$$= (4.87\times 10^4\,\mathrm{m}^{-3}\,\mathrm{K}^{-1})R^2 \tag{7.25}$$

The expansion will be easiest to observe in a crystalline solid, because the distance between neighbouring ions can then be measured accurately, as a function of temperature, by x-ray diffraction. An example for which this has been done is potassium chloride KCl, for which $\langle r\rangle$ has the value 0.315 nm at room temperature. According to (7.25) the spacing should increase by 4.8×10^{-6} nm for every 1 K rise in temperature. This is only about a factor of 4 smaller than the observed rate of expansion; it turns out that most of the discrepancy can be taken care of by modifying the potential to include the contributions from the more distant ions in the solid structure.

Summary. An asymmetric non-linear return force introduces both even and odd harmonics into the vibration, and shifts the average position of the mass slightly to one side of its equilibrium position. There is, however, no first-order frequency shift.

The thermal expansion of a solid can be understood in terms of the non-linear force acting between the atoms. Because the force is 'hard' for compression and 'soft' for stretching, the average distance between two vibrating atoms increases with the amplitude, and therefore with the temperature of the solid.

7.3. Forced vibrations of non-linear systems

We can get some idea of how a non-linear system responds to a harmonic driving force by considering a lightly damped system at driving frequencies well below the resonance frequency ($\omega \ll \omega_0$). We know that ψ then depends mainly on the stiffness, and hardly at all on the mass or the resistance (5.9). The effects of non-linearity in the spring may be expected to be most pronounced under these conditions. (By contrast, we can presume that non-linearity is unimportant in the mass controlled region $\omega \gg \omega_0$.)

For any slightly non-linear spring, we could express the displacement ψ as a power series in the return force. (See for example problem 7.7.) We can therefore write immediately

$$\psi \approx aF + bF^2 + cF^3 + \dots \tag{7.26}$$

where $a, b, c, \dots$ are constants, and F is the *driving force* which, for stiffness controlled motion, is at all times nearly equal and opposite to the spring force. If F is a single harmonic force, we put

$$F = F_0 \cos \omega t$$

in (7.26). We can see straight away that second, third and higher harmonics, as well as a constant term, will appear in ψ since

$$\cos^2 \omega t = \tfrac{1}{2}(1 + \cos 2\omega t)$$

$$\cos^3 \omega t = \tfrac{1}{4} \cos 3\omega t + \tfrac{3}{4} \cos \omega t$$

and there are equivalent identities for higher powers of $\cos \omega t$. As with free vibrations, these new contributions to ψ will become increasingly important as the amplitude is increased.

Sub-harmonic resonances. If we begin with the driving frequency at the resonance frequency of the system and then gradually reduce it, ω will successively pass through the values $\omega_0/2, \omega_0/3, \dots$, at which the harmonics in the driven motion are close to the resonance frequency, i.e. $2\omega \approx \omega_0, 3\omega \approx \omega_0, \dots$. At one of these frequencies, we would expect the relevant component to go through a resonance, so that the dominant motion will be at the resonance frequency (not the driving frequency!). Subsidiary resonance peaks will appear at each of these driving frequencies, which are *sub-harmonics* of the resonance frequency (fig. 7.7). Of course, our starting assumption of stiffness controlled motion will become invalid at these frequencies, and so we shall not go into the details.

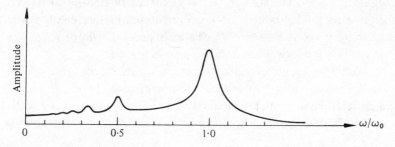

Fig. 7.7 In forced vibration of a lightly damped non-linear system, subsidiary resonances occur at driving frequencies which are sub-harmonics of the resonance frequency.

Combination frequencies. Something quite new emerges when we consider the effect of two coherent driving forces with different frequencies. We now write

$$F = F_1 \cos \omega_1 t + F_2 \cos \omega_2 t \tag{7.27}$$

and assume that $\omega_2 > \omega_1$. (We could include phase constants for the two forces, but they would make no difference to the results.)

For simplicity we assume that only the linear and quadratic terms in (7.26) are significant. Substituting (7.27) then gives

$$\psi \approx a (F_1 \cos \omega_1 t + F_2 \cos \omega_2 t) + b (F_1 \cos \omega_1 t + F_2 \cos \omega_2 t)^2$$

The first half of this expression is already familiar from the linear case (section 5.2); it will give rise to beats if $\omega_1 \approx \omega_2$. To investigate the new behaviour, we expand

$$(F_1 \cos \omega_1 t + F_2 \cos \omega_2 t)^2 = F_1^2 \cos^2 \omega_1 t + F_2^2 \cos^2 \omega_2 t$$
$$+ 2F_1 F_2 \cos \omega_1 t \cos \omega_2 t$$

The first two terms on the right are also familiar: they will give rise to second harmonics of both driving frequencies, and to a constant term.

Finally, we look at the term

$$2F_1 F_2 \cos \omega_1 t \cos \omega_2 t = F_1 F_2 [\cos (\omega_1 + \omega_2)t + \cos (\omega_2 - \omega_1)t]$$

Two completely new components have now appeared: one at the sum frequency $(\omega_1 + \omega_2)/2\pi$, and one at the difference frequency $(\omega_2 - \omega_1)/2\pi$. Like the harmonics, these *combination frequencies*† will become increasingly important as the driving force amplitudes F_1 and F_2 are increased. We also expect a subsidiary resonance whenever the driving forces have a combination frequency equal to the resonance frequency.

Examples. The most characteristic consequence of non-linearity in a forced vibrator is the production of components (harmonics, combination frequencies, or a constant term) which are not present in the driving force or forces. This is nearly always undesirable in practice.

For example, a sound reproduction system which 'invents' frequencies is clearly unsatisfactory. The forced vibrator involved here is the loudspeaker. It is difficult to make a loudspeaker in which the mounting of the cone (the moving part whose forced vibration produces the sound)

---

† Keeping the F^3 term in (7.26) would have led to additional combination frequencies of the form $2\omega_1 \pm \omega_2$ and $2\omega_2 \pm \omega_1$.

provides a strictly linear return force.† A common solution is to over-design the unit and in use to give it only displacements which are small compared with the design range. We have seen that non-linear effects are always most pronounced at large amplitudes.

The ear itself is a non-linear forced vibrator. The eardrum responds quite linearly to harmonic pressure variations produced by sound reaching it, but these vibrations are transmitted to the sensor organ (the basilar membrane) through a fluid in a coiled tube known as the cochlea, and this part of the system appears to be non-linear. If the pressure variations at the eardrum have frequencies ν and $\frac{3}{2}\nu$, for example, the brain receives sound sensations corresponding to notes of these frequencies, but also to the difference frequency‡ $\frac{1}{2}\nu$. The musical interval corresponding to any two frequencies in the ratio $3:2$ is a perfect fifth, and a note of frequency $\frac{1}{2}\nu$ lies an octave below one of frequency ν. A violin player listens for this octave and other combination frequencies when tuning two strings to an exact fifth.

Summary. By discussing a lightly damped non-linear system in the stiffness controlled region of driving frequencies, we have discovered the following characteristics of forced vibrations in non-linear systems:

(1) the forced vibration, like the free vibration, contains harmonics (and, in the case of an asymmetric system, a constant term) that are not present in the driving force;

(2) subsidiary resonances occur at driving frequencies which are subharmonics of the resonance frequency;

(3) when two harmonic driving forces act coherently, the vibration contains not only harmonics of both driving frequencies but also combination frequencies.

Problems

7.1 Is the first-order formula (7.11) for ω_f accurate enough to predict the frequency of a pendulum to within about 0.1 per cent, if the amplitude is (a) 5°, (b) 50°?

7.2 (a) A simple pendulum is set into vibration with an amplitude of exactly 45°. Calculate its frequency in terms of ν_0.

---

† In addition the electromagnetic driving force may itself be anharmonic, either because it is not strictly proportional to the signal current, or because the current is not a true reflection of the source signal.

‡ Combination frequencies due to the cubic term in (2.26) also occur but, for psychophysical reasons not yet understood, the sum frequency $\nu + \frac{3}{2}\nu$ cannot be heard.

(b) The amplitude gradually decreases as a consequence of light damping by the air, and so the frequency gradually rises towards the value ν_0. Calculate the amplitude when the frequency is $0.990\nu_0$. (Assume that the damping affects the frequency *only* by changing the amplitude.)

(c) If the damping width γ is 0.0240 s^{-1}, calculate the time taken for the amplitude to fall to the value found in (b).

7.3   A rubber string of negligible mass has a relaxed length a_0. When it is stretched, the tension increases linearly with the extension. The string is stretched between two anchor points which are separated by a distance $a > a_0$. A mass is attached at the mid point of the string and set into *transverse* vibration, along a straight line perpendicular to the string.

(a) Show that the third-harmonic admixture in the vibration is

$$\varepsilon \approx \frac{a_0}{a - a_0}\left(\frac{A}{4a}\right)^2$$

where A is the amplitude. (Neglect gravity.)

(b) If a_0 is 1.0 m, a is 1.1 m, and ε must not exceed 0.01, find the maximum permissible amplitude.

7.4   For a vibration governed by the quadratic force (7.12), show that the mass swings farther out on the 'soft' side than on the 'hard' side, by an amount $\frac{2}{3}\beta A^2$.

7.5   Assume an inter-ionic potential of the form

$$V(r) = -\frac{e^2}{4\pi\varepsilon_0 r} + \frac{B}{r^p}$$

where B and p are positive constants. Show that

$$\beta R = -(p^2 + 3p - 4)/2(p - 1)$$

where β and R have the same meanings as in the text.

Find the numerical values of βR when (a) $p = 8$, and (b) $p = 10$. Compare them with the value found in equation (7.24) for $p = 9$. (This shows that the exact form of the repulsion term in the potential does not have a sensitive effect on the anharmonicity.)

7.6   The force between two *neutral* atoms or molecules is sometimes represented by a potential of the form

$$V(r) = w\left[-2\left(\frac{R}{r}\right)^6 + \left(\frac{R}{r}\right)^{12}\right]$$

where R is the bond length and w is a constant energy (the well depth).

Show that $sR^2 = 72w$ and $\beta R = -10.5$ for this potential, where s and β have the same meanings as in the text (7.12).

Show that the average distance between the two atoms should increase by a fraction $0.15k/w$ for every 1 K rise in temperature, where k is the Boltzmann constant.

The coefficient of linear expansion of a certain solid is 0.001 K^{-1}. Estimate the well depth w for a pair of atoms of the same element.

7.7   For the quadratic return force (7.12) show that

$$-\psi \approx [1+(\beta/s)F_s](F_s/s)$$

7.8   An asymmetric non-linear vibrator is driven by coherent harmonic forces whose frequencies (50 Hz and 60 Hz) lie well below the resonance frequency. Write down all the frequencies present in the forced vibration. (Neglect third and higher powers of the force in the series expansion for ψ.)

7.9   A certain cheap and nasty loudspeaker has a resonance frequency of 120 Hz and is highly non-linear. It is driven simultaneously by two alternating voltage supplies: one of variable frequency and one whose frequency is fixed at 50 Hz. At what values of the variable frequency between 25 and 100 Hz would you expect to find peaks in the response?

8
Two-coordinate vibrations

Systems which can be described in terms of a single coordinate are exceptional, and we must expect the vibrational possibilities to become more numerous and more complex as the required number of coordinates increases.

Most of the new features can be revealed by thinking about a system with only two coordinates. Even a two-coordinate system can exhibit very complicated vibrational behaviour, but we shall find that there are two very simple basic motions, known as modes, and that any other possible vibration can be treated as a superposition of these.

In the first two sections of this chapter we shall discover, without undue rigour, the physical nature of the modes and the way in which the mathematics can be simplified by the choice of a special coordinate system. The third section outlines the technique for finding these 'mode coordinates' for a given system, and the technique is applied to a physical example in section 8.4. Only sections 8.1 and 8.2 are strictly necessary to enable you to understand the rest of the book.

8.1. Modes and mode coordinates

We start with the simple system shown in fig. 8.1. It consists of two identical masses, connected to each other and to two fixed points by a symmetric arrangement of three springs, of which the outer ones each have stiffness s, and the central one has stiffness S. We ignore all forces other than the spring forces, and we consider one-dimensional motion of each mass along the line of the springs. Thus the situation at any given time can be completely specified by two coordinates: one for each mass. We suppose for simplicity that the springs exert no forces when the

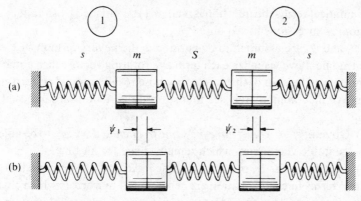

Fig. 8.1 (a) A symmetric two-coordinate system in equilibrium. (b) The masses are instantaneously displaced distances ψ_1 and ψ_2 to the right of their respective equilibrium positions.

system is in equilibrium,† and we use ψ_1 and ψ_2 to denote the displacements of the masses to the right of their individual equilibrium positions.

The behaviour of the system after the masses are disturbed will be governed by two equations of motion,

$$m\ddot{\psi}_1 = -s\psi_1 - S(\psi_1 - \psi_2)$$

$$m\ddot{\psi}_2 = -s\psi_2 - S(\psi_2 - \psi_1)$$

which may be rearranged to read

$$\ddot{\psi}_1 + [(s+S)/m]\psi_1 - (S/m)\psi_2 = 0$$

$$\ddot{\psi}_2 + [(s+S)/m]\psi_2 - (S/m)\psi_1 = 0$$

(8.1)

A disquieting feature of these equations is immediately apparent. Because the tension in the central spring depends on the positions of *both* masses, each equation contains terms in both ψ_1 and ψ_2. Thus, before we could solve the first equation for $\psi_1(t)$, we should need to know $\psi_2(t)$, and conversely for the second equation. The functions $\psi_1(t)$ and $\psi_2(t)$ must in fact be obtained as simultaneous solutions of these *coupled equations*.

Modes. When there is no coupling between the masses ($S = 0$) each mass moves harmonically, independently of the other. It seems reasonable to ask whether the coupled system can also undergo a vibration in which both masses retain the elementary harmonic kind of motion. In

† Even if they did the vibration would be unaffected, provided all the force–extension relationships were linear.

mathematical terms the question is: can (8.1) be satisfied by simultaneous harmonic expressions for ψ_1 and ψ_2?

If ψ_1 and ψ_2 are assumed to be harmonic, then every term in (8.1) will be harmonic. If we visualize each term as a rotating vector, each equation is represented by a set of three vectors whose resultant must be zero at all times. We may immediately see that several restrictions are imposed on ψ_1 and ψ_2.

(1) Obviously *they must have the same frequency* if we are to be able to form a suitable vector sum which remains fixed for all time.

(2) They must also be in a simple phase relationship. We know that the first two terms in each equation are represented by vectors which always point in opposite directions, so that a balance will be possible only if the third vector lies along the same line (fig. 8.2). This means that ψ_1 *and* ψ_2 *must be either in phase or in antiphase with each other*, and so always go through their equilibrium positions at the same instant.

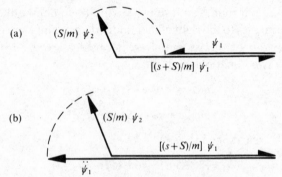

Fig. 8.2 A vector diagram representing the first equation in (8.1). Because the ψ_1 and $\ddot{\psi}_1$ vectors always point in opposite directions, the ψ_2 vector must lie either (a) in the ψ_1 direction, or (b) in the $-\psi_1$ direction. The vector diagram for the second equation is similar.

(3) Their amplitudes must also be related. Since ψ_1 and ψ_2 have the same frequency, and keep in phase or in antiphase, we may write

$$\psi_1 = A_1 \cos(\omega t + \phi)$$

$$\psi_2 = \pm A_2 \cos(\omega t + \phi)$$

and define a constant ratio

$$r \equiv \psi_2/\psi_1 = \pm A_2/A_1$$

Substitution into (8.1) gives

$$\omega^2 - (s+S)/m + rS/m = 0$$

$$\omega^2 - (s+S)/m + S/rm = 0$$

(8.2)

which can be simultaneously satisfied only if $1/r = r$, or

$$r = \pm 1$$

Thus ψ_1 *and* ψ_2 *must have equal amplitudes.* We can have a vibration in which $\psi_1 = \psi_2$ at all times, or one in which $\psi_1 = -\psi_2$ at all times, but there are no other harmonic possibilities.

We call these two special states of motion the *modes* (or *normal modes*) of the system. We can find the *mode frequencies* by substituting r in either member of (8.2). For the 'in-phase' mode, which we shall call mode 1, we have $r = +1$, giving an angular frequency

$$\omega_1 \equiv (s/m)^{1/2} \qquad (8.3)$$

Mode 2 (the 'in-antiphase' mode, with $r = -1$) has a higher frequency

$$\omega_2 \equiv [(s + 2S)/m]^{1/2} \qquad (8.4)$$

We may summarize these findings by writing

$$\text{(mode 1)} \quad \psi_2 = \psi_1 = A_1 \cos(\omega_1 t + \phi_1)$$
$$\text{(mode 2)} \quad \psi_2 = -\psi_1 = A_2 \cos(\omega_2 t + \phi_2) \qquad (8.5)$$

We are now using the pair of numbers 1 and 2 for a new purpose, to label the mode angular frequencies ω_1 and ω_2, and the mode phase constants ϕ_1 and ϕ_2. It is important to be clear that there is no particular association of ω_1 with ψ_1, for example, or ω_2 with ψ_2: the frequencies are properties of *the system as a whole.* When we use the numbers in the old way, as labels for parts of the system, reference to the parent symbol (m, ψ or A) should prevent confusion.

Mode coordinates. We may notice from (8.5) that, when the system is vibrating in mode 1, the quantity $(\psi_2 - \psi_1)$ is always zero, and $(\psi_1 + \psi_2)$ varies harmonically; in mode 2 the reverse is true. We are led to take a closer look at these variables: we shall do this by defining the closely related quantities

$$q_1 \equiv (m/2)^{1/2}(\psi_1 + \psi_2)$$
$$q_2 \equiv (m/2)^{1/2}(-\psi_1 + \psi_2) \qquad (8.6)$$

Since only one of them is active in a mode, we call them the *mode coordinates* (or *normal coordinates*) of the system. The reason for the factors $(m/2)^{1/2}$ on the right will appear later. The quantities $\dot{q}_1$ and $\dot{q}_2$ are connected to $\dot{\psi}_1$ and $\dot{\psi}_2$ by similar equations, as are higher derivatives.

We shall need corresponding expressions for the ψ's in terms of the q's; by adding and subtracting the two members of (8.6) we obtain

$$\psi_1 = (1/2m)^{1/2}(q_1 - q_2)$$
$$\psi_2 = (1/2m)^{1/2}(q_1 + q_2)$$
(8.7)

The change of coordinates has a remarkable effect on the equations of motion. Substituting (8.7) into (8.1) gives *uncoupled* equations

$$\ddot{q}_1 + \omega_1^2 q_1 = 0$$
$$\ddot{q}_2 + \omega_2^2 q_2 = 0$$
(8.8)

Their solutions are harmonic, as we have already seen. We may write them in the form

$$q_1 = \mathscr{A}_1 \cos(\omega_1 t + \phi_1)$$
$$q_2 = \mathscr{A}_2 \cos(\omega_2 t + \phi_2)$$
(8.9)

observing that the *mode amplitudes* $\mathscr{A}_1$ and $\mathscr{A}_2$ must have the same units $(\text{kg}^{1/2}\text{m})$ as the mode coordinates themselves.

Since the new equations of motion (8.8) have exactly the form of (1.4), they may be treated by the methods developed in section 1.1. We are thus encouraged to solve a two-coordinate problem in terms of mode coordinates whenever we can. Although q_1 and q_2 do not measure the positions of single points of the system in the way usually associated with simple coordinates, they nevertheless provide a complete specification of the system at any instant: we can use (8.6) and (8.7) to go freely back and forth between the two sets of coordinates without sacrificing information.

Since the equations of motion are second-order partial differential equations, the two special pairs of solutions we have found may be taken in any linear combination to represent the general motion of the system. Combining (8.7) and (8.9) gives such a general solution

$$\psi_1 = (1/2m)^{1/2}[\mathscr{A}_1 \cos(\omega_1 t + \phi_1) - \mathscr{A}_2 \cos(\omega_2 t + \phi_2)]$$
$$\psi_2 = (1/2m)^{1/2}[\mathscr{A}_1 \cos(\omega_1 t + \phi_1) + \mathscr{A}_2 \cos(\omega_2 t + \phi_2)]$$

having the required four arbitrary constants ($\mathscr{A}_1$, $\mathscr{A}_2$, ϕ_1 and ϕ_2) which will accommodate any given set of initial conditions (two displacements and two velocities). When only mode 1 is 'excited' we have $\mathscr{A}_2 = 0$; when only mode 2 is excited we have $\mathscr{A}_1 = 0$. In all other states of motion the vibration of each mass depends on *both* mode frequencies, and so is no longer harmonic. We investigate one such situation below.

Examples. We now take a more detailed look at the vibrations resulting from three specific sets of initial conditions. If, first, we give the two masses equal displacements A_0 to the right and then release them, the initial conditions are

$$\psi_1(0) = A_0 \qquad \psi_2(0) = A_0$$
$$\dot{\psi}_1(0) = 0 \qquad \dot{\psi}_2(0) = 0$$

Transforming these conditions to mode coordinates with the aid of (8.6) we obtain

$$q_1(0) = \mathscr{A}_0 \qquad q_2(0) = 0$$
$$\dot{q}_1(0) = 0 \qquad \dot{q}_2(0) = 0$$

where

$$\mathscr{A}_0 \equiv (2m)^{1/2} A_0 \qquad (8.10)$$

By starting the vibration in this way we have excited mode 1 on its own; from these initial conditions we can immediately write down the mode solution

$$q_1 = \mathscr{A}_0 \cos \omega_1 t$$

To see what this means in terms of the two vibrating masses, we transform back to the ψ's by means of (8.7), when we find

$$\psi_1 = A_0 \cos \omega_1 t$$
$$\psi_2 = A_0 \cos \omega_1 t$$

A cycle of the vibration is illustrated in fig. 8.3(a). The masses are always moving in the same direction, and are always at equal distances from their respective equilibrium positions. The centre spring is never stretched or compressed, and might as well be removed since it has no effect on the motion. This is the physical reason for the fact that the frequency (8.3) does not depend on S.

As a second example, we displace mass 1 a distance A_0 to the right and mass 2 an equal distance *to the left*, and then release them, so that

$$\psi_1(0) = A_0 \qquad \psi_2(0) = -A_0$$
$$\dot{\psi}_1(0) = 0 \qquad \dot{\psi}_2(0) = 0$$

These transform to

$$q_1(0) = 0 \qquad q_2(0) = -\mathscr{A}_0$$
$$\dot{q}_1(0) = 0 \qquad \dot{q}_2(0) = 0$$

with $\mathscr{A}_0$ again given by (8.10).

(a)　(b)　(c)

In this case we have mode 2 only, and

$$q_2 = \mathscr{A}_0 \cos{(\omega_2 t + \pi)}$$

which is equivalent to

$$\psi_1 = A_0 \cos{\omega_2 t}$$

$$\psi_2 = A_0 \cos{(\omega_2 t + \pi)} = -A_0 \cos{\omega_2 t}$$

This mode is illustrated in fig. 8.3(b). The masses again vibrate with equal amplitudes and frequencies, but are now always moving in *opposite* directions. The centre spring is stretched once and compressed once in each cycle, and, because of the extra return force which it contributes, the angular frequency ω_2 (8.4) is greater than ω_1.

Finally we consider the situation when mass 1 is initially moved a distance $2^{1/2} A_0$ to the right but mass 2 *is held fixed*, and both are then released. (The choice of an initial displacement $2^{1/2} A_0$ leads to amplitudes of A_0 when the energy is shared equally between the masses.) The initial conditions are

$$\psi_1(0) = 2^{1/2} A_0 \qquad \psi_2(0) = 0$$

$$\dot{\psi}_1(0) = 0 \qquad \dot{\psi}_2(0) = 0$$

In terms of mode coordinates these conditions are

$$q_1(0) = \mathscr{A}_0/2^{1/2} \qquad q_2(0) = -\mathscr{A}_0/2^{1/2}$$

$$\dot{q}_1(0) = 0 \qquad \dot{q}_2(0) = 0$$

with the same $\mathscr{A}_0$ as in the previous examples (8.10).

This time both modes are excited, with equal amplitudes; we readily obtain the solutions

$$q_1 = (\mathscr{A}_0/2^{1/2}) \cos{\omega_1 t}$$

$$q_2 = (\mathscr{A}_0/2^{1/2}) \cos{(\omega_2 t + \pi)} = -(\mathscr{A}_0/2^{1/2}) \cos{\omega_2 t}$$

which, after transformation back to the ψ's, become

$$\psi_1 = (A_0/2^{1/2})(\cos{\omega_1 t} + \cos{\omega_2 t})$$

$$\psi_2 = (A_0/2^{1/2})(\cos{\omega_1 t} - \cos{\omega_2 t}) \tag{8.11}$$

Fig. 8.3 (a) and (b) The two modes of the system shown in fig. 8.1. (c) Superposition of the two modes, excited equally in antiphase with each other, with $S/s = 20/81$. In this vibration the masses are in quadrature. During the cycle shown they have equal amplitudes, but this does not persist.

If S is small the mode angular frequencies ω_1 and ω_2 will be nearly equal (8.4). The motion of each mass will then take the form of beats. We have previously met beats between two driving forces (p. 71), and beats between a driving force and a free vibration (p. 74). Now we have an example of beats between the modes of a system consisting of two weakly coupled vibrators. The resulting vibration provides one of the most striking spectacles in physics.

Using the notation introduced previously (5.28) we can rewrite (8.11) in the form

$$\psi_1 = 2^{1/2} A_0 \cos \Omega_- t \cos \Omega_+ t$$
$$\psi_2 = 2^{1/2} A_0 \sin \Omega_- t \sin \Omega_+ t$$

Now we can see that the two masses vibrate in quadrature with each other, and with angular frequencies that are both close to ω_0; the modulating factors $\cos \Omega_- t$ and $\sin \Omega_- t$ are also in quadrature. As the motion develops from time $t = 0$, the vibration of mass 1 dies away to nothing, while that of mass 2 builds up from nothing to a peak amplitude of $2^{1/2} A_0$. The trends are then reversed until, after approximately s/S periods, mass 2 is again at rest. Both $\psi_1(t)$ and $\psi_2(t)$ are plotted in fig. 8.4 for the case in which $S/s = 20/81$.

If the coupling is so weak that we may always neglect the energy stored in the coupling spring, we can think of the whole process as the perpetual exchange of energy between vibrator 1 (which has all the energy originally) and vibrator 2. The vibrator whose amplitude is increasing at any

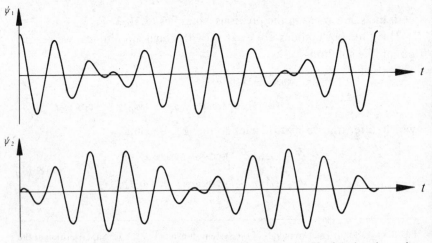

Fig. 8.4 Beating modes: vibrations of the two masses in the system of fig. 8.1 when the modes are excited equally. In this example $S/s = 20/81$, giving $\omega_2/\omega_1 = 11/9$ and $\Omega_+ = 10\Omega_-$. The diagram shows a cycle of the vibration when the amplitudes of the masses are equal.

given instant is the one that is lagging in phase by 90°, as we expect for a vibrator absorbing power from a driving force at resonance. The two vibrators in our example alternately act as 'driver' and as 'driven vibrator', and this relationship is always maintained until the driver has given up all of its energy. When it comes to rest, the driver acquires a sudden phase lag of 180° (as its modulating factor changes sign), thereby exchanging roles with the driven vibrator. The masses will have approximately equal amplitudes when $\Omega_- t \approx \frac{1}{4}\pi$, giving

$$\psi_1 \approx A_0 \cos \Omega_+ t$$

$$\psi_2 \approx A_0 \sin \Omega_+ t$$

Figure 8.3(c) shows a cycle of the motion under these conditions.

We are now in a position to appreciate, in figs. 8.3(a), 8.3(b) and 8.3(c), the characteristic phase differences *between the masses* in the three examples we have discussed. The masses are *in phase* with each other in the first example, and *in antiphase* in the second. In the third example they are *in quadrature*, the leadership being exchanged between the two vibrators at intervals $\pi/\Omega_- \approx s/S$.

Variation of mode frequencies with coupling. We have seen that, if the stiffness of the centre spring is very small, the system behaves to some extent like two simple vibrators with their own separate frequencies (which happen to be equal in our symmetric system). Their vibrations are, however, profoundly modified by even a small degree of coupling, which produces the spectacular interchange of energy between the two vibrators. If we could continuously 'wind up' the value of S, we should see the separate identities of the two vibrators becoming merged in that of the system as a whole, as a result of the increasing difference between the

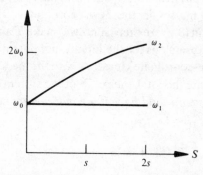

Fig. 8.5 Dependence of the mode frequencies on the coupling strength, for the system in fig. 8.1.

mode frequencies. The variation of ω_2 as S changes is governed by (8.4) and is plotted in fig. 8.5. The lower mode frequency ω_1 is, as we have already noted, independent of S; this is a rather special property of our simple system, but the increasing separation between ω_1 and ω_2 with increasing coupling strength is typical of all systems.

Energy. Neither of the features of the mode coordinates so far mentioned (the uncoupling of the equations of motion, and the vanishing of one coordinate in a mode) depends on the particular scale factor included in the definition (8.6). The reason for the choice we made there emerges when we write down the energy of the system in terms of the mode coordinates.

The kinetic and potential energies are respectively

$$T = \tfrac{1}{2}m\dot{\psi}_1^2 + \tfrac{1}{2}m\dot{\psi}_2^2$$

$$V = \tfrac{1}{2}s\psi_1^2 + \tfrac{1}{2}s\psi_2^2 + \tfrac{1}{2}S(\psi_2 - \psi_1)^2$$

$$= \tfrac{1}{2}(s+S)\psi_1^2 + \tfrac{1}{2}(s+S)\psi_2^2 - S\psi_1\psi_2$$

To turn T and V into functions of the q's we use (8.7) and obtain

$$T = \tfrac{1}{2}\dot{q}_1^2 + \tfrac{1}{2}\dot{q}_2^2$$

$$V = \tfrac{1}{2}(s/m)q_1^2 + \tfrac{1}{2}[(s+2S)/m]q_2^2 \tag{8.12}$$

$$= \tfrac{1}{2}\omega_1^2 q_1^2 + \tfrac{1}{2}\omega_2^2 q_2^2$$

The first thing we notice is that V now contains *no cross-product term* in q_1q_2; moreover, the coefficients of q_1^2 and q_2^2 depend only on the respective mode frequencies. Neither T nor V explicitly contains any masses or stiffnesses, which are properties of the individual parts of the system, and not of the modes. We could not, for example, relate 'the mass of a mode' to any identifiable physical object, and (8.6) has avoided this difficulty by incorporating the masses in the new coordinates. The factors $1/2^{1/2}$ which also appear in (8.6) are there simply to make T and V in (8.12) look like the familiar expressions for the kinetic and potential energies (1.10) and (1.11) of single-coordinate vibrators, with the mass factors removed.

We may now write the total energy W of the system as a simple *sum of the energies the system would have if the modes were excited separately*,

$$W = T + V$$

$$= (\tfrac{1}{2}\dot{q}_1^2 + \tfrac{1}{2}\omega_1^2 q_1^2) + (\tfrac{1}{2}\dot{q}_2^2 + \tfrac{1}{2}\omega_2^2 q_2^2)$$

Here we see once more how the modes of the system have a unique status as the basic components of all the other possible vibrations.

Summary. Our symmetric two-coordinate vibrator is capable of two special kinds of vibration, called modes, in which the two masses move harmonically with equal frequencies and amplitudes, vibrating in phase with each other in one mode, and in antiphase in the other. Increasing the degree of coupling between the two vibrators from which the system is constructed increases the separation between the two mode frequencies.

The most suitable pair of coordinates for describing any free vibration of the system are the mode coordinates, which have the following special properties:

(1) they satisfy uncoupled equations of motion;

(2) it takes only one mode coordinate to describe a mode;

(3) the expressions for the system kinetic and potential energies T and V both become simple sums of squares, the two coefficients in V being fixed by the two mode frequencies.

8.2. Forced vibrations

In this section we investigate very briefly how the two-coordinate system (fig. 8.1) responds to a long-established driving force. We assume that the two masses are identically damped, and that the drag forces are of the standard type described by (3.1).

Finding the mode recipe. We are now familiar with the idea that the vibration of each mass in the system can be described as a superposition of its motion in the two modes. For free vibrations, the amplitudes and phase constants of the modes are determined by the initial conditions. In a steady-state forced vibration the mode 'recipe' depends, as we shall see, on the driving frequency.

We shall investigate the particular problem of finding the recipe when the driving force is applied to the left-hand mass, and has the form $F_0 \cos \omega t$. As we saw in section 5.1, we can if we wish arrange this by moving the left-hand anchor point so that its displacement is $(F_0/s) \cos \omega t$.

The equations of motion are

$$m\ddot{\psi}_1 = -s\psi_1 - S(\psi_1 - \psi_2) - b\dot{\psi}_1 + F_0 \cos \omega t$$
$$m\ddot{\psi}_2 = -s\psi_2 - S(\psi_2 - \psi_1) - b\dot{\psi}_2$$

It is not difficult to verify that our previous mode coordinate transformation (8.7), when applied to these equations, produces a pair of 'forced' mode equations

$$\ddot{q}_1 + \gamma\dot{q}_1 + \omega_1^2 q_1 = (1/2m)^{1/2} F_0 \cos \omega t$$

$$\ddot{q}_2 + \gamma\dot{q}_2 + \omega_2^2 q_2 = -(1/2m)^{1/2} F_0 \cos \omega t \qquad (8.13)$$

$$= (1/2m)^{1/2} F_0 \cos (\omega t + \pi)$$

in which the symbols γ, ω_1 and ω_2 all have their previous meanings. The equations are now in essentially the same form as the equations of motion for a forced single-coordinate vibrator (5.3). Of course we are now dealing with q's, and so all the terms in (8.13) have the appropriate units ($\mathrm{kg}^{1/2}\,\mathrm{m\,s}^{-2}$). We notice, however, that there is an additional phase constant of π which puts the mode 2 driving 'force' in antiphase with mode 1 driving 'force'; this new feature can be handled easily, by shifting the phase of the solution by $180°$.

We deduce immediately that there will be steady-state harmonic solutions for the modes, similar to (5.4),

$$q_1 = \mathscr{A}_1 \cos (\omega t + \phi_1)$$

$$q_2 = \mathscr{A}_2 \cos (\omega t + \phi_2)$$

From these we can obtain, in the usual way, expressions for the mass coordinates

$$\psi_1 = (1/2m)^{1/2}[\mathscr{A}_1 \cos (\omega t + \phi_1) - \mathscr{A}_2 \cos (\omega t + \phi_2)]$$

$$\psi_2 = (1/2m)^{1/2}[\mathscr{A}_1 \cos (\omega t + \phi_1) + \mathscr{A}_2 \cos (\omega t + \phi_2)] \qquad (8.14)$$

We are reminded here of what happens when two driving forces of the same frequency are applied simultaneously to a single-coordinate vibrator; the steady-state motion in that case was a superposition of the separate motions (5.23) and (5.24) produced by the two forces acting on their own. The amplitudes of the two masses in the present example can be found from formulas

$$A_1^2 = (1/2m)[\mathscr{A}_1^2 + \mathscr{A}_2^2 - 2\mathscr{A}_1\mathscr{A}_2 \cos (\phi_2 - \phi_1)]$$

$$A_2^2 = (1/2m)[\mathscr{A}_1^2 + \mathscr{A}_2^2 + 2\mathscr{A}_1\mathscr{A}_2 \cos (\phi_2 - \phi_1)] \qquad (8.15)$$

which are similar to (5.25).

The essential difference becomes apparent when we write down the mode amplitudes $\mathscr{A}_1$ and $\mathscr{A}_2$; these can be found from diagrams similar to fig. 5.2, and are

$$\mathscr{A}_1 = \frac{F_0}{(2m)^{1/2}}\left[\frac{1}{(\omega_1^2-\omega^2)^2+\gamma^2\omega^2}\right]^{1/2}$$

$$\mathscr{A}_2 = \frac{F_0}{(2m)^{1/2}}\left[\frac{1}{(\omega_2^2-\omega^2)^2+\gamma^2\omega^2}\right]^{1/2}$$

Because $\omega_1 \neq \omega_2$ in general, $\mathscr{A}_1$ and $\mathscr{A}_2$ do not keep in step as they rise and fall with changing frequency. They are plotted together in fig. 8.6(a) for comparison.

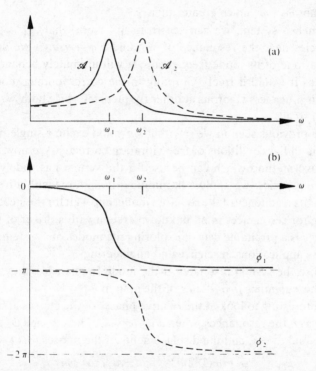

Fig. 8.6 Variation with driving frequency of (a) the mode amplitudes, and (b) the mode phase constants, for forced vibrations of the system in fig. 8.1. Phases are measured relative to the driving force.

The mode phase difference $\phi_2 - \phi_1$ in (8.15) is also dependent on the driving frequency. The mode 1 phase constant ϕ_1 varies with ω like the curve shown in fig. 5.4, with the swing from 0 to $-\pi$ occurring at frequencies near ω_1; for ϕ_2 the graph has a similar shape, but swings over near ω_2 and is also shifted vertically by $-\pi$ to allow for the antiphase driving 'force'.

Curves showing this behaviour are given in fig. 8.6(b). They show that q_1 and q_2 are in antiphase with each other at driving frequencies $\omega \ll \omega_1$ and $\omega \gg \omega_2$. There is not generally such a simple phase relationship between the modes at other frequencies; if we wish to calculate ϕ_1 or ϕ_2 at some given frequency, we can use (5.6) with ω_1 or ω_2, as appropriate, in place of ω_0.

Resonances in a two-coordinate system. Without having to do any calculations, we can discuss qualitatively the response of the system at different driving frequencies. We assume that the modes are separated by an amount $\omega_2 - \omega_1$ much greater than γ.

For such a system, we can see from fig. 8.6(a) that $\mathcal{A}_1$ is small at frequencies near the resonance of $\mathcal{A}_2$, and *vice versa*. If we drive the system at one of its mode frequencies, it will ultimately behave almost exactly as it would if freely vibrating in the corresponding mode. The motion is much less vigorous at other frequencies, since both $\mathcal{A}_1$ and $\mathcal{A}_2$ are then small.

In the previous section we saw that we could excite a single mode by fixing the initial conditions of free vibration correctly. We now have an alternative method which can be used if the system has widely spaced mode frequencies: we apply a harmonic driving force with a frequency equal to the frequency of the required mode, and wait for the steady state. Looking for resonances in an unknown system with a driver of variable frequency is a profitable way of exploring the modes of a system; it finds countless applications in science and engineering.

Because the system resonates at the mode frequencies, they are also called the *resonance frequencies* of the system.

It is interesting to look at the relative phases of the masses at frequencies *between* the resonances, when $\omega_1 \ll \omega \ll \omega_2$. Then ϕ_1 and ϕ_2 are both nearly equal to $-\pi$, and the displacements of the masses (8.14) are

$$\psi_1 \approx -(1/2m)^{1/2}(\mathcal{A}_1 - \mathcal{A}_2) \cos \omega t$$

$$\psi_2 \approx -(1/2m)^{1/2}(\mathcal{A}_1 + \mathcal{A}_2) \cos \omega t$$

It is obvious that $|\psi_2|$ is always greater than $|\psi_1|$ in this frequency range.

We know that $\mathcal{A}_1$ and $\mathcal{A}_2$ are small at these frequencies; they will not generally be equal. At the lower frequencies, $\mathcal{A}_1$ will be larger than $\mathcal{A}_2$ and mass 1 will move in antiphase with the driving force. At the higher frequencies $\mathcal{A}_1$ is less than $\mathcal{A}_2$ and mass 1 moves in phase with the driving force. Mass 2 is in antiphase with the force at all frequencies in the range.

The motion of the system is in fact similar to that in mode 1 (masses in phase with each other) at angular frequencies near ω_1 and to that in mode

2 (masses in antiphase) at angular frequencies near ω_2. Naturally there will be a value of ω at which mass 1 is stationary, with $\mathscr{A}_1$ and $\mathscr{A}_2$ equal. Its phase constant therefore swings through 180° while the mass is hardly moving: very little power is absorbed, in accord with our knowledge that significant power absorption occurs only near a resonance.

Figure 8.7 shows how the amplitude and the phase constant of mass 2 vary as we take the driving frequency through the entire range $\omega \ll \omega_1$ to $\omega \gg \omega_2$.

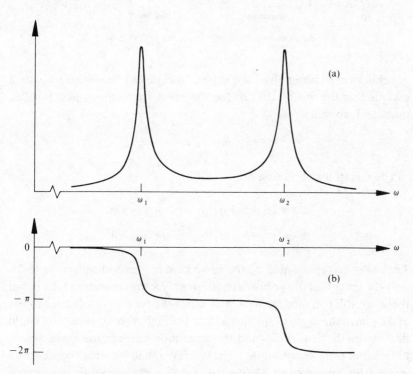

Fig. 8.7 (a) The amplitude of mass 2 as a function of driving frequency, for the system in fig. 8.1, when the driving force is applied to mass 1. (b) The phase lag of mass 2, relative to the driving force, as a function of driving frequency. In this example $\omega_2 - \omega_1 = 20\gamma$.

Summary. When a two-coordinate system is acted on by a harmonically varying driving force, the steady-state vibration at any frequency is one in which both masses vibrate harmonically. If the mode frequencies for the system are many widths apart, the system resonates whenever the driving frequency is close to a mode frequency. At a resonance, the motion is almost exactly the same as the free vibration in the corresponding mode. Far from resonance, the motion is less vigorous, but the masses still move nearly in phase or in antiphase with each other.

8.3. How to find the mode coordinates

A more general system than that of fig. 8.1 is the one shown in fig. 8.8. In this system the symmetry of the previous example has been abandoned, which means that, if the centre spring were removed, one mass would in

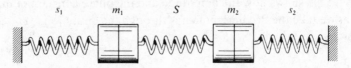

Fig. 8.8 A more general version of the system in fig. 8.1.

general vibrate faster than the other. In order to be definite we shall assume that the one on the left has the lower frequency, and is labelled number 1, so that

$$s_1/m_1 < s_2/m_2$$

The equations of motion are

$$\ddot{\psi}_1 + [(s_1 + S)/m_1]\psi_1 - (S/m_1)\psi_2 = 0$$
$$\ddot{\psi}_2 + [(s_2 + S)/m_2]\psi_2 - (S/m_2)\psi_1 = 0$$

They may be represented by the same kind of vector diagrams as (8.1): only the lengths of the vectors are different. We may therefore take it that there are still two modes, in each of which the harmonic solutions for ψ_1 and ψ_2 have the same frequency and go through zero together. We could find the mode frequencies and the amplitude ratios in the same way as before, but the algebra would be heavier. We shall therefore concentrate instead on the problem of how to find the mode coordinate system. Discovering the mode coordinates will give us not only the mode frequencies and amplitude ratios, but also the all-important uncoupled equations of motion, and will clear the way for the solution of two-coordinate problems by one-coordinate methods.

A theorem about the mode coordinate transformation. We have already seen that the transformation (8.6) which previously uncoupled the equations of motion also removed the masses from the kinetic energy and the cross-products from the potential energy (8.12). Now we shall prove the converse: *the transformation that gives T and V the forms of (8.12) will automatically produce uncoupled equations of motion like (8.8).*

First we rewrite the equations of motion in a new form derived from the kinetic and potential energies, which are

$$T = \tfrac{1}{2}m_1\dot{\psi}_1^2 + \tfrac{1}{2}m_2\dot{\psi}_2^2$$

$$V = \tfrac{1}{2}s_1\psi_1^2 + \tfrac{1}{2}s_2\psi_2^2 + \tfrac{1}{2}S(\psi_2 - \psi_1)^2 \tag{8.16}$$

$$= \tfrac{1}{2}(s_1 + S)\psi_1^2 + \tfrac{1}{2}(s_2 + S)\psi_2^2 - S\psi_1\psi_2$$

It is easy to show, by direct substitution, that the equations

$$\frac{\mathrm{d}}{\mathrm{d}t} \cdot \frac{\partial T}{\partial \dot{\psi}_1} + \frac{\partial V}{\partial \psi_1} = 0$$

$$\frac{\mathrm{d}}{\mathrm{d}t} \cdot \frac{\partial T}{\partial \dot{\psi}_2} + \frac{\partial V}{\partial \psi_2} = 0 \tag{8.17}$$

are equivalent to (8.1).

The most important property of the last equations† is that *their form is preserved under any linear coordinate transformation*. Take, for example, the transformation

$$\psi_1 = a_{11}q_1 + a_{12}q_2$$

$$\psi_2 = a_{21}q_1 + a_{22}q_2 \tag{8.18}$$

in which the four a's are constants. We now have

$$\frac{\mathrm{d}}{\mathrm{d}t} \cdot \frac{\partial T}{\partial \dot{q}_1} = \frac{\mathrm{d}}{\mathrm{d}t}\left(\frac{\partial T}{\partial \dot{\psi}_1} \cdot \frac{\partial \dot{\psi}_1}{\partial \dot{q}_1} + \frac{\partial T}{\partial \dot{\psi}_2} \cdot \frac{\partial \dot{\psi}_2}{\partial \dot{q}_2}\right)$$

$$= a_{11}\left(\frac{\mathrm{d}}{\mathrm{d}t} \cdot \frac{\partial T}{\partial \dot{\psi}_1}\right) + a_{21}\left(\frac{\mathrm{d}}{\mathrm{d}t} \cdot \frac{\partial T}{\partial \dot{\psi}_2}\right) \tag{8.19}$$

and, similarly,

$$\frac{\partial V}{\partial q_1} = a_{11}\left(\frac{\partial V}{\partial \psi_1}\right) + a_{21}\left(\frac{\partial V}{\partial \psi_2}\right) \tag{8.20}$$

Combining (8.19) and (8.20), and using (8.17), we find

$$\frac{\mathrm{d}}{\mathrm{d}t} \cdot \frac{\partial T}{\partial \dot{q}_1} + \frac{\partial V}{\partial q_1} = 0$$

$$\frac{\mathrm{d}}{\mathrm{d}t} \cdot \frac{\partial T}{\partial \dot{q}_2} + \frac{\partial V}{\partial q_2} = 0$$

which are identical in form to (8.17).

† They are commonly written as the *Lagrange equations*

$$\frac{\mathrm{d}}{\mathrm{d}t} \cdot \frac{\partial L}{\partial \dot{\psi}_i} - \frac{\partial L}{\partial \psi_i} = 0$$

where the function $L \equiv T - V$ is known as the Lagrangian.

If, now, the q coordinates in (8.18) are the ones that put T and V into their special forms (8.12), we must have

$$\frac{\mathrm{d}}{\mathrm{d}t} \cdot \frac{\partial T}{\partial \dot{q}_1} = \ddot{q}_1 \qquad \frac{\mathrm{d}}{\mathrm{d}t} \cdot \frac{\partial T}{\partial \dot{q}_2} = \ddot{q}_2$$

$$\frac{\partial V}{\partial q_1} = \omega_1^2 q_1 \qquad \frac{\partial V}{\partial q_2} = \omega_2^2 q_2$$

The uncoupled equations (8.8) follow immediately, and the theorem is proved.

Transforming T. Having convinced ourselves that the coordinates we are looking for are simply those in which T and V have the tidy forms of (8.12), we are left with the task of finding the particular values of a_{11}, a_{12}, a_{21} and a_{22} in (8.18) which give them those forms. We shall conduct the search in two stages: first we look for a transformation which puts T into the required form, and then we do the same for V, taking care that the second transformation does not undo the work of the first.

It is helpful to use diagrams to illustrate the various transformations involved. Since the kinetic energy, for example, is a function of two variables ($\dot{\psi}_1$ and $\dot{\psi}_2$ in the initial coordinate system), the problem of showing pictorially how it depends on them is the same one that we face when we want to show the height of land as a function of latitude and longitude on a map, and we adopt the same solution: the use of contours.

Figure 8.9(a) shows kinetic energy contours in the $\dot{\psi}_1\dot{\psi}_2$ plane. All points at which the kinetic energy is T_0, for example, are joined together

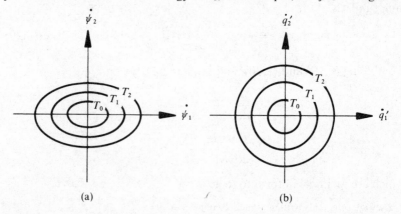

(a)   (b)

Fig. 8.9 Kinetic energy contours. The coordinates of every point on a given contour correspond to the kinetic energy value indicated, and $T_0 < T_1 < T_2$. (a) When mass coordinates $\dot{\psi}_1$ and $\dot{\psi}_2$ are used the contours are elliptical. (b) With a suitable change to new coordinates $\dot{q}_1'$ and $\dot{q}_2'$ the contours become circular.

to give the curve

$$\tfrac{1}{2}m_1\dot{\psi}_1^2 + \tfrac{1}{2}m_2\dot{\psi}_2^2 = T_0 \tag{8.21}$$

This curve is an ellipse whose principal axes lie along the $\dot{\psi}_1$ and $\dot{\psi}_2$ axes, intersecting them at the points $\pm(2T_0/m_1)^{1/2}$ and $\pm(2T_0/m_2)^{1/2}$ respectively.

In the final system of coordinates the kinetic energy contours are circles. The $T = T_0$ contour, for example, is

$$\tfrac{1}{2}\dot{q}_1^2 + \tfrac{1}{2}\dot{q}_2^2 = T_0$$

which has a radius $(2T_0)^{1/2}$. As a first step, therefore, we decide to squeeze the ellipses of fig. 8.9(a) into the circles of fig. 8.9(b) by *changing the scales of both axes*, going to an intermediate system of coordinates q_1' and q_2' defined by

$$q_1' \equiv m_1^{1/2}\psi_1$$
$$q_2' \equiv m_2^{1/2}\psi_2 \tag{8.22}$$

The q' coordinates have the same units ($\mathrm{kg}^{1/2}\,\mathrm{m}$) as the q's, and the kinetic energy has the special form required. We are not entitled to call q_1' and q_2' mode coordinates, however, since V is not yet in the correct form.

Transforming V. We must first find out what our scale changes have done to the potential energy contours. With the help of (8.22) we see from (8.16) that

$$V = \tfrac{1}{2}[(s_1+S)/m_1]q_1'^2 + \tfrac{1}{2}[(s_2+S)/m_2]q_2'^2 - (S/m_1^{1/2}m_2^{1/2})q_1'q_2'$$
$$\equiv aq_1'^2 + bq_2'^2 + cq_1'q_2' \tag{8.23}$$

The new coefficients a, b and c are introduced as a temporary measure, merely to simplify the notation.

A contour map for V in the q' coordinate system is drawn in fig. 8.10(a). Because of the cross-product term with the negative coefficient in (8.23) each contour is pushed out in the first and third quadrants. (We have to travel farther out from the origin to reach a given value of V.) Conversely, it is pulled in in the second and fourth quadrants. The curves are in fact ellipses whose major axes lie at some common angle θ to the q_1' axis; the $V = V_0$ contour has the equation

$$aq_1'^2 + bq_2'^2 + cq_1'q_2' = V_0$$

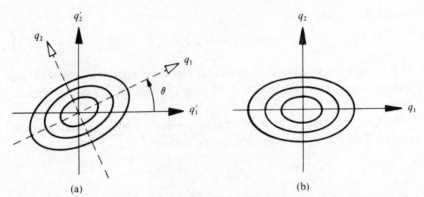

Fig. 8.10 Potential energy contours. (a) When the coordinates q'_1 and q'_2 are used, the major axes of the ellipses make an angle θ with the coordinate axes. (b) By rotating the coordinate axes through θ, the axes of the ellipses can be made to lie along the coordinate axes.

The situation we wish to achieve is the one in which the cross-product has vanished from V, and the $V = V_0$ ellipse, for example, has become

$$\tfrac{1}{2}\omega_1^2 q_1^2 + \tfrac{1}{2}\omega_2^2 q_2^2 = V_0$$

This ellipse, like that of (8.21), has its principal axes lying along the coordinate axes; it intersects the q_1 axis at $\pm(2V_0/\omega_1^2)^{1/2}$ and the q_2 axis at $\pm(2V_0/\omega_2^2)^{1/2}$. What we require, therefore, is a simple anticlockwise *rotation of the coordinate axes* to the positions q_1 and q_2 shown in fig. 8.10(a); plotted on these axes the contour map will look like fig. 8.10(b). An axis rotation will satisfy the requirement that the form of T should be unimpaired, since we have already turned the kinetic energy contours into circles, which can be rotated about their centres without affecting our results.

It only remains to discover the size of the angle, shown as θ in fig. 8.10(a), through which we must rotate. The relation between the coordinates of a given point before and after a rotation θ is (fig. 8.11)

$$q'_1 = q_1 \cos \theta - q_2 \sin \theta$$

$$q'_2 = q_1 \sin \theta + q_2 \cos \theta$$

(8.24)

Using this relation to transform (8.23), we find that the coefficient of $q_1 q_2$ will be zero if

$$2(b-a)\sin \theta \cos \theta + c(\cos^2 \theta - \sin^2 \theta) = 0$$

$$(b-a)\sin 2\theta + c \cos 2\theta = 0$$

(8.25)

$$\tan 2\theta = -\frac{c}{b-a} = \frac{2m_1^{1/2}m_2^{1/2}S}{m_1 s_2 - m_2 s_1 + (m_1 - m_2)S}$$

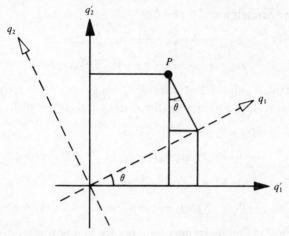

Fig. 8.11 Rotation of coordinate axes. Coordinates q_1 and q_2 of P are the lengths of the long sides of the two shaded triangles; coordinates q'_1 and q'_2 are the lengths of the sides of the large rectangle. Subtract the horizontal sides of the triangles to find q'_1 in terms of q_1 and q_2; add the vertical sides of the triangles to find q'_2. The answers are given in equation (8.24).

The complete transformation. We combine our two steps (scale changes followed by a rotation of axes) to obtain the over-all transformation

$$\psi_1 = [(1/m_1^{1/2})\cos\theta]q_1 - [(1/m_1^{1/2})\sin\theta]q_2$$
$$\psi_2 = [(1/m_2^{1/2})\sin\theta]q_1 + [(1/m_2^{1/2})\cos\theta]q_2 \tag{8.26}$$

We can now quickly find an expression for the ratio of the displacements in each mode: since $q_2 = 0$ when only mode 1 is excited, and $q_1 = 0$ when only mode 2 is excited, we have

$$\text{(mode 1)} \quad \psi_2/\psi_1 = (m_1/m_2)^{1/2}\tan\theta$$
$$\text{(mode 2)} \quad \psi_2/\psi_1 = -(m_1/m_2)^{1/2}\cot\theta \tag{8.27}$$

Since $\tan\theta$ is positive, the masses are again in phase in mode 1, and in antiphase in mode 2. In any particular example in which the various masses and stiffnesses are given, the two values of ψ_2/ψ_1 can be obtained from (8.25) and (8.27).

Finally we use (8.24) to transform (8.23) once more, looking this time for the coefficients of q_1^2 and q_2^2 which we know (8.12) to be $\frac{1}{2}\omega_1^2$ and $\frac{1}{2}\omega_2^2$ respectively. We find

$$\tfrac{1}{2}\omega_1^2 = a\cos^2\theta + b\sin^2\theta + c\cos\theta\sin\theta$$

$$\tfrac{1}{2}\omega_2^2 = a\sin^2\theta + b\cos^2\theta - c\cos\theta\sin\theta$$

From these we obtain, with the aid of standard relations connecting trigonometric functions of θ and 2θ,

$$\omega_1^2 = a + b + (a - b + c \tan 2\theta) \cos 2\theta$$

$$\omega_2^2 = a + b - (a - b + c \tan 2\theta) \cos 2\theta$$

Taking $\tan 2\theta$ and $\cos 2\theta$ from (8.25), we get the mode frequencies in terms of the system parameters (the masses and stiffnesses),

$$2\omega_1^2 = (s_1 + S)/m_1 + (s_2 + S)/m_2$$

$$-\{[(s_1 + S)/m_1 - (s_2 + S)/m_2]^2 + 4S^2/m_1 m_2\}^{1/2}$$

$$2\omega_2^2 = (s_1 + S)/m_1 + (s_2 + S)/m_2$$

$$+\{[(s_1 + S)/m_1 - (s_2 + S)/m_2]^2 + 4S^2/m_1 m_2\}^{1/2}$$

Having found the mode frequencies, we are in a position to write down the uncoupled equations of motion (8.8), leaving the way clear for the solution of any specific problem we may be given.

Figure 8.12 shows how the two mode frequencies vary with the coupling strength, with values chosen to give $s_2/m_2 = 6s_1/m_1$.

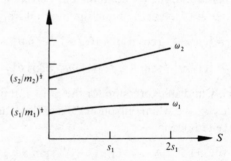

Fig. 8.12 Dependence of the mode frequencies on the coupling strength, for the system in fig. 8.8 with $m_1 = 3m_2$ and $2s_1 = s_2$ (giving $s_2/m_2 = 6s_1/m_1$).

We are now in a position to see where our original transformation (8.6) came from, since the results in this section may all be applied to the symmetric system by putting $m_1 = m_2 = m$ and $s_1 = s_2 = s$. We then have $a = b$ (8.23), so that $\theta = \frac{1}{4}\pi$ (8.25), a result we might have foreseen by looking at fig. 8.10(a) and remembering that the symmetry has obliterated all means of telling which coordinate is q_1' and which is q_2'. Putting $\cos \theta = \sin \theta = 1/2^{1/2}$ in (8.26) leads directly to (8.7).

Summary. This example has been worked through in complete detail to illustrate the method of solving two-coordinate vibration problems by means of mode coordinate transformations.

The mode coordinates are the ones which give T and V the special forms (8.12) corresponding to kinetic energy contours which are circles of radius $(2T)^{1/2}$ and potential energy contours which are ellipses lined up on the coordinate axes. The necessary manipulation of the coordinate axes comprised, in this example, two independent scale changes and a rotation. The transformation to mode coordinates is the key to the whole problem, for, having found it, we may:

(1) read off the mode frequencies by inspecting the coefficients of q_1^2 and q_2^2 in the new V;

(2) deduce the relative amplitudes and the phase constants of the two masses in either mode, by setting the other mode coordinate to zero in the transformation;

(3) use the straightforward techniques of chapter 1 to solve the equations of motion under any given initial conditions, these equations now being of the tractable uncoupled variety.

8.4. Coupled circuits

One of the most important examples of a two-coordinate vibrator arises through the inductive coupling of two simple circuits of the kind discussed in section 2.5. We shall use the general method developed in section 8.3 to analyze the free oscillations of such a system (fig. 8.13).

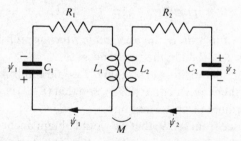

Fig. 8.13 Coupled circuits. Clockwise currents and e.m.f.'s are taken as positive. The figure shows the relative signs of the charges on the capacitor plates when ψ_1 and ψ_2 are positive. The results derived in the text assume that the coupling is such that a clockwise current in one circuit induces an anticlockwise e.m.f. in the other circuit. (Results for the opposite situation can be obtained by making the mutual inductance M negative.)

The circuits are coupled because the current oscillating in one circuit sets up an oscillating magnetic flux through the other circuit, and so induces an oscillating e.m.f. in that circuit. If we adopt the familiar convention that clockwise currents and e.m.f.'s are labelled positive, the equations of 'motion' are

$$L_1\ddot{\psi}_1 + M\ddot{\psi}_2 + (1/C_1)\psi_1 + R_1\dot{\psi}_1 = 0$$
$$L_2\ddot{\psi}_2 + M\ddot{\psi}_1 + (1/C_2)\psi_2 + R_2\dot{\psi}_2 = 0 \tag{8.28}$$

The terms coupling these equations are linear in $\ddot{\psi}$ rather than ψ. To solve them, we must first extend (8.17) to include the terms in $\dot{\psi}_1$ and $\dot{\psi}_2$ that also appear. These new terms are very similar to the existing terms in ψ_1 and ψ_2, and we ought to be able to rewrite them in the form $\partial U/\partial \dot{\psi}_1$ and $\partial U/\partial \dot{\psi}_2$, just as the ψ terms became $\partial V/\partial \psi_1$ and $\partial V/\partial \psi_2$ in (8.17). The complete equations will then be

$$\frac{\mathrm{d}}{\mathrm{d}t} \cdot \frac{\partial T}{\partial \dot{\psi}_1} + \frac{\partial V}{\partial \psi_1} + \frac{\partial U}{\partial \dot{\psi}_1} = 0$$
$$\frac{\mathrm{d}}{\mathrm{d}t} \cdot \frac{\partial T}{\partial \dot{\psi}_2} + \frac{\partial V}{\partial \psi_2} + \frac{\partial U}{\partial \dot{\psi}_2} = 0 \tag{8.29}$$

The behaviour of the system is now described in terms of the three functions T, V and U. If we write down T and V for our coupled circuit, it is easy to invent a suitable U by noticing how the ψ's and $\dot{\psi}$'s appear in (8.28). This gives us the set

$$T = \tfrac{1}{2}L_1\dot{\psi}_1^2 + \tfrac{1}{2}L_2\dot{\psi}_2^2 + M\dot{\psi}_1\dot{\psi}_2$$
$$V = \tfrac{1}{2}(1/C_1)\psi_1^2 + \tfrac{1}{2}(1/C_2)\psi_2^2 \tag{8.30}$$
$$U = \tfrac{1}{2}R_1\dot{\psi}_1^2 + \tfrac{1}{2}R_2\dot{\psi}_2^2$$

We call the new function U the *dissipation function*; it is half the rate at which power is absorbed in the resistances.

Arguments similar to those of section 8.3 would confirm that (8.29) and (8.30) are together equivalent to (8.28), and that (8.29) is invariant under any linear coordinate transformation.

We can also see from (8.29) that we would obtain uncoupled equations of motion for damped vibrations if we were able to remove the cross-product terms from T, V and U. Figure 8.14 shows, however, that this is not possible in general: we cannot put all three sets of ellipses simultaneously into the required form just by stretching and rotating the coordinate axes.

If, however, we happen to be dealing with a particular circuit for which

$$R_1C_1 = R_2C_2 \tag{8.31}$$

then the V and U ellipses in fig. 8.14 will have equal eccentricities, and a suitable sequence of steps can be found. We first rotate to bring the T

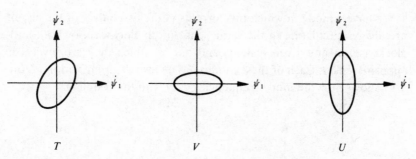

Fig. 8.14 It is in general impossible to line up the T, V and U contours along the same coordinate axes. Scale changes which make V circular will not make U circular, and a rotation to align T will turn U out of alignment.

contours on to the axes, and change scales to turn the ellipses into circles; then, because the V and U contours have the same shape, one more rotation will bring them simultaneously on to the axes. In general, however, *only certain simple damped systems can be treated by the method of mode coordinates.*†

Example. The completely symmetric circuit with $L_1 = L_2 \equiv L$, $C_1 = C_2 \equiv C$ and $R_1 = R_2 \equiv R$ satisfies the condition (8.31). Application of the method to this example leads to uncoupled equations

$$\ddot{q}_1 + [R/(L+M)]\dot{q}_1 + [1/C(L+M)]q_1 = 0$$
$$\ddot{q}_2 + [R/(L-M)]\dot{q}_2 + [1/C(L-M)]q_2 = 0 \tag{8.32}$$

These show that both q_1 and q_2 will vary in a damped harmonic manner. (We assume that R is small enough to make the damping light.) Identification of the coefficients in (8.32) yields the mode frequencies, given by

$$\omega_1^2 = 1/C(L+M)$$
$$\omega_2^2 = 1/C(L-M) \tag{8.33}$$

and a corresponding pair of widths

$$\gamma_1 = R(L+M)$$
$$\gamma_2 = R(L-M) \tag{8.34}$$

The frequency of a free oscillation in either mode can now be found from a formula like (3.7) or (3.9).

---

† It turns out that, even in cases for which no mode coordinates exist, there are always two states of vibration in which ψ_1 and ψ_2 execute damped harmonic motion at the same frequency, but *do not go through zero at the same times*.

We have a mode in which the currents in the two halves of the circuit are always circulating in the same sensè (both clockwise, or both anti-clockwise, at any given instant) and one in which they are always in opposite senses. Each of these modes has its own damping width. Figure 8.15 shows how the mode frequencies vary with the coupling.

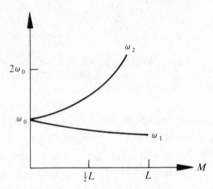

Fig. 8.15 How the mode frequencies vary with the mutual inductance, for identical coupled circuits.

As usual the simplest behaviour is seen when only one mode is excited. With both modes excited the oscillation appears complicated at first but, because $\gamma_1 \neq \gamma_2$, one mode will die out faster than the other and there will come a time when only one mode is appreciable. Here then is a third way of making a system vibrate in a single mode: we can start an arbitrary vibration and wait for one of the modes to be damped out.

Summary. The procedure for finding mode coordinates can be extended to damped examples with the help of the dissipation function, but it will only work for certain special systems. Only these special systems possess modes in which both coordinates go through zero together.

Problems

8.1   For the symmetric system (fig. 8.1) with weak coupling ($S \ll s$) show that the fraction by which the larger mode frequency exceeds the smaller mode frequency is approximately S/s.

8.2   At a certain instant during the vibration of the symmetric system (fig. 8.1) the mode coordinates have the values $q_1 = 1.36 \times 10^{-3} \text{ kg}^{1/2} \text{ m}$ and $q_2 = -0.34 \times 10^{-3} \text{ kg}^{1/2} \text{ m}$. Calculate ψ_1 and ψ_2 at the same instant, if $m = 0.020$ kg.

8.3   The symmetric system of fig. 8.1 is set into forced vibration by moving the mid point of the centre spring harmonically with angular frequency ω. Show that, in the steady state, the 'in-antiphase' mode is not excited at any driving frequency. (This illustrates the general point that the mode mixture depends on how the driving force is applied to a system.)

8.4 If the system in fig. 8.8 has $m_1 = 3m_2$ and $s_1 = s_2 = 2S$, find (a) the amplitude ratio A_2/A_1 for the 'in-phase' mode; (b) the amplitude ratio A_2/A_1 for the 'in-antiphase' mode; (c) the ratio of the mode frequencies.

8.5 Obtain expressions for the mode coordinates in terms of ψ_1, ψ_2 and m for the system of fig. 8.8, when $m_1 = m_2 \equiv m$ and $2S = 3^{1/2}(s_2 - s_1)$.

8.6 Two identical simple pendulums are connected by a light spring attached to the bobs. Each bob has a mass of 1.00 kg, and the stiffness of the coupling spring is 0.800 N m^{-1}. When one pendulum is clamped, the period of the other is found to be 1.25 s. Find the periods of the two modes of the system with the clamp removed.

8.7 Figure 8.16 shows 100 mJ kinetic and potential energy contours for a two-coordinate system in a certain system of coordinates (x_1, x_2). (a) Express the mode coordinates in terms of x_1 and x_2. (b) Calculate the mode frequencies of the system.

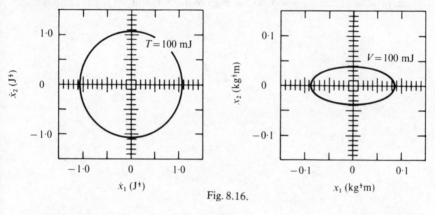

Fig. 8.16.

8.8 Consider the forced vibrations of the *asymmetric* system of fig. 8.8, with $s_2 = 0$. The harmonic driving force is applied to mass 1. Show that, in the steady state, mass 1 will have zero amplitude at the driving frequency for which $\omega^2 = S/m_2$.

8.9 Consider the resistance-coupled circuit shown in fig. 8.17. Find (a) the mode

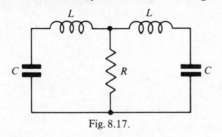

Fig. 8.17.

frequencies, and (b) the mode widths. (c) Explain physically why one of the modes is not damped out but continues indefinitely.

8.10   Most instruments of the violin family possess a 'wolf note'. We can understand the wolf note in a simplified way by thinking of the instrument playing it as a pair of weakly coupled vibrators consisting of the stopped string and the main body resonance of the instrument. The throbbing sound heard when the wolf note is played may be identified with the beating modes of this system.

The author's son has a cello with a wolf note at a frequency of 175 Hz (F below middle C) which throbs about 10 times per second. Estimate the mode frequencies.

8.11   A friend of the author was puzzled by the behaviour of a recently inherited grandfather clock. The suspended weight which drives the clock was wound up every Sunday morning. The clock should then have kept going for over a week, but it always stopped during the night between Thursday and Friday. The stoppage was accompanied by the sound of the weight knocking against the inside of the case. When the clock was examined after one of these stoppages, the distance between the centre of gravity of the weight and its point of suspension was found to be 0.99 m.

Explain the phenomenon with the help of fig. 8.4, and suggest a remedy.

9

Non-dispersive waves

We turn our attention now to *continuous, extended systems*, in which there is no demarcation between parts that contribute mass and those that contribute stiffness. We noted in passing that the air in a flask (section 2.2) was such a system, though we were able to make a notional separation into a mass part and a stiffness part for the purposes of an approximate analysis. As our model system we take a flexible, elastic string, initially stretched along the z axis. We consider *transverse* motion of points on the string in a fixed plane containing the z axis.

Our starting point must be an equation of motion suitable for any point on the string. We apply Newton's second law to a short segment initially lying between the points z and $z + \Delta z$ (fig. 9.1(a)). If the mass per unit length of the string is μ, the mass of the segment will be $\mu \Delta z$. We assume that each point on the string moves strictly at right angles to the z axis. At some instant our segment may appear as in fig. 9.1(b). The segment has been stretched by an amount which depends on the

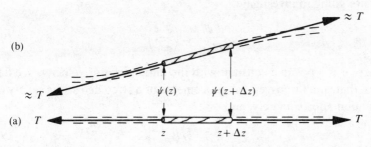

Fig. 9.1 A segment of the stretched string (a) in its equilibrium position, and (b) displaced to some new position. It is assumed that every point is displaced strictly at right angles to the original line of the string.

angle it makes with the axis, but this stretching will have had no effect on its mass.

For small displacements, we may assume that the tension at all points in the string remains close to its equilibrium value T. Newton's second law then says

$$\mu \, \Delta z \left(\frac{\partial^2 \psi}{\partial t^2}\right) \approx -T\left(\frac{\partial \psi}{\partial z}\right)_z + T\left(\frac{\partial \psi}{\partial z}\right)_{z+\Delta z} \tag{9.1}$$

This equation contains partial derivatives with respect to z and t, since the displacement $\psi(z, t)$ depends on both.

Clearly the return force on the right of (9.1) will be zero unless $\partial \psi/\partial z$ varies from point to point along the string: the tension therefore provides a net transverse force on the segment only when the string is *curved*. For a curved string, the change in $\partial \psi/\partial z$ on going from z to $z + \Delta z$ is

$$\Delta\left(\frac{\partial \psi}{\partial z}\right) = \left(\frac{\partial^2 \psi}{\partial z^2}\right) \Delta z$$

and the equation of motion (9.1) becomes

$$\mu \Delta z \left(\frac{\partial^2 \psi}{\partial t^2}\right) \approx T\left(\frac{\partial^2 \psi}{\partial z^2}\right) \Delta z$$

Cancelling Δz, we obtain finally

$$\frac{\partial^2 \psi}{\partial t^2} \approx \frac{T}{\mu}\left(\frac{\partial^2 \psi}{\partial z^2}\right) \tag{9.2}$$

Any motion of the string must satisfy (9.2). The string is, of course, only a prototype of other, more interesting systems. As in the case of the prototype vibrator, we can make the results more general by using more general notation. We therefore write, as a standard form of the equation we are going to investigate,

$$\frac{\partial^2 \psi}{\partial t^2} = c^2\left(\frac{\partial^2 \psi}{\partial z^2}\right) \tag{9.3}$$

where c is a positive constant with the dimensions of velocity. We have seen that small transverse displacements of a stretched string satisfy this equation approximately, and that

$$c = |(T/\mu)^{1/2}| \tag{9.4}$$

for that particular system. We shall meet systems which obey similar equations in chapter 10.

9.1. Travelling waves

The partial differential equation (9.3) which we now investigate is linear
and homogeneous, and has constant coefficients. In section 1.2 we solved
the standard vibrator equation (1.4), which also has these properties, by
the method of substituting an exponential trial solution whose exponent
was linear in the variables involved: in that case simply t. We found then
that a purely imaginary exponent produced harmonic vibrations of ψ
and that we should add two terms which were complex conjugates of
each other (1.16), in order to keep ψ real.

We now have two variables t and z, and so we may try the form C
solution

$$\psi = C \exp\left[i(\omega t - kz)\right] + C^* \exp\left[-i(\omega t - kz)\right] \qquad (9.5)$$

The minus sign in front of the constant k is there purely for convenience,
and does not restrict the solution because k can be positive or negative.
The most general solution will allow for both possibilities; here we
investigate the motion when k is *either* positive *or* negative. A solution of
this form, if it is acceptable, will correspond to a motion in which every
point on the string vibrates harmonically with angular frequency ω
(presumed to be positive), since fixing the value of z simply leads to (1.16)
with an extra term $-kz$ in the phase constant.

By putting (9.5) into (9.3) we find immediately that the solution is
acceptable provided that ω and k satisfy

$$\omega^2 = c^2 k^2$$
$$\omega = \pm ck \qquad (9.6)$$

We can write (9.5) in other ways, just as we did when discussing
vibrations (1.22). The four available forms are

$$\psi = A \cos(\omega t - kz + \phi)$$
$$\psi = B_p \cos(\omega t - kz) + B_q \sin(\omega t - kz)$$
$$\psi = C \exp\left[i(\omega t - kz)\right] + C^* \exp\left[-i(\omega t - kz)\right] \qquad (9.7)$$
$$\psi = \text{Re}\{D \exp\left[i(\omega t - kz)\right]\}$$

The various constants are related to each other exactly as before (1.23).

Sinusoidal travelling waves. We concentrate on form A to investigate the
salient features of the motion.

We have already observed that each point on the string vibrates
harmonically, and that all points have the same frequency; they also have

a common amplitude. The phase constant, however, increases or decreases linearly with distance z along the string, and it is this feature which is most characteristic of the motion.

At any given moment (fixed t) the shape of the string is sinusoidal, and repeats itself in a distance $2\pi/|k|$. If, however, we concentrate our attention on a particular point on the string (fixed z) while the time increases slightly from t to $t + \Delta t$, we shall see the displacement change from $\psi(z, t)$ to a new value $\psi(z, t + \Delta t)$. The displacements of other points nearby change also, and there will be one of these points, at $z + \Delta z$ say, where the displacement $\psi(z + \Delta z, t + \Delta t)$ is equal to the original displacement $\psi(z, t)$ at point z. These displacements $\psi(z, t)$ and $\psi(z + \Delta z, t + \Delta t)$ will be equal if the corresponding phase angles are equal, which means

$$\omega t - kz = \omega(t + \Delta t) - k(z + \Delta z)$$

$$\omega \Delta t - k \Delta z = 0$$

$$\frac{\Delta z}{\Delta t} = \frac{\omega}{k}$$

For vanishingly small values of Δz and Δt we may write

$$\frac{dz}{dt} = \frac{\omega}{k} \equiv v_\phi \tag{9.8}$$

Since our argument is equally valid for all values of z, (9.8) tells us that the whole sinusoidal profile moves along to left or to right at a speed v_ϕ, which is known as the *phase velocity* because it denotes the velocity of a point of fixed phase angle. Our original decision (9.5) to write the z term in the phase angle as $-kz$ makes the sign of v_ϕ the same as the sign of k: the profile moves in the positive z direction if k is positive, and *vice versa*. It is essential to realise that the vibration of the string itself is entirely transverse: no point on the string ever moves forward with the travelling profile.

The motion is known as a *sinusoidal travelling wave*. Any differential equation, such as (9.3), which is satisfied by sinusoidal travelling wave solutions (9.7) is called a *wave equation*. There are other forms of wave equation, but (9.3) is the simplest. We shall meet some others in later chapters.

The actual value of the phase velocity v_ϕ for any particular system depends on how ω and k are related to each other in that system (9.8). For a stretched string the relevant equation is (9.6), which gives

$$v_\phi = \pm c \tag{9.9}$$

Since the value of c is fixed by (9.4) the magnitude of v_ϕ depends only on T and μ for the string, and not on any property of the wave, such as its frequency. As we have seen, T and μ respectively determine the size of the return force acting on any segment of the string, and its inertia; they thus have similar functions to the stiffness and the mass of a simple vibrator. Since sending a wave along the string involves the acceleration of masses, it is reasonable to find that high phase velocities are associated with large forces (controlled by T) and a small mass per unit length. We shall find, however, that v_ϕ is *not* independent of the wave properties in all systems.

The distance after which the pattern on the string starts to repeat itself is known as the *wavelength*

$$\lambda \equiv 2\pi/|k| \tag{9.10}$$

In terms of the wavelength we can write

$$c = \nu\lambda \tag{9.11}$$

where ν is the frequency with which each point on the string vibrates (the *wave frequency*). The last equation relates the speed, the wave frequency and the wavelength for *any* sinusoidal travelling wave on the string, but it says nothing about the direction since c, ν and λ were all defined as positive quantities.

We have not so far given a name to the all-important quantity k that characterizes a sinusoidal travelling wave. In chapter 17 we shall meet it in its three-dimensional form as a vector $\mathbf{k}$. We anticipate, therefore, calling k the *wavevector*. Other names in use are circular wavevector (in parallel with the term angular frequency for ω) and propagation vector.

Other travelling waves. We have found that, when each point on the string vibrates harmonically according to (9.7), constant values of the phase angle move along the string to right or to left at speed c. By rearranging the form A expression for ψ with the help of (9.6) to read

$$\psi(z, t) = A \cos\left[(\omega/c)(ct \mp z + c\phi/\omega)\right]$$

we highlight a significant fact: the phase angle will remain constant as long as the special variable

$$Z \equiv ct \mp z$$

does so. The fact that Z does not contain either of the 'harmonic' quantities ω and k suggests that harmonic vibration may not be an essential feature of a travelling wave.

In fact we may easily verify that *any* function $\psi(z, t)$ in which z and t appear only in the form $ct \mp z$ will satisfy (9.3). We have

$$\left(\frac{\mathrm{d}\psi}{\mathrm{d}Z}\right)\left(\frac{\partial Z}{\partial t}\right) = c\left(\frac{\mathrm{d}\psi}{\mathrm{d}Z}\right)$$

$$\left(\frac{\mathrm{d}\psi}{\mathrm{d}Z}\right)\left(\frac{\partial Z}{\partial z}\right) = \mp\left(\frac{\mathrm{d}\psi}{\mathrm{d}Z}\right)$$

and thus

$$\frac{\partial\psi}{\partial t} \pm c\left(\frac{\partial\psi}{\partial z}\right) = 0 \tag{9.12}$$

must hold for any function $\psi(ct \mp z)$.

Since $\partial\psi/\partial t$ and $\partial\psi/\partial z$ also depend on $ct \mp z$ only, it follows that

$$\frac{\partial^2\psi}{\partial t^2} = \mp c\left[\mp c\left(\frac{\partial^2\psi}{\partial z^2}\right)\right] = c^2\left(\frac{\partial^2\psi}{\partial z^2}\right)$$

and the wave equation is satisfied.

We have discovered something very general indeed. At time $t = 0$ the shape of the string is described by a function $\psi(\mp z)$, which can be anything at all: we are not restricted to sinusoidal shapes. Furthermore, at position $z = 0$ the motion is described by a function $\psi(ct)$, which can also be anything we like: we are not restricted to harmonic vibration, and *we are not even restricted to periodic motion*. Any 'disturbance' of the string from its (straight) equilibrium shape may be propagated to left or to right at a speed c. The propagated shape is known as the *waveform*.

This property is unique to the wave equation (9.3) under discussion. A system obeying this wave equation is said to be *non-dispersive*, and we shall call the equation itself the *non-dispersive wave equation*. This chapter and the next are about non-dispersive waves.

All possible disturbances of the string obey (9.3), but only travelling waves obey (9.12). For convenience we shall invent the name *travelling wave equation* for (9.12). It says that the transverse speed of any point on the string and the slope of the string at that point are proportional to each other, in a travelling wave. The plus sign in (9.12) corresponds to a disturbance moving in the positive z direction, and the minus sign to one moving in the negative z direction. If we are thinking only about travelling waves in a given system, we can use a travelling wave equation in preference to the general wave equation for the system.

One example of an aperiodic disturbance is shown in fig. 9.2. Points on the string which are initially to the right of the disturbance remain at rest until the disturbance arrives. As it passes they undergo some kind of

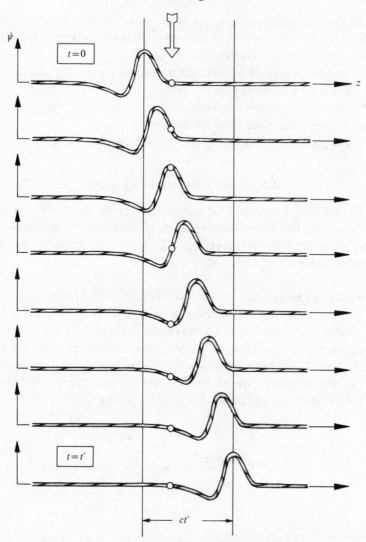

Fig. 9.2 A pulse travelling along a string from left to right. (Time increases from top to bottom.) The specimen point on the string moves only in the transverse direction. Other points move in exactly the same way, but earlier or later than the point marked.

transverse motion, returning to rest once more at $\psi = 0$ when the disturbance has passed. Every point is disturbed in exactly the same way, but different points are disturbed at different times. A disturbance which is over in a short time is called a *pulse*.

Summary. A non-equilibrium profile of any shape can be propagated along a stretched string as a travelling wave. The speed at which the profile travels depends only on the mass per unit length of the string and its tension. Although the disturbance moves along the string, the actual motion of the string is entirely transverse.

In the special case of a sinusoidal travelling wave, all points on the string vibrate harmonically, and

(1) all points have the same frequency;

(2) all points have the same amplitude;

(3) the phase constant varies linearly with distance along the string.

9.2. Reflection of travelling waves

So far we have imagined that the string extends indefinitely in both directions. In this section we investigate the situation at one end of the string when the other end is still very far away. In section 9.3 we shall consider both ends of the string at once.

Characteristic impedance. We choose the end of the string as the point $z = 0$ and assume that the other end is at $z = +\infty$. A disturbance $\psi(t, z)$, where ψ is a function of $ct - z$, travels at a speed c in the positive z direction. Something must be causing this disturbance, and we may invoke some kind of driving device (fig. 9.3) to give the end of the string the required transverse movement. We wish to know what kind of force the driver must apply to produce a given travelling wave.

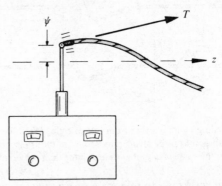

Fig. 9.3 A travelling wave is sent along the string from left to right by means of a driver which moves the end of the string in the transverse direction.

At any given moment the force must balance the transverse component of the string tension at $z = 0$. This requires a driving force

$$-T\left(\frac{\partial\psi}{\partial z}\right)_{z=0} = +\frac{T}{c}\left(\frac{\partial\psi}{\partial t}\right)_{z=0} = (T\mu)^{1/2}\left(\frac{\partial\psi}{\partial t}\right)_{z=0} \qquad (9.13)$$

This force acts outwards from $\psi = 0$ if it is positive, and inwards if it is negative. We have used the travelling wave equation (9.12) with the plus sign, since we are dealing with a disturbance moving in the positive z direction.

We see that the applied force must always be proportional to the instantaneous transverse velocity of the end of the string. If the travelling wave is sinusoidal, the driving force must lead ψ by $90°$ at all times, since only then will it remain in phase with $\partial\psi/\partial t$. This is just the kind of force that would be required to balance a 'standard' drag force (3.1). Thus the end of the string acts not like a mass being accelerated or a spring being stretched, but like a resistance being dragged against. This is in spite of the fact that we have so far ignored ordinary drag forces. Looking ahead, we can appreciate that the damping action is associated with the fact that energy is being carried away along the string by the wave.

The constant of proportionality connecting the force and the velocity is called the *characteristic impedance*

$$Z_0 = |(T\mu)^{1/2}| \tag{9.14}$$

It is a property of the string and its tension only, and does not depend on the particular form of travelling disturbance involved. (The term impedance is used rather than resistance, to embrace other systems in which Z_0 may be complex: see section 9.5.)

Whenever possible we shall write formulas in terms of c and Z_0, rather than T and μ, to make them applicable to other systems. If we wish to find T or μ, we can use one of the relations

$$T = Z_0 c$$
$$\mu = Z_0/c \tag{9.15}$$

derived by solving (9.4) and (9.14). Similar equations will enable us to transform to analogous parameters belonging to other systems.

A correctly terminated string. Now we consider the other end of the string. For convenience we place this at $z = 0$, so that the driver is now at $z = -\infty$. We expect the motion of the string to be affected by the mechanical conditions at $z = 0$; these are very easy to discuss if we use the concept of characteristic impedance.

We can, for example, be sure that a travelling wave (in the positive z direction) can exist, provided the end of the string is attached to *any* device which presents a transverse force equal to the instantaneous transverse velocity multiplied by $-Z_0$. (The minus sign appears because we are now calculating the force exerted on the string by the device, instead of the other way round.) A second piece of identical string, attached to the first at $z = 0$ and extending to $z = +\infty$, will react on the first with just such a force, and the first string must respond identically in the two cases. In either case the first string is said to be *correctly terminated*.

A suitable terminating device for a stretched string is shown in an idealized form in fig. 9.4. It is a damper consisting of a lubricated piston

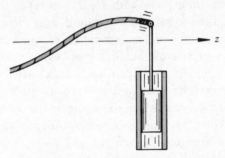

Fig. 9.4 The string may be correctly terminated by attaching the end to a damper whose resistance is equal to the characteristic impedance of the string. (The mass of the piston is negligible.)

whose mass is negligible, free to slide in a cylinder at right angles to the z axis. By suitable choice of lubricant we can arrange for the damper impedance to be 'matched' to the characteristic impedance of the string in use. Conditions at any point on the string will then be the same as they would be if the string extended to infinity in the positive z direction.

Incorrect termination: reflection. It is easy to see that the motion must be more complicated if the damper impedance is not equal to the string impedance. By arguing as we did in arriving at (9.13), we can find that the string (characteristic impedance Z_1, say) will exert on the damper a transverse force approximately given by

$$-T\left(\frac{\partial \psi}{\partial z}\right)_{z=0} = Z_1\left(\frac{\partial \psi}{\partial t}\right)_{z=0}$$

when a wave travels in the positive z direction. The force produced at the end of the string by the damper (impedance Z_2) will be $-Z_2(\partial \psi/\partial t)_{z=0}$ and, if $Z_2 \neq Z_1$, we shall have an unbalanced transverse force

$$(Z_1 - Z_2)\left(\frac{\partial \psi}{\partial t}\right)_{z=0} \neq 0$$

acting at the *point* $z = 0$. This will produce a very large acceleration there, since the damper piston has negligible mass. Such an acceleration is not consistent with a simple travelling wave, and so we are obliged to consider something more elaborate.

We turn therefore to the *general* solution of the wave equation: a combination of arbitrary disturbances travelling both to the right and to the left. We shall denote these by $\psi_i(t, z)$ and $\psi_r(t, z)$ respectively. (We shall find that the suffixes i and r stand for 'incident' and 'reflected'.) The displacement of the string is then given by the superposition

$$\psi(t, z) = \psi_i(t, z) + \psi_r(t, z)$$

We know that ψ_i and ψ_r travel at the same speed c. We now wish to discover what further relation must exist between them if the boundary conditions at $z = 0$ are to be satisfied.

The essential condition is that the transverse force exerted by the string at $z = 0$ should balance the drag force produced by the damper. The string force is approximately

$$-T\left(\frac{\partial \psi_i}{\partial z}\right)_{z=0} - T\left(\frac{\partial \psi_r}{\partial z}\right)_{z=0} = Z_1\left(\frac{\partial \psi_i}{\partial t} - \frac{\partial \psi_r}{\partial t}\right)_{z=0}$$

We have used the travelling wave equation (9.12) once again, choosing the plus sign for ψ_i, and the minus sign for ψ_r. Equating this force to the drag force gives

$$Z_1\left(\frac{\partial \psi_i}{\partial t} - \frac{\partial \psi_r}{\partial t}\right)_{z=0} = Z_2\left(\frac{\partial \psi_i}{\partial t} + \frac{\partial \psi_r}{\partial t}\right)_{z=0}$$

$$\left(\frac{\partial \psi_r}{\partial t}\right)_{z=0} = \frac{Z_1 - Z_2}{Z_1 + Z_2}\left(\frac{\partial \psi_i}{\partial t}\right)_{z=0}$$

Integrating both sides with respect to time, we find

$$\psi_r(t, 0) = \left(\frac{Z_1 - Z_2}{Z_1 + Z_2}\right)\psi_i(t, 0) \tag{9.16}$$

(No constant of integration is necessary: obviously ψ_r must be zero if ψ_i is zero.)

The last equation shows how ψ_r must be related to ψ_i at the point $z = 0$, at any time t. From it we can deduce a corresponding relation connecting ψ_i and ψ_r at any other point on the string. We choose the point $z = -l$ where l is some positive length.

Since ψ_i is a function of $ct - z$, we may write

$$\psi_i(t - l/c, -l) = \psi_i(t, 0)$$

This says simply that ψ_i has the same value at $z = 0$ as it had at $z = -l$, at a time l/c in the past. Similarly for ψ_r, a function of $ct + z$, we have

$$\psi_r(t + l/c, -l) = \psi_r(t, 0)$$

Since ψ_r travels to the left, its value at $z = 0$ will be reproduced at $z = -l$, at a time l/c in the future.

Putting these relationships into (9.16) we find

$$\psi_r(t + l/c, -l) = \left(\frac{Z_1 - Z_2}{Z_1 + Z_2}\right)\psi_i(t - l/c, -l)$$

which says the following: the displacement produced at $z = -l$ by the passage of ψ_r is proportional to the displacement produced there at a time $2l/c$ in the past, by the passage of ψ_i. This can only remain true for all values of l if the shape of ψ_r copies the shape of ψ_i, but in the reverse direction along the z axis. It must also be scaled down in the transverse direction by a factor

$$R = \frac{Z_1 - Z_2}{Z_1 + Z_2} \tag{9.17}$$

Since the time delay $2l/c$ is just the time it would take for ψ_i to travel to the end of the string and back again, and because ψ_r is a scaled down mirror image of ψ_i, we describe the process occurring at the end of the string as *partial reflection*. The quantities ψ_i and ψ_r are known respectively as the *incident disturbance* and the *reflected disturbance*, and R is called the *reflection coefficient*.

The superposition of ψ_i and ψ_r at various times to produce the resultant string shape is illustrated in fig. 9.5(a) for the improbable† but pictorial case of a triangular pulse, arriving at a termination for which $Z_2 = \frac{1}{2}Z_1$ (i.e. $R = \frac{1}{3}$). For greater clarity the mathematical forms of ψ_i and ψ_r are plotted for both positive and negative values of z, although they have physical significance only when $z < 0$.

---

† A travelling disturbance with absolutely sharp corners could not occur in practice, since a sudden change of slope would imply a sudden change of transverse velocity (9.12), calling for an infinitely large force.

Fig. 9.5 (a) Reflection of a travelling disturbance at an impedance equal to half the characteristic impedance of the string. To illustrate how the string shape is constructed, the incident disturbance and the reflected disturbance are shown on both sides of the junction, although they have a physical meaning only on the left. (b) Transmission of a travelling disturbance across a junction between two strings. The characteristic impedance of the second string (on the right) is half the characteristic impedance of the first string. The transmitted disturbance is shown approaching the junction from the left (at twice the speed of the incident disturbance), but it has a physical meaning only on the right of the junction.

(a)   (b)

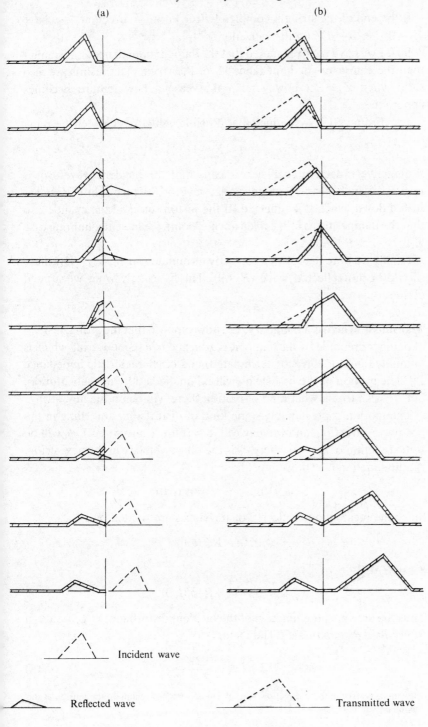

△ —————— Incident wave

◁▱⟍ —————— Reflected wave   ⟍⟋⟍ —————— Transmitted wave

If the end of the string is completely free in the ψ direction,[†] we have $Z_2 = 0$ (i.e. $R = 1$) and the reflection is 'total'.

So far we have tacitly assumed that the damper impedance Z_2 is smaller than the characteristic impedance Z_1 of the string. The results are also valid when $Z_2 \geqslant Z_1$, however. In that case a new feature becomes apparent.

From (9.17) we deduce that possible values of R cover the range

$$-1 \leqslant R \leqslant 1$$

A negative reflection coefficient means that the incident waveform is inverted (which means, for a sinusoidal wave, a $180°$ phase shift) as well as scaled down, when it is reflected. If the piston seizes in the cylinder, so that the damper turns into a fixed anchor point, we have the limiting case $Z_2 = \infty$ (i.e. $R = -1$).

Finally we note that, with a correctly terminated string ($Z_2 = Z_1$) there should be no reflected wave ($R = 0$). This is exactly what we argued earlier.

Transmission across a boundary. We now suppose that the string, instead of being terminated by the damper, is attached to a second string which is infinitely long (or correctly terminated) and of characteristic impedance Z_2. The motion of the first string will be unaffected by the substitution, but we wish to ask what kind of motion there will be on the new string.

The motion is presumably some kind of disturbance travelling in the positive z direction, and we may call it ψ_t (t for 'transmitted'); ψ_t will be some function of $c_2 t - z$, where c_2 is the phase velocity in the new string. At the junction ($z = 0$) we must have

$$\psi_i(t, 0) + \psi_r(t, 0) = \psi_t(t, 0)$$

Since the incident and reflected disturbances are related by

$$\psi_r(t, 0) = R\psi_i(t, 0)$$

we may write

$$\psi_t(t, 0) = (1 + R)\psi_i(t, 0)$$

Thus the arrival at the junction of the incident disturbance ψ_i generates a *transmitted disturbance* ψ_t. The factor

$$T = 1 + R = \frac{2Z_1}{Z_1 + Z_2} \tag{9.18}$$

---

† It cannot be free in the z direction, since we must somehow maintain the tension at the assumed value T.

by which ψ_t is magnified in the ψ direction relative to ψ_i is called the *transmission coefficient.*

In this case there is also a change of scale in the z direction, due to the difference in the phase velocities on the two strings. If the velocity for the original string is c_1, the transmitted waveform will be stretched or compressed by a factor c_2/c_1.

The formation of the transmitted wave is illustrated in fig. 9.5(b), for the same incident disturbance and the same relative values of Z_1 and Z_2 as in the previous case (termination by a damper); only values of ψ_t for $z > 0$ correspond to physical displacements of the string. The reflected wave is of course identical in the two cases.

The transmission coefficient (9.18) can have any value in the range

$$0 \leqslant T \leqslant 2$$

Since T is always positive, inversion of the transmitted wave never occurs. If Z_2 is larger than the matching value Z_1, the height of the transmitted disturbance is less than that of the incident disturbance; if Z_2 is smaller than Z_1 the transmitted disturbance is the larger.

The fact that the transmitted disturbance is sometimes larger than the incident disturbance does not carry any awkward implications such as non-conservation of energy. If, for example, the two string tensions are equal (the most common situation, since any other condition would require some sophisticated arrangement at the junction) the condition $Z_2 < Z_1$ simply means that the second string is lighter than the first (9.14). The maximum transmission coefficient of 2 occurs when Z_2 is zero: in practice a very small value of Z_2 could be obtained by making the second string very light. (The velocity would be correspondingly large; this would lead to a large longitudinal stretching factor, so that the second string would move bodily up and down, while remaining almost straight.)

For the case of equal tensions we have

$$\frac{Z_1}{Z_2} = \frac{c_2}{c_1} \tag{9.19}$$

and the reflection and transmission coefficients may be written

$$R = \frac{c_2 - c_1}{c_1 + c_2}$$

$$T = \frac{2c_2}{c_1 + c_2} \tag{9.20}$$

Summary. A device which is to send a travelling wave down an infinitely long string must apply a force that is always proportional to the transverse velocity of the end it is driving. The constant of proportionality, known as the characteristic impedance of the string, is a convenient quantity for describing events at the ends of the string.

If the string is terminated with an impedance equal to its own characteristic impedance, the simple travelling wave motion we found on an infinitely long string can still be supported. Otherwise, a reflected disturbance coexists with the incident disturbance; this reflected wave is inverted if the terminating impedance is greater than the critical value. If the end is either rigidly fixed, or free in the transverse direction, the incident and reflected waves have equal heights.

When the terminating device is a second string, an upright transmitted disturbance is found. Depending on the relative values of the characteristic impedances, the transmitted wave may be smaller or larger (by up to a factor of 2) than the incident wave. Depending on the relative values of the phase velocities, the transmitted wave may be stretched or compressed in the propagation direction.

9.3. Standing waves

It is not always possible to visualize the incident and reflected disturbances as separate entities, as we can in fig. 9.5 for example. If they are continuous wave trains, we shall only be aware of the total motion resulting from their superposition.

Let us consider what happens when a sinusoidal travelling wave

$$\psi_i = A' \cos(\omega t - kz + \phi')$$

reaches a fixed end at $z = 0$. To be definite we shall assume that the wave comes from the left ($k > 0$) though the direction does not affect the results.

Because the end of the string is fixed, the reflected wave is inverted, and we write

$$\psi_r = -A' \cos(\omega t + kz + \phi') \tag{9.21}$$

The complete motion is given by

$$\psi = A'[\cos(\omega t - kz + \phi') - \cos(\omega t + kz + \phi')]$$
$$= 2A' \sin kz \, \sin(\omega t + \phi') \tag{9.22}$$

If we now define the new quantities

$$A \equiv 2A'$$

$$\phi \equiv \phi' - \tfrac{1}{2}\pi$$

(9.23)

we can write (9.22) in the form

$$\psi = A \sin kz \cos(\omega t + \phi)$$

(9.24)

The most striking thing about this ψ is the separation of the variables t and z. At any given moment the string profile is sinusoidal, as in a travelling wave. The motion of every point on the string is harmonic, and all points have the same frequency; but *they now have a common phase constant ϕ*. All points for which $\sin kz$ is positive vibrate in phase with each other and in antiphase with the points for which $\sin kz$ is negative. There is no advance of the profile along the string such as occurs in a travelling wave.

The disturbance described by (9.24) is known as a *standing wave*. The amplitude of the vibration varies from point to point along the string, and the pattern of the vibration repeats itself in a distance λ, given by (9.10). That is, the wavelength of the standing wave is the same as that of the travelling waves. The pattern on the string, although it is now stationary, depends on the wave frequency ν and the phase velocity c in the same way as before (9.11).

At those points on the string for which

$$kz = m\pi \qquad (m = 0, \pm 1, \pm 2, \ldots)$$

(9.25)

we have $\psi = 0$ at all times. These points, which are called *nodes*, never move at all; they are spaced out at intervals $\tfrac{1}{2}\lambda$ along the string. Midway between them are the *antinodes*: the points where the amplitude is greatest.

Standing waves on a finite string. In practice the string is most likely to be of such a length that we must take *both* ends into account. For example, the standing waves on a string of length L, fixed at both ends, must have nodes at the two anchor points. The standing wave in (9.24) automatically has a node at $z = 0$, but there will be another one at $z = L$ only if $|k|$ has one of the uniformly spaced values given by

$$\sin k_n L = 0$$

$$k_n = n\pi/L \qquad (n = 1, 2, 3, \ldots)$$

(9.26)

This can be seen to restrict the wavelength to the values

$$\lambda_n = 2\pi/k_n = 2L/n \qquad (9.27)$$

The number of loops (or antinodes) fitted between the anchor points increases as k_n increases. We can see from (9.25) and (9.26) that nodes are to be found where

$$z = (m/n)L$$

In standing wave n there must therefore be $n - 1$ nodes (not counting the anchor points) and n antinodes.

The discussion so far has been about a string which is fixed at both ends. If the ends are 'free' (in the usual sense of being capable of free transverse movement) the results look slightly different. For the reflected wave at $z = 0$ we have, in place of (9.21),

$$\psi_r = +A' \cos(\omega t + kz + \phi')$$

This changes the form of the standing wave (9.22) to

$$\begin{aligned} \psi &= A' \left[\cos(\omega t - kz + \phi') + \cos(\omega t + kz + \phi') \right] \\ &= 2A' \cos kz \cos(\omega t + \phi') \end{aligned} \qquad (9.28)$$

Making the same substitutions as before (9.23) we obtain

$$\psi = A \cos kz \sin(\omega t + \phi) \qquad (9.29)$$

This solution has the necessary antinode at $z = 0$, and will have another at $z = L$ if $|k|$ has one of the values k_n given by

$$\cos k_n L = 1$$

$$k_n = n\pi/L$$

These are the same values as before (9.26) although the standing waves have different shapes.

Figure 9.6(a) shows 'photographs' of some of the standing waves in the fixed-ends case. These string profiles are plotted at heights which are proportional to their respective k_n values. In fig. 9.6(b) we compare the profiles when the ends are free.

If the string is fixed at $z = 0$ but free at $z = L$, we can use the original standing wave (9.24) and the various wavevectors are found from the condition

$$\sin k_n L = 1$$

$$k_n = (2n - 1)\pi/2L \qquad (n = 1, 2, 3, \ldots)$$

This gives an antinode at $z = L$ as required.

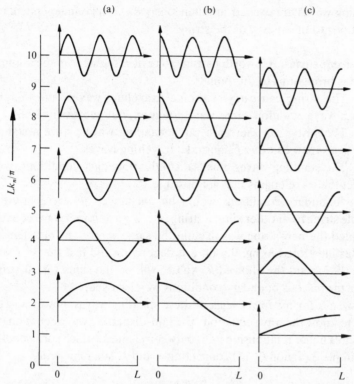

Fig. 9.6 The first five standing waves on a string of length L (a) fixed at both ends, (b) free at both ends, and (c) fixed at the left but free at the right. The plots are placed at heights proportional to their k_n values.

In this case the lowest k_n value (the one corresponding to $n = 1$) is half its previous size, but the interval between successive k_n's is the same as before. The string shapes are shown in fig. 9.6(c).

We can bring all these results together by writing

$$\psi_n(z, t) = A_n u_n(z) \cos{(\omega_n t + \phi_n)} \tag{9.30}$$

to represent *any* standing wave on a finite string. The space functions $u_n(z)$, known as the *eigenfunctions* of the system, are all sinusoidal (harmonic in space).† They have wavelengths (or k_n values) which are fixed by the boundary conditions at the ends of the string.

The nature of the motion in the standing waves, with all the parts vibrating in phase or in antiphase, reminds us of the modes of a two-coordinate system. In fact we shall see in section 11.1 that the standing waves possess the most characteristic feature of modes: when a number of

† A string which is not uniform along its length has non-sinusoidal eigenfunctions.

standing waves are excited simultaneously they make independent contributions to the energy of the string.

Eigenfrequencies. It is worth pausing to review the steps in the argument that has brought us to this point.

(1) The string can carry sinusoidal travelling waves, and therefore obeys a wave equation – *any* wave equation, not necessarily (9.3).

(2) The string can therefore carry standing waves, since a standing wave is a superposition of sinusoidal travelling waves.

(3) The standing waves must satisfy the boundary conditions, which restrict the wavelengths that are possible.

The boundary conditions would be unchanged if we replaced the flexible string by another kind of string obeying a different wave equation, provided the new string was fixed in the same way as the old. Thus the profiles illustrated in fig. 9.6 do not depend on the fact that our string obeys (9.3), and the allowed k_n values will be the same for all strings, under the various boundary conditions we discussed.

If we ask for the *frequencies* of these standing waves, however, we do need to know the wave equation. The non-dispersive wave equation (9.3) leads to a linear relationship (9.6) between ω and k, which corresponds to the simple equation (9.11) connecting ν and λ. We may write

$$\nu_n \doteq (c/2\pi)k_n$$

for the *eigenfrequency* of the mode with wavevector of magnitude k_n. For other strings obeying other wave equations we would find different relationships.

For a flexible string which obeys (9.3), therefore, the eigenfrequencies form a series similar to that of the k_n values for the appropriate boundary conditions. A string with both ends fixed, for example, will have

$$\nu_n = nc/2L \qquad (n = 1, 2, 3, \ldots) \tag{9.31}$$

These particular eigenfrequencies are *harmonics*: we may write

$$\nu_n = n\nu_1$$

where ν_1 is the *fundamental* frequency $c/2L$. Harmonic eigenfrequencies are exceptional.

Summary. The sinusoidal waves formed on a long string with one end either fixed or free are standing waves, in which all points on the string vibrate harmonically with the same phase constant.

When the length of the string is finite, only a certain series of standing waves is possible; they are characterized by uniformly spaced k_n values which depend on the boundary conditions. If the string obeys the non-dispersive wave equation the eigenfrequencies are also uniformly spaced.

9.4. Energy propagation

The stretched string provides a means of conveying energy from one point to another. The motion of the damper in fig. 9.4 continuously generates heat, and the necessary power must be provided by the driver at the other end of the string. In this section we examine the process by which power is transmitted along the string by the wave.

Energy density. A segment of elastic string can store both kinetic energy (due to its motion) and potential energy (due to its stretching). The instantaneous kinetic energy per unit length at a point on the string will be given by the expression $\frac{1}{2}\mu(\partial\psi/\partial t)^2$ which we may call the instantaneous *kinetic energy density* at that point.

The potential energy density at any point depends on how much stretching has occurred there, and the amount of stretching depends in turn on the slope of the string. Referring to fig. 9.1, we see that the length of a displaced segment whose equilibrium length was Δz is

$$\Delta z\left[1+\left(\frac{\partial\psi}{\partial z}\right)^2\right]^{1/2} \approx \Delta z\left[1+\tfrac{1}{2}\left(\frac{\partial\psi}{\partial z}\right)^2\right]$$

if the displacement is small. The extra length has been gained by pulling out against the tension T, and so the stored potential energy density at that point, due to the wave, is approximately $\frac{1}{2}T(\partial\psi/\partial z)^2$.

We combine these two quantities in the *total energy density* †

$$
\begin{aligned}
w(z,t) &= \tfrac{1}{2}\mu\left(\frac{\partial\psi}{\partial t}\right)^2 + \tfrac{1}{2}T\left(\frac{\partial\psi}{\partial z}\right)^2 \\
&= \frac{Z_0}{2c}\left[\left(\frac{\partial\psi}{\partial t}\right)^2 + c^2\left(\frac{\partial\psi}{\partial z}\right)^2\right]
\end{aligned}
\tag{9.32}
$$

This expression holds for any disturbance at all; but if the travelling wave equation (9.12) is also satisfied, the two terms on the right of (9.32) will be

---

† This equation and subsequent ones are approximate for the string, valid for small displacements; but we write them as exact, and go quickly from T and μ to Z_0 and c, to make them more general.

equal, everywhere and always, and we can say that *the instantaneous kinetic and potential energy densities are equal at any point on a string carrying a travelling wave.*

Energy flow. We now wish to find the rate at which energy passes a point z on the string, from left to right. We know that the string immediately to the right of z experiences an outward force, due to the string on the left, approximately equal to $-T(\partial\psi/\partial z)$. (Compare the situation at the driver in fig. 9.3.) This force will do work if the point z moves outwards from $\psi = 0$: that is, if $\partial\psi/\partial t$ there is positive at that instant. The rate at which work is done is

$$P(z, t) = -T\left(\frac{\partial\psi}{\partial z}\right)\left(\frac{\partial\psi}{\partial t}\right) = -Z_0 c\left(\frac{\partial\psi}{\partial z}\right)\left(\frac{\partial\psi}{\partial t}\right) \tag{9.33}$$

This is the power delivered from left to right past the point z.

Again, this expression is quite general; now we apply it to a travelling wave. Using the travelling wave equation (9.12), we get

$$P(z, t) = \pm\frac{T}{c}\left(\frac{\partial\psi}{\partial t}\right)^2 = \pm Z_0\left(\frac{\partial\psi}{\partial t}\right)^2 \tag{9.34}$$

Remembering that the terms on the right of (9.32) are equal for a travelling wave, we may rewrite the last result in terms of the total energy density,

$$P(z, t) = \pm cw(z, t) \tag{9.35}$$

The power $P(z, t)$ is positive for a wave travelling to the right, and negative for a wave travelling to the left, and so the energy travels in the same direction as the wave.

Example: a sinusoidal travelling wave. For illustration we turn once more to sinusoidal waves. The energy density (9.32) in a travelling wave is, in form A,

$$w(z, t) = (Z_0/2c)(\omega^2 + c^2 k^2)A^2 \sin^2(\omega t - kz + \phi) \tag{9.36}$$

The terms in the second bracket on the right are equal, by (9.6). Figure 9.7 shows the instantaneous distribution of energy along a section of the string. Peaks of energy density occur at those points where the string momentarily has its greatest speed and slope: these are in fact the points where the displacement is zero. As (9.35) suggests, the energy pattern keeps in step with the wave profile as it moves along the string to left or right.

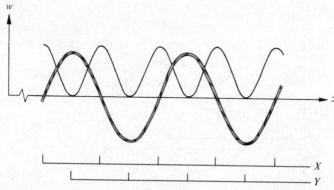

Fig. 9.7 Energy distribution along a string carrying a sinusoidal travelling wave. At points X the string has maximum transverse speed and maximum slope; at points Y the string is stationary and has zero slope.

Standing waves. In a standing wave the variables z and t appear in separate factors, and so the transverse force $-T(\partial\psi/\partial z)$ at any point on the string is in phase with the displacement. The transverse velocity $\partial\psi/\partial t$ is, of course, in quadrature with ψ, as with any other harmonic vibration. The product of these two quantities, which determines the power flow (9.33), thus averages to zero in a standing wave: no net power flows along the string. Figure 9.8 shows how the energy flowing to the right past any

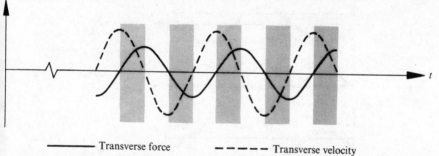

——— Transverse force   − − − − − Transverse velocity

Fig. 9.8 Energy flow in a standing wave. The graph shows the force and the velocity at a fixed point z on the string, as a function of time. At times in the shaded regions the string is moving in the direction in which the force is acting; at other times the velocity and the force are in opposite directions. The power delivered past z is proportional to the product of the two quantities plotted, and therefore averages to zero in a complete half cycle.

point in one quarter-cycle is cancelled by an equal amount flowing to the left in the next.

What happens is that energy surges to and fro between the nodes and the antinodes. An expression for the energy density at z and t is easily found by substituting a standing wave formula for ψ into (9.32). As an example we see in fig. 9.9 how the energy is distributed along the string at

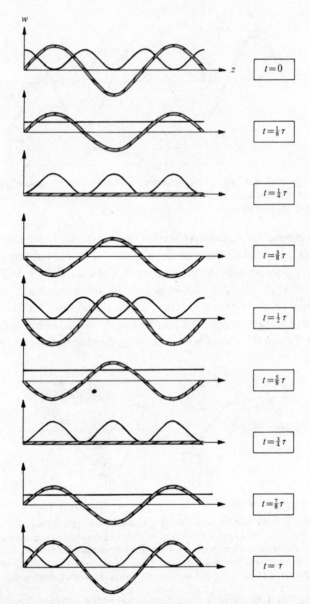

Fig. 9.9 Energy distribution along a string fixed at both ends, during a cycle of vibration in a standing wave with three antinodes. At $t = 0$, $\frac{1}{2}\tau$ and τ the string is stationary; the energy is entirely potential, and is concentrated near the nodes. At $t = \frac{1}{4}\tau$ and $\frac{3}{4}\tau$ the string is straight; the energy is entirely kinetic, and is concentrated near the antinodes.

times separated by one-eighth of a period, for a standing wave with three antinodes. At only four instants during the cycle (the times half way between those for maximum and minimum displacements) is $w(z, t)$ the same at all points on the string.

Summary. In a travelling wave on a string, the kinetic and potential energy densities at any point are equal, and the energy distribution pattern travels along the string at the same speed as the wave profile. In a standing wave, energy passes to and fro between the nodes and the antinodes, but there is no net flow of energy along the string.

9.5. Attenuation

Most real wave systems are affected by damping forces which cause a travelling wave to lose its energy as it propagates through the medium. We can discuss the process by imagining our stretched string to be immersed in a treacly fluid. There would then be added to the return force acting on a segment of length Δz a new force $-\beta(\partial \psi / \partial t)\Delta z$, where β is the resistance per unit length. This force puts a new term into the wave equation, which now reads

$$\frac{\partial^2 \psi}{\partial t^2} \approx \frac{T}{\mu}\left(\frac{\partial^2 \psi}{\partial z^2}\right) - \frac{\beta}{\mu}\left(\frac{\partial \psi}{\partial t}\right)$$

We shall discuss this equation in the standardized form

$$\frac{\partial^2 \psi}{\partial t^2} + \Gamma\left(\frac{\partial \psi}{\partial t}\right) = c^2\left(\frac{\partial^2 \psi}{\partial z^2}\right) \tag{9.37}$$

where

$$\Gamma \equiv \beta/\mu$$

and the other symbols have their previous meanings.

Is (9.37) a wave equation? That is, can it be satisfied by the expressions listed in (9.7)? The easiest version to test is form D, because it is, as we have seen, the easiest to differentiate. We have

$$\frac{\partial^2 \psi}{\partial t^2} = \text{Re}\,\{-\omega^2 D \exp[i(\omega t - kz)]\}$$

$$\frac{\partial^2 \psi}{\partial z^2} = \text{Re}\,\{-k^2 D \exp[i(\omega t - kz)]\}$$

The condition that the complex expressions in curly brackets (and therefore their real parts) satisfy (9.37) is

$$\omega^2 - i\Gamma\omega = c^2 k^2 \qquad (9.38)$$

This replaces (9.6), the effect of the damping being to add an imaginary term, linear in ω. Since ω is real it follows that k^2, and therefore k, must be complex. We shall meet a number of examples of complex wavevectors; we shall always write them in the form

$$k = K - i\kappa \qquad (9.39)$$

where K and κ are both real. (It will also turn out that they always have the same sign.) With this form of k, a sinusoidal travelling wave in form D becomes

$$\psi = \text{Re} \{D \exp [i(\omega t - Kz + i\kappa z)]\}$$
$$= \exp(-\kappa z) \, \text{Re} \{D \exp [i(\omega t - Kz)]\} \qquad (9.40)$$

To find how K and κ depend on the frequency, we could separate (9.38) into its real and imaginary parts

$$\omega^2 = c^2(K^2 - \kappa^2)$$
$$\Gamma\omega = 2c^2 K\kappa \qquad (9.41)$$

which we could then solve for K and κ. (See problem 9.17.) Clearer physical insight can be gained, however, if we concentrate instead on the approximate solutions which hold when the ratio Γ/ω is either very small or very large. A surprisingly large number of physical situations belong to one or other of these categories. For a given wave frequency, Γ/ω measures the degree of damping in much the same way as γ/ω_0 does for a vibrator (chapter 3).

Very light damping ($\Gamma/\omega \ll 1$). When Γ is much smaller than ω we have

$$\omega^2(1 - i\Gamma/\omega) = c^2 k^2$$
$$\omega(1 - i\Gamma/2\omega) \approx \pm c(K - i\kappa)$$

Separating the real and imaginary parts gives

$$K \approx \pm\omega/c$$
$$\kappa \approx \pm\Gamma/2c \qquad (9.42)$$

At this extreme the wave (9.40) differs from a non-dispersive sinusoidal travelling wave only in the factor $\exp(-\kappa z)$, in which κ is virtually constant. Every point on the string vibrates harmonically, and points separated by a distance $2\pi c/\omega$ vibrate in phase with each other. Thus a point of fixed phase angle travels along the string at a speed c, regardless of the frequency.

The new feature is *attenuation*: points which are in phase with each other no longer have the same amplitude. The amplitude falls off exponentially as we go along the string, the amplitude at any particular point being a factor e smaller than that of the point lying a distance $2c/\Gamma$ upstream. Since the ratio

$$\kappa/K \approx \Gamma/2\omega$$

is small, one attenuation length $1/|\kappa|$ contains many wavelengths. Two examples are shown in fig. 9.10.

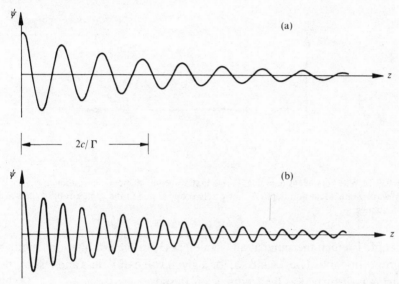

Fig. 9.10 Two attenuated travelling waves on the same string, with very light damping ($\Gamma/\omega \ll 1$); only the frequencies are different in the two cases. The waves are damped out in a fixed distance, not in a fixed number of wavelengths. (a) $\omega = 10\Gamma$; (b) $\omega = 20\Gamma$.

This is a convenient point at which to comment on the choice of a minus sign in front of the imaginary term on the right of (9.39). A wave travelling in the positive z direction has $K > 0$, and κ must be positive if the wave is to be attenuated rather than 'explode' unphysically; the reverse is true for a wave travelling in the negative z direction. Thus K and κ, as we defined them in (9.39), will always have the same sign.

Very heavy damping ($\Gamma/\omega \gg 1$). In this case (9.38) takes the approximate form

$$-i\Gamma\omega \approx c^2(K - i\kappa)^2$$

We can take the square root on the left if we remember that

$$(-i)^{1/2} = \pm(1-i)/2^{1/2}$$

We then find

$$K - i\kappa \approx \pm(1-i)|(\Gamma\omega/2c^2)^{1/2}|$$

which may be separated to give

$$K \approx \kappa \approx \pm|(\Gamma\omega/2c^2)^{1/2}| \qquad (9.43)$$

Now the attenuation is so great that the wave is hardly recognizable at all (fig. 9.11). Since K and κ are nearly equal, the *shape* of the string

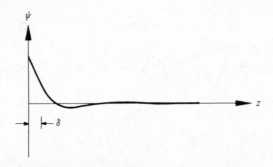

Fig. 9.11 With very heavy damping ($\Gamma/\omega \gg 1$) all waves are damped out in about a quarter of a 'wavelength'. The skin depth δ is inversely proportional to the square root of the wave frequency.

profile is much the same at all frequencies; but the attenuation length is proportional to $1/\omega^{1/2}$ and so, for a given value of Γ, less and less of the string is disturbed as the frequency is raised.

The analogous phenomenon in electricity, in which high-frequency alternating currents are damped out within a short distance of the surface of a conductor, is known as the *skin effect*, and the limiting value $(2c^2/\Gamma\omega)^{1/2}$ of the attenuation length is called the *skin depth*. We shall adopt these useful terms when discussing any waves for which $\Gamma/\omega \gg 1$.

Complex characteristic impedances. In section 5.1 we introduced two complex functions, the compliance and the impedance, which could be used as response functions expressing the amplitude *and the phase lag* of a forced vibrator at any driving frequency. In a similar way we can extend the concept of characteristic impedance to cover cases, such as the present one, in which there is a phase difference between the transverse force and the transverse velocity in a travelling wave.

For any sinusoidal travelling wave in form D, the transverse force

$$-T\left(\frac{\partial \psi}{\partial z}\right) = \text{Re}\,\{ikTD\,\exp\,[i(\omega t - kz)]\}$$

has a complex amplitude $ikTD$. We know that the complex amplitude of the velocity is $i\omega D$. From these two quantities we can form the complex characteristic impedance

$$Z_0(\omega) = Tk/\omega = T(K - i\kappa)/\omega \qquad (9.44)$$

When the damping is very light, our previous approximations (9.42) lead to

$$Z_0(\omega) \approx (T/c)(1 - i\Gamma/2\omega) = (T\mu)^{1/2}(1 - i\Gamma/2\omega) \qquad (9.45)$$

which has only a small imaginary part. The corresponding expression for the very heavy damping case is, by (9.43),

$$Z_0(\omega) \approx (T\mu)^{1/2}(1 - i)(\Gamma/2\omega)^{1/2} \qquad (9.46)$$

Reflection of travelling waves at a fluid boundary. If we form a reflection coefficient by putting complex impedances into an expression like (9.17) we shall get a complex function which we can denote by $R(\omega)$. This quantity is the ratio of the complex amplitudes of the reflected and incident waves, and it will tell us of any phase changes that occur as part of the reflection process.

There are many important examples in which waves are reflected at the boundaries of damping media. We can find prototype equations to use in these problems by seeing what would happen if a wave travelling on an undamped string came to a boundary at which the string entered a damping fluid – a situation easier to imagine than to engineer! We know that at least some of the power is absorbed in the fluid, causing attenuation; any that is not absorbed is presumably reflected.

The characteristic impedance of the undamped string is $(T\mu)^{1/2}$, and we use (9.45) or (9.46) for the string in the fluid, depending on whether Γ/ω is very small or very large. In the former case the complex reflection coefficient is

$$R(\omega) \approx i\Gamma/4\omega \qquad (9.47)$$

Since Γ/ω is small $|R(\omega)|$ is small, and most of the power is absorbed by the fluid. Thus under these conditions the second string acts as an almost correct termination for the first: the energy is gradually absorbed as the wave travels along the second string, but hardly any is reflected at the join.

When the damping is very heavy we have

$$R(\omega) \approx \frac{1 - (1-i)(\Gamma/2\omega)^{1/2}}{1 + (1-i)(\Gamma/2\omega)^{1/2}} \qquad (9.48)$$

Since $\Gamma/2\omega \gg 1$ now, $R(\omega)$ is nearly equal to -1. This means that the second string, because it is so heavily damped, acts almost like a fixed anchor point and produces almost total reflection, accompanied by a phase change of almost $180°$.

Summary. Damping leads to the attenuation of travelling waves. When the damping is very light ($\Gamma/\omega \ll 1$) the attenuation length has the same value $2c/\Gamma$ for all wave frequencies. When the damping is very heavy ($\Gamma/\omega \gg 1$) all waves are damped out in a fraction of a wavelength; the attenuation length in this case (the 'skin depth') is inversely proportional to the square root of the wave frequency.

A travelling wave coming up to a very lightly damped string suffers hardly any reflection, most of its power being absorbed by the fluid. A travelling wave coming up to a very heavily damped string is almost totally reflected.

Problems

9.1 Calculate (a) the amplitude, (b) the phase constant, and (c) the complex amplitude, for the sinusoidal travelling wave given by

$$\psi = (10 \text{ mm}) \cos(\omega t - kz) + (17 \text{ mm}) \sin(\omega t - kz)$$

9.2 Calculate (a) the frequency, (b) the wavelength, and (c) the phase velocity, for the wave given by

$$\psi = (0.0010 \text{ m}) \cos[(15 \text{ s}^{-1})t + (7.5 \text{ m}^{-1})z]$$

(d) What is the distance between two adjacent points on the string whose displacements have a phase difference of $45°$?

9.3 For the wave in problem 9.2, find (a) the greatest transverse speed reached by each point on the string, and (b) the greatest percentage stretching produced at each point on the string.

9.4 Figure 9.12 shows a 'photograph' of a string carrying a travelling wave moving from left to right. For each of the points marked, state whether the string was moving upwards or downwards when the photograph was taken. Was point A or point B moving faster? In each case check your answer against the travelling wave equation (9.12).

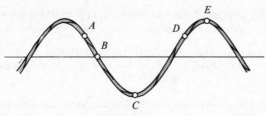

Fig. 9.12.

9.5 Calculate (a) the tension and (b) the mass per unit length of a stretched string on which waves travel at 30 m s^{-1}, and which has a characteristic impedance of 3.0 kg s^{-1}.

9.6 A stretched string is terminated by a damper, as in fig. 9.4. When a travelling disturbance is sent along the string, an upright reflected disturbance is obtained. The height of the reflected disturbance is half the height of the incident disturbance. How could reflection be eliminated by adjustment of the string tension?

9.7 If the picture in fig. 9.12 were a photograph of a string carrying a standing wave, how much could you say about the motion of each of the points marked, at the instant when the photograph was taken?

9.8 A long uniform string of linear density 0.1 kg m^{-1} is stretched with tension 40 N. Pulses are generated by moving both ends upwards at a constant speed of 10 m s^{-1} for 5 ms, holding them at rest for 5 ms, and then lowering them to their original positions at 10 m s^{-1}. Draw diagrams to show the following: (a) the displacement of the string as a function of position at a particular instant, for one of the pulses, and the velocity distribution for points on the string within this pulse; (b) the shape, and the velocity distribution, at the instant when the two pulses are partly overlapped, with corresponding points on the two pulses separated by 100 mm; (c) the shape, and the velocity distribution, when the pulses overlap completely. Draw corresponding diagrams for the case in which one of the pulses is inverted.

9.9 A single pulse of the shape described in problem 9.8 approaches a junction where the mass per unit length of the string changes abruptly. The tension remains 40 N. Draw the shapes of the reflected and transmitted pulses for the cases in which the linear density beyond the junction is (a) 0.2 kg m^{-1}, and (b) 0.05 kg m^{-1}.

9.10 A travelling pulse is transmitted from the left along a very long stretched elastic string. Figure 9.13 shows the string at some later time. The mass per unit

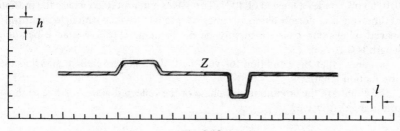

Fig. 9.13.

length to the left of point Z has a different value from the mass per unit length to the right of Z. Describe the original pulse (polarity, height and length).

9.11   Show that the average power carried along a stretched string by a sinusoidal travelling wave is $\frac{1}{2}Z_0\omega^2 A^2$, where the symbols have the same meanings as in the text.

9.12.   For a standing wave on a stretched string, show that the maximum kinetic energy density at an antinode is $\frac{1}{2}\mu\omega^2 A^2$, and the maximum potential energy density at a node is $\frac{1}{2}Tk^2 A^2$, where the symbols have the same meanings as in the text.

9.13   The E string of a violin has a mass per unit length of 4.0×10^{-4} kg m^{-1}, and must be tuned to a frequency of 660 Hz. The vibrating length of the string is 0.33 m. Calculate the tension when the string is in tune.

9.14   If a string is incorrectly terminated, superposition of incident and reflected sinusoidal waves gives rise to a combination of a standing wave and a travelling wave, no point on the string having zero amplitude. Show that the ratio of the maximum amplitude to the minimum amplitude (the 'standing wave ratio') is Z_1/Z_2 where Z_1 is the characteristic impedance of the string and Z_2 is the terminating impedance. (Measuring the standing wave ratio is a convenient way of finding the impedance of a termination, in acoustic and electromagnetic wave systems.)

9.15   A sinusoidal wave of amplitude A_1 approaches and travels through a region of the string in which the mass per unit length changes gradually from μ_1 to μ_2. It can be shown that, in such a situation, there is no reflection provided the distance over which the change occurs is large compared with the wavelength. On this assumption, show that the final amplitude A_2 is given by

$$A_2/A_1 = (\mu_1/\mu_2)^{1/4}$$

9.16   Show that, with very light damping, the amplitude of an attenuated travelling wave of frequency ν is reduced by a factor e in $2\nu/\Gamma$ wavelengths.

Calculate Γ for a wave of frequency 5.0 Hz, if the attenuation length is 100 wavelengths.

9.17   Show that the real and imaginary parts of k in (9.38) are given exactly by

$$cK = \pm\{\tfrac{1}{2}\omega[\omega + (\omega^2 + \Gamma^2)^{1/2}]\}^{1/2}$$
$$c\kappa = \pm\Gamma\omega/\{2\omega[\omega + (\omega^2 + \Gamma^2)^{1/2}]\}^{1/2}$$

9.18   Sinusoidal waves of frequency 5.0 Hz travel on a string of linear density 0.10 kg m^{-1} under a tension of 40 N. The waves enter a region where the motion of the string is uniformly damped. (Imagine it to be threaded with light, uniformly spaced, ping-pong balls which provide air damping.) The resistance per unit length is 0.10 N m^{-2} s.

(a) Show that the condition for very light damping is fulfilled, and find the attenuation length.

(b) What are the amplitude and phase of the reflected wave, relative to those of the incident wave?

10

Non-dispersive waves in physics

Before going on to discuss other wave equations, we pause to examine a few physical systems which obey (exactly or approximately) the non-dispersive form of wave equation (9.3), or its damped version (9.37).

10.1. Longitudinal waves

Our prototype waves were transverse. Any transverse disturbance, even one that travels along the string, makes each part of the string move sideways, and sideways only. Longitudinal waves disturb the system by producing actual motion along the direction in which the waves travel. We shall investigate longitudinal disturbances of a stretched string or wire; the results will be applicable to rigid objects such as cylindrical rods, provided they are not too thick.

We consider small distortions involving local compressions and extensions of the string along its own axis, which we take as the z axis. Under such a distortion, which is illustrated (and exaggerated) in fig. 10.1, a segment contained between the planes z and $z + \Delta z$ when the string is in equilibrium will be deformed into another cylindrical segment which is also bounded by planes perpendicular to the z axis. We suppose that the left boundary plane is displaced from z to a new position $z + \psi$, and that the right boundary plane moves from $z + \Delta z$ to $(z + \Delta z) + (\psi + \Delta \psi)$ where

$$|\Delta \psi| \ll |\Delta z| \tag{10.1}$$

In the deformation process the segment is both stretched and shifted. Its cross-section is also slightly decreased, but we neglect this effect. We denote by F the magnitude of the new stress force stretching the

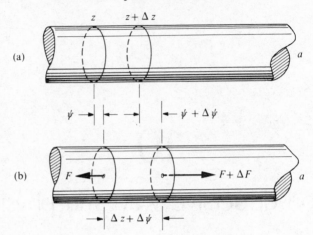

Fig. 10.1 (a) Segment of string or wire in equilibrium. (b) The same segment, shifted a distance ψ to the right and stretched by an amount $\Delta\psi$.

segment,† and by ΔF the excess force accelerating the whole segment to the right.

Since the length of the segment has increased from Δz to $\Delta z + \Delta\psi$, the corresponding *strain* is $\Delta\psi/\Delta z$. This is related to the *stress* F/a by the equation

$$\frac{F}{a} = E\left(\frac{\Delta\psi}{\Delta z}\right) \tag{10.2}$$

where a is the cross-sectional area and E is Young's modulus.

For a vanishingly thin segment we have

$$F = aE\left(\frac{\partial\psi}{\partial z}\right) \tag{10.3}$$

Thus the excess force responsible for accelerating the segment is

$$\Delta F = aE\left(\frac{\partial^2\psi}{\partial z^2}\right)\Delta z$$

The mass of the segment is approximately $\rho a\,\Delta z$ where ρ is the density, and the acceleration of the mass centre is approximately $\partial^2\psi/\partial t^2$. Newton's second law therefore becomes

$$\rho a\left(\frac{\partial^2\psi}{\partial t^2}\right)\Delta z \approx aE\left(\frac{\partial^2\psi}{\partial z^2}\right)\Delta z$$

† The initial stretching due to the equilibrium tension, if any, will be uniform along the string and can be ignored.

We may cancel $a\Delta z$ to obtain the wave equation

$$\frac{\partial^2 \psi}{\partial t^2} \approx \frac{E}{\rho}\left(\frac{\partial^2 \psi}{\partial z^2}\right) \tag{10.4}$$

We can now say that small-displacement longitudinal waves satisfy the non-dispersive wave equation (9.3), with

$$c = (E/\rho)^{1/2} \tag{10.5}$$

Longitudinal travelling waves. Equation (10.5) gives the speed of longitudinal travelling waves on a string or a wire made from a material whose density and Young's modulus are known. Taking a steel wire as an example, we shall have a value in the region of 8000 kg m^{-3} for the density and about $2 \times 10^{11} \text{ N m}^{-2}$ for Young's modulus. These figures give approximately 5 km s^{-1} for c. Since E tends to be large in high density materials and small in low density materials, phase velocities of several km s^{-1} are typical.

Such values are much higher than most transverse wave speeds. To produce a transverse wave with a phase velocity equal to that of a longitudinal wave on the same string, one would have to apply an equilibrium stress T/a equal to Young's modulus for the material; no ordinary string could withstand such a stress, which would double its unstretched length.

Calculating the speed of longitudinal waves on a wire made of steel, say, gives a reasonable rough estimate of the speed of longitudinal waves through bulk steel. We can imagine the wire to be thickened indefinitely, so that we end up with an infinite *plane wave*. The wave is called 'plane' because we continue to assume that nothing varies with any space coordinate other than z: thus identical events occur simultaneously all over the plane defined by z.

Our analysis is not strictly valid for infinite plane waves because it uses Young's modulus, which is applicable to situations in which the material is free to contract sideways as it is stretched. We can expect c to be somewhat higher in the bulk material than in wires or thin rods, because of the extra stiffness due to the rigidity in the transverse direction.

Characteristic impedance. The travelling wave equation (9.12) will be approximately obeyed by longitudinal travelling waves. We may therefore rewrite the stress force (10.3) at any point, in terms of the longitudinal velocity at that point,

$$F = \frac{aE}{c}\left(\frac{\partial \psi}{\partial t}\right) = a\rho c\left(\frac{\partial \psi}{\partial t}\right)$$

Hence we obtain the characteristic impedance (stress force per unit velocity at any point along the string, in a travelling wave)

$$Z_0 = a\rho c = a(E\rho)^{1/2} \tag{10.6}$$

A steel wire of cross-section 1 mm^2 will have a characteristic impedance of about 40 kg s^{-1} for longitudinal travelling waves. Because E and ρ are multiplied together in the characteristic impedance, values for Z_0 in different materials vary more than the corresponding values for c. For a given string the ratio of the characteristic impedances for longitudinal and transverse waves is the same as the ratio of the phase velocities, namely $(aE/T)^{1/2}$.

Longitudinal standing waves. For a finite string with both ends fixed (or a thin rod with both ends free) the eigenfrequencies of the longitudinal standing waves will be given by (9.31) with the appropriate wave speed (10.5). A string on a violin has a vibrating length of 0.33 m. Thus the eigenfrequency of the fundamental longitudinal mode ($n = 1$) is about 7.5 kHz for a violin string made of steel. If longitudinal standing waves are excited the instrument will produce sounds of this frequency and higher. The ear is extremely sensitive to such sounds, which are high pitched and unpleasant. Successful violinists excite transverse standing waves only.

Summary. The values of c and Z_0 for longitudinal waves on a string or a wire depend on the elasticity and the density of the material from which it is made. The phase velocities of longitudinal waves are typically several km s^{-1}.

10.2. Acoustic waves

As a simple introduction to acoustic waves, we consider longitudinal disturbances in a gas within a long cylindrical pipe. The geometry of this situation (fig. 10.2) is similar to that of the previous section. Since a gas cannot support shear stresses, we need not place any restriction on the thickness of the pipe.

In equilibrium the gas is at some uniform, steady pressure p. Under these conditions a thin slice of gas bounded by the planes z and $z + \Delta z$ must have a force ap acting on each of its end faces, a being the cross-sectional area of the pipe.

We choose the axis of the pipe as the z axis, and assume as before that nothing varies with any other space coordinate. When the gas is dis-

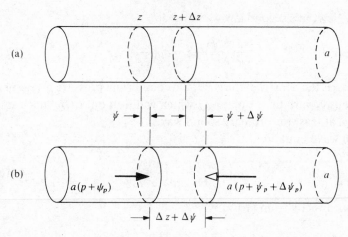

Fig. 10.2 (a) Slice of gas in a pipe, in equilibrium. (b) The same slice of gas, shifted a distance ψ to the right and expanded in volume by $a\Delta\psi$.

turbed, our slice may be transformed into one of width $\Delta z + \Delta\psi$, with its left face at a new position $z + \psi$. The displacement is assumed to be 'small': that is, (10.1) holds.

The disturbance will be accompanied by pressure changes. We call the new pressure on the left face of the slice $p + \psi_p$, and that on the right face $p + \psi_p + \Delta\psi_p$.

To find the wave equation we consider in turn the stiffness and inertia properties of the gas. First we obtain a relation connecting ψ_p and $\Delta\psi$, by thinking about the compression of the slice by the forces on its opposite faces. Then we write down Newton's second law, which governs the acceleration of the slice by the unbalanced force $a\Delta\psi_p$.

The first process involves the compressibility of the gas (2.6). The slice has had its volume changed by the fractional amount $\Delta\psi/\Delta z$, as a result of a pressure change ψ_p, and

$$\Delta\psi/\Delta z = -\kappa\psi_p \tag{10.7}$$

For an infinitely thin slice, we have

$$\psi_p = -\frac{1}{\kappa}\left(\frac{\partial\psi}{\partial z}\right) \tag{10.8}$$

To describe the acceleration we require not ψ_p but $\Delta\psi_p$, found by differentiating (10.8) with respect to z,

$$\Delta\psi_p = -\frac{1}{\kappa}\left(\frac{\partial^2\psi}{\partial z^2}\right)\Delta z$$

From Newton's second law we now get

$$\frac{a}{\kappa}\left(\frac{\partial^2\psi}{\partial z^2}\right)\Delta z \approx a\rho\left(\frac{\partial^2\psi}{\partial t^2}\right)\Delta z \tag{10.9}$$

where ρ is the density of the gas at the equilibrium pressure p. (We neglect the density variations due to ψ_p, which will turn out to be much smaller than p, at least for ordinary sound waves.)

The wave equation

$$\frac{\partial^2\psi}{\partial t^2} \approx \frac{1}{\kappa\rho}\left(\frac{\partial^2\psi}{\partial z^2}\right) \tag{10.10}$$

follows directly from (10.9).

Acoustic pressure. The variable ψ denotes a rather intangible quantity: the displacement of a thin slice of gas. There is actually no way in which we can identify any given 'piece' of a gas, far less fix its position, since the molecules which make it up are constantly being substituted by different ones in the course of their thermal motion.

The pressure variations ψ_p, on the other hand, are measurable. Since ψ_p is proportional to $\partial\psi/\partial z$ (10.8), and because $\partial\psi/\partial z$ must satisfy the same wave equation as ψ, it follows that ψ_p satisfies (10.10). It is customary therefore to use ψ_p rather than ψ as the variable to describe an acoustic wave. We can use the phrase 'pressure wave' to make it quite clear that this is what we are doing.

It should be remembered that ψ_p is the instantaneous pressure *excess* above (or below) the steady value p prevailing before the wave came along. The wave equation

$$\frac{\partial^2\psi_p}{\partial t^2} \approx \frac{1}{\kappa\rho}\left(\frac{\partial^2\psi_p}{\partial z^2}\right) \tag{10.11}$$

does not actually mention the equilibrium pressure, but p is implicitly involved through the compressibility κ and the density ρ. We refer to ψ_p as the *acoustic pressure*.

The speed of sound. Before we can use (10.11) to calculate the phase velocity of acoustic travelling waves, we need to know whether to use the isothermal or the adiabatic compressibility (or some value in between) for κ. We were faced with the same uncertainty in section 2.2, and you should now re-read the previous discussion (p. 19). The speed of acoustic waves along a pipe is something which can easily be measured with the help of an ultrasonic transducer and a small microphone, and the choice of the correct compressibility is ultimately a matter for experiment. It turns out that the adiabatic compressibility gives the closest prediction for the phase velocity of acoustic waves. The implication is that negligible amounts of heat are able to flow from regions of positive ψ_p (where the gas is compressed) to regions of negative ψ_p (where it is expanded) in the time available (half a period).

For a perfect gas we have therefore

$$c = (\rho\kappa_S)^{-1/2} = (\gamma p/\rho)^{1/2} = (\gamma RT/M)^{1/2} \tag{10.12}$$

where we have used our previous expression (2.8) for $\rho\kappa_S$ in a perfect gas. Like ω_0 for an acoustic vibration, c is independent of the gas pressure, and increases as $T^{1/2}$.

If the gas in the pipe is air at room temperature (table 2.1) we find $c \approx 350 \text{ m s}^{-1}$. Other gases at similar temperatures will give similar values: about an order of magnitude smaller than the corresponding values for longitudinal waves in solids (section 10.1).

The range of wavelengths in air corresponding to *audible* acoustic waves (somewhere in the frequency range 20 Hz to 20 kHz for human beings) is, by (9.11), about 17.5 mm to 17.5 m. Thus sound waves have wavelengths comparable with the dimensions of objects they are likely to encounter in the atmosphere (books, people, cars, houses, etc.): a fact of great significance, as we shall see when we discuss diffraction.

Characteristic impedance. We have seen (10.8) that the acoustic pressure at any point along the pipe is proportional to $\partial\psi/\partial z$ at that point. We know also that $\partial\psi/\partial z$ is proportional to $\partial\psi/\partial t$ in a travelling wave (9.12). Thus, in a travelling wave, the acoustic pressure per unit speed is the same at all points, and

$$\psi_p = \pm\frac{1}{c\kappa}\left(\frac{\partial\psi}{\partial t}\right) = \pm\frac{\gamma p}{c}\left(\frac{\partial\psi}{\partial t}\right) \tag{10.13}$$

Since the force which must be applied to produce the pressure increase ψ_p is $a\psi_p$, the characteristic impedance is

$$Z_0 = a\gamma p/c \tag{10.14}$$

For most purposes the characteristic impedance *per unit area* (the acoustic pressure per unit speed†)

$$\mathcal{Z}_0 \equiv Z_0/a = \gamma p/c = \rho c \tag{10.15}$$

---

† Workers in acoustics usually call this the *acoustic impedance*. Sometimes, however, acoustic impedance is defined as $\mathcal{Z}_0/a$ (acoustic pressure per unit volume velocity).

is a more useful quantity. In the case of a perfect gas we have

$$\mathcal{Z}_0 = (\gamma M/RT)^{1/2} p \qquad (10.16)$$

For air at room temperature (table 2.1) and atmospheric pressure ($p \approx 1.0 \times 10^5 \, \mathrm{N \, m^{-2}}$) $\mathcal{Z}_0$ is approximately $400 \, \mathrm{kg \, m^{-2} \, s^{-1}}$.

Intensity. The basic formula for the power carried in a travelling wave is (9.34). For an acoustic travelling wave we may express $\partial\psi/\partial t$ in terms of ψ_p with the help of (10.13), thereby obtaining

$$P(z, t) = \pm(a^2/Z_0)\psi_p^2 = \pm(a/\mathcal{Z}_0)\psi_p^2 \qquad (10.17)$$

The power *per unit area*

$$P(z, t)/a = \pm\psi_p^2/\mathcal{Z}_0 \qquad (10.18)$$

fluctuates with the changing pressure produced by the disturbance. Its average value is known as the *intensity* of the wave; the intensity of a sinusoidal travelling wave with a pressure amplitude A_p is

$$I(z, t) \equiv \langle P(z, t)\rangle/a = A_p^2/2\mathcal{Z}_0 \qquad (10.19)$$

It is noteworthy that the intensity, when expressed in terms of the pressure amplitude, does not depend on the frequency or the wavelength of the wave.

The pressure amplitude in a sound wave. We may use (10.19) to test the validity of the underlying 'small displacement' assumption (10.1). In pressure terms that assumption is, by (10.7),

$$|\psi_p| \ll 1/\kappa$$

If the gas obeyed Boyle's law, κ would be simply $1/p$, and we could write

$$|\psi_p| \ll p \qquad (10.20)$$

We know that a gas carrying an acoustic wave does *not* obey Boyle's law, but the difference is negligible from the present point of view. What we have established is that (10.11) will become a better approximation to the true wave equation, and our deductions from it will become more reliable, as the pressure fluctuations in the wave become smaller relative to the equilibrium pressure.

We now investigate the validity of (10.20) for sound waves in the atmosphere. The human ear is adapted to respond, without discomfort, to sound intensities over an enormous range: roughly from $1 \, \mathrm{pW \, m^{-2}}$ to $1 \, \mathrm{W \, m^{-2}}$. Taking the upper figure, and using the value $\mathcal{Z}_0 \approx 400 \, \mathrm{kg \, m^{-2} s^{-1}}$

obtained previously, we find from (10.19) that $A_p \approx 30$ N m^{-2}, or about 3×10^{-4} atmosphere. Clearly our assumption is a good one for ordinary sound waves.

We may note in passing that the pain produced by very loud sounds whose intensities exceed 1 W m^{-2} cannot be simply explained in terms of the pressure amplitudes to which the ear drum is being subjected. Very much greater, if slower, variations of atmospheric pressure can be experienced just by driving up and down quite small hills.

Standing waves in pipes. We can obtain an expression for ψ_p in an acoustic standing wave which has a *displacement* (ψ) node at $z = 0$, by substituting (9.24) into (10.8). The result is

$$\psi_p = -(kA/\kappa) \cos kz \cos (\omega t + \phi)$$
$$= -A_p \cos kz \cos (\omega t + \phi)$$

(10.21)

where

$$A_p \equiv kA/\kappa$$

is the acoustic pressure amplitude: in this case the maximum pressure excess achieved at any pressure antinode.

The standing wave described by (10.21) has a pressure antinode at $z = 0$, where $\partial \psi / \partial z$ is a maximum *but ψ is zero always*, corresponding to the string fixed at $z = 0$.

Conversely, we expect a pressure node to occur at an *open* end (where ψ is a maximum), since the pressure outside the pipe has the steady value of the surrounding atmosphere. In practice some of the air just outside the pipe will take part in the vibration, and the node will be found slightly beyond the open end.

From these arguments it follows that the standing waves in a completely closed pipe of length L will be those illustrated in fig. 9.6(b), with the string profiles in that diagram re-interpreted as graphs showing the instantaneous value of ψ_p along the axis of the pipe. Because acoustic waves are non-dispersive, the eigenfrequencies are uniformly spaced, like the k_n values.

Figure 9.6(a) gives the same information for a pipe with two open ends, the length of the pipe being slightly less than L. A flute behaves like such a pipe: by exciting the standing wave of frequency ν_2 instead of the fundamental, the player can, without changing his fingering, produce a note an octave higher in pitch.

Figure 9.6(c) applies to a pipe with one end open (at $z = 0$) and the other end closed, its length again being slightly less than L. This situation is exemplified by a clarinet. Since the second eigenfrequency is not twice but three times the frequency of the fundamental mode, the result of 'overblowing' a clarinet is that the pitch jumps, not by an octave but by an octave plus a fifth.

As a numerical example, we estimate the frequency of the fundamental standing wave in a pipe with one end open to the atmosphere (at room temperature) and the other end closed, when it has an effective length of 1 m. The required frequency is c/λ where $c \approx 350 \text{ m s}^{-1}$ and $\lambda = 4 \text{ m}$, giving an approximate value of 90 Hz.

If we think about the motion of the air in this standing wave (maximum longitudinal movement at the open end, no movement at the closed end) we are reminded of the assumptions on which we based our estimate of the vibration frequency of the air in a flask, in section 2.2. These were that only the air in the neck of the flask moves, while the air in the bulb is at rest and merely provides stiffness. Our estimates of the frequency are comparable, and we now recognize the second factor on the right of (2.9) as our expression (10.12) for the wave speed. What we estimated in the case of the flask was in fact the eigenfrequency of the standing wave with $n = 1$. Higher standing waves will also occur, though we must expect their eigenfrequencies to form a more complicated series than those for a cylindrical pipe.

Summary. We have obtained a wave equation for acoustic waves in terms of the acoustic pressure as variable. The values of c and Z_0 for acoustic waves depend on the adiabatic compressibility and the density of the gas. The phase velocities of acoustic waves in gases at ordinary temperatures are several hundred m s^{-1}.

In acoustic standing waves in pipes, pressure antinodes occur at closed ends and pressure nodes occur just outside open ends.

10.3. Cable waves

We wish to find the wave equation that governs electrical disturbances on a length of uniform, two-conductor cable such as a feeder cable used to carry the signal from the aerial (antenna) to a television set. We shall make no assumptions as to what the cross-section of the cable looks like, although the example shown in, fig. 10.3 is simply two parallel wires separated by an air space. We use the coordinate z to measure distance along the cable from one end.

We assume that, before the disturbance arrives, the potential difference between the conductors is zero everywhere, and that no current is

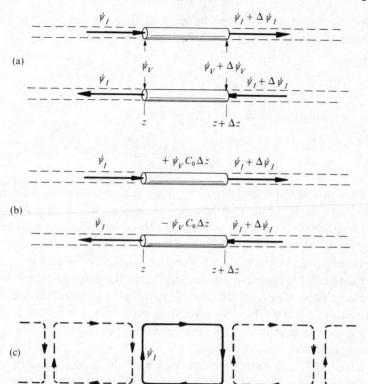

Fig. 10.3 (a) Instantaneous potential differences and currents at points z and $z + \Delta z$ along a cable. (b) The cable segment of length Δz can be regarded as a capacitor whose plates carry charges $\pm\psi_V C_0 \Delta z$. (c) The induced voltages can be derived by imagining the cable to consist of a large number of short loops, each carrying a circulating current.

flowing. (It would not be difficult to include steady voltages and currents in the equations, but the results would be unchanged.)

Figure 10.3(a) shows the state of affairs at some instant during the disturbance. The voltage across the cable at z has become ψ_V. At the same moment currents ψ_I flow in the two conductors, in the directions shown; their magnitudes are the same because the cable is uniform along its length. At position $z + \Delta z$ and the same instant, the corresponding voltage and current values are $\psi_V + \Delta\psi_V$ and $\psi_I + \Delta\psi_I$.

Wherever a potential difference exists across the cable there must also be charges of opposite sign on the two conductors. If we think of the cable as a large number of very short capacitors connected in parallel, the instantaneous charge on the capacitor between z and $z + \Delta z$ is $(C_0\Delta z)\psi_V$, where C_0 is the capacitance per unit length of the cable: see fig. 10.3(b). The currents reaching and leaving each 'plate' of the capacitor are unequal, however: thus the charge, and therefore the voltage, must change with time. In fact, the charge increases at the rate

$$\Delta \psi_I = -\frac{\partial}{\partial t}(C_0 \, \Delta z \, \psi_V) = -C_0 \, \Delta z \left(\frac{\partial \psi_V}{\partial t}\right)$$

For vanishingly small values of Δz we have

$$\frac{\partial \psi_I}{\partial z} = -C_0 \left(\frac{\partial \psi_V}{\partial t}\right) \tag{10.22}$$

The current too will change with time, and we know that fluctuating currents give rise to induced voltages. The easiest way to discuss these is to imagine that the cable is actually a chain of very short loops of length Δz, insulated from each other. The loop at z in fig. 10.3(c) carries an instantaneous current ψ_I. If Δz is small enough, the imaginary currents perpendicular to the cable will be cancelled by the perpendicular currents in the neighbouring loops, the net effect being the same as that due to the currents in the real cable. We know how to calculate the voltage induced in a closed loop; in this way we can understand the induction of voltages along the cable.

If the self inductance per unit length of the cable is L_0, then the imaginary loop between z and Δz has inductance $L_0 \, \Delta z$. When the current increases at rate $\partial \psi_I / \partial t$, an induced voltage $-L_0 \, \Delta z \, (\partial \psi_I / \partial t)$ will appear round this loop. Part of the voltage increment $\Delta \psi_V$ between z and $z + \Delta z$ comes from this induced voltage. The remainder is simply the Ohm's law contribution due to the resistance of the conductors. Combining these two terms gives

$$\Delta \psi_V = -L_0 \, \Delta z \left(\frac{\partial \psi_I}{\partial t}\right) - R_0 \, \Delta z \, \psi_I$$

where R_0 is the resistance per unit length. For a vanishingly short loop we have

$$\frac{\partial \psi_V}{\partial z} = -L_0 \left(\frac{\partial \psi_I}{\partial t}\right) - R_0 \psi_I \tag{10.23}$$

The wave equation for ψ_V is now quickly found by differentiating (10.22) with respect to t, differentiating (10.23) with respect to z, and eliminating derivatives of ψ_I. The result,

$$\frac{\partial^2 \psi_V}{\partial t^2} + \frac{R_0}{L_0}\left(\frac{\partial \psi_V}{\partial t}\right) = \frac{1}{L_0 C_0}\left(\frac{\partial^2 \psi_V}{\partial z^2}\right) \tag{10.24}$$

TABLE 10.1
Analogous quantities in a stretched string and an air-insulated cable

String		Cable	
Displacement	ψ		
Transverse velocity	$\dfrac{\partial \psi}{\partial t}$	Current	$\dfrac{\partial \psi}{\partial t}$
Slope of string	$\dfrac{\partial \psi}{\partial z}$	Charge density	$\dfrac{\partial \psi}{\partial z}$
Mass per unit length	μ	Inductance per unit length	L_0
Tension	T	Length per unit capacitance	$1/C_0$
Resistance per unit length	β	Resistance per unit length	R_0
Transverse component of tension	$T\left(\dfrac{\partial \psi}{\partial z}\right)$	Voltage across cable	$\dfrac{1}{C_0}\left(\dfrac{\partial \psi}{\partial z}\right)$
Return force on segment	$T\left(\dfrac{\partial^2 \psi}{\partial z^2}\right)\Delta z$	Voltage increment along segment	$\dfrac{1}{C_0}\left(\dfrac{\partial^2 \psi}{\partial z^2}\right)\Delta z$
Drag force on segment	$\beta\left(\dfrac{\partial \psi}{\partial t}\right)\Delta z$	Resistive voltage drop along segment	$R_0\left(\dfrac{\partial \psi}{\partial t}\right)\Delta z$
Force accelerating segment	$\mu\left(\dfrac{\partial^2 \psi}{\partial t^2}\right)\Delta z$	Induced voltage	$L_0\left(\dfrac{\partial^2 \psi}{\partial t^2}\right)\Delta z$
Potential energy density	$\tfrac{1}{2}T\left(\dfrac{\partial \psi}{\partial z}\right)^2$	Electric energy density	$\dfrac{1}{2C_0}\left(\dfrac{\partial \psi}{\partial z}\right)^2$
Kinetic energy density	$\tfrac{1}{2}\mu\left(\dfrac{\partial \psi}{\partial t}\right)^2$	Magnetic energy density	$\tfrac{1}{2}L_0\left(\dfrac{\partial \psi}{\partial t}\right)^2$

is a damped, non-dispersive wave equation like (9.37) with

$$c = (1/L_0 C_0)^{1/2}$$
$$\Gamma = R_0/L_0$$

(10.25)

Mechanical analogies. We exploit our knowledge of the prototype system (the stretched string) by establishing a set of analogies (table 10.1) similar to that of table 2.2. To make the new analogies consistent with the old, we should obviously pair L_0 (the inductance per unit length of the cable) with μ (the mass per unit length of the string). Similarly, we connect $1/C_0$ with the tension T, just as $1/C$ took the place of the stiffness s. The electrical resistance per unit length R_0 goes naturally with the mechanical resistance per unit length β.

Pairing the variables involved is not quite so straightforward, since the charge on the capacitor has no obvious counterpart on the cable. We have previously met the charge *per unit length* $C_0\psi_V$, however. If we write this as $\partial\psi/\partial z$, every quantity involving ψ_V can be readily identified with an analogous quantity on the left of the table. The remaining quantities, involving ψ_I, can be similarly treated by writing the current as $\partial\psi/\partial t$.

To identify the charge ψ itself would mean integrating $C_0\psi_V$ with respect to z, or ψ_I with respect to t. In either case the result would contain an arbitrary constant term. Clearly ψ_V is a more convenient variable than ψ for cable waves, and we shall continue to use it. We do not need to be specific about the identity of the charge ψ, and the right-hand side of table 10.1 contains no entry for ψ.

It is worth noticing here that ψ_I would be an equally satisfactory variable for the description of cable waves. The wave equation for the current, found by differentiating (10.22) with respect to z and (10.23) with respect to t,

$$\frac{\partial^2 \psi_I}{\partial t^2} + \frac{R_0}{L_0}\left(\frac{\partial\psi_I}{\partial t}\right) = \frac{1}{L_0 C_0}\left(\frac{\partial^2\psi_I}{\partial z^2}\right)$$

is identical with the equation (10.24) for the voltage; but obviously a different set of analogies would be needed. It is a good idea to talk about 'voltage waves' or 'current waves', to make it quite clear whether the variable in use is ψ_V or ψ_I.

Coaxial cables. A widely used type of cable for high frequency signals consists of a copper wire symmetrically surrounded by a hollow cylinder of copper braid (fig. 10.4). For mechanical reasons the conductors must be supported with the help of some insulating material, but for simplicity we shall initially pretend that they have only air between them.

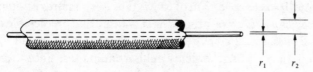

Fig. 10.4 A coaxial cable.

The capacitance and the inductance of a coaxial cable can be calculated quite simply. It is shown in electricity textbooks that

$$1/C_0 = (1/2\pi\varepsilon_0) \ln (r_2/r_1)$$
$$L_0 = (\mu_0/2\pi) \ln (r_2/r_1)$$

(10.26)

for a cable with inner and outer conductors of respective radii r_1 and r_2, in a vacuum. The value of $\ln (r_2/r_1)$ is unlikely to differ much from 1; consequently C_0 will be several tens of pF m^{-1}, and L_0 somewhat under 1 μH m^{-1}, for almost any air-cored coaxial cable.

The exact values of C_0 and L_0 do not, however, influence the phase velocity. Equations (10.25) and (10.26) give

$$c = (\mu_0\varepsilon_0)^{-1/2} = 2.998 \times 10^8 \text{ m s}^{-1}$$

(10.27)

and this value is obtained for a cable of *any* cross-section, since it turns out that L_0 always depends on the cable geometry in the same way as $1/C_0$.

The calculated value is for a vacuum-insulated cable. The permittivity of air is sufficiently close to ε_0 to allow (10.27) to be used with air insulation. In practice the conductors are of course separated not by air but by a solid insulator such as polyethylene ('polythene') or poly-tetrafluorethylene (PTFE), possibly containing air spaces. The phase velocity is reduced because the permittivity in the space between the conductors is significantly greater than ε_0.

Polyethylene has permittivity $2.3\varepsilon_0$, and the value for PTFE is about $2.1\varepsilon_0$; thus c may be as small as two-thirds of the above value (10.27). (Both materials have permeabilities very close to μ_0.)

Clearly the voltage fluctuations have to be very rapid indeed before we become aware of waves on the cable. For an alternating current supply working at 50 Hz, the wavelength (9.11) can be as high as 6000 km. For any reasonable length of cable, the instantaneous potential difference will be effectively the same all along the cable. At 50 MHz, on the other hand, the wavelength would be only 6 m.

Similarly, if there is to be an observable interval between the arrival of a voltage *pulse* at two points on a cable, the duration of the pulse must be small in comparison with the travel time between the points in question. It takes about 10 ns to cover 3 m of cable; thus a few metres of simple cable can be used as a *delay line*, producing significant and predictable delays in the arrival of nanosecond pulses at a specified point in a circuit. Different techniques must be used to delay millisecond pulses, since these would require thousands of kilometres of cable.

Attenuation. The parameter Γ/ω whose size tells us whether the damping is light or heavy is given in this case by $R_0/\omega L_0$. For an air-insulated coaxial cable R_0 might typically be $\sim 0.1\ \Omega\ \mathrm{m}^{-1}$; taking $L_0 \sim 1\ \mu\mathrm{H}\ \mathrm{m}^{-1}$ gives $\Gamma \sim 10^5\ \mathrm{s}^{-1}$. Most applications in which these cables are used have $\omega \gg 10^5\ \mathrm{s}^{-1}$, giving very light damping ($\Gamma/\omega \ll 1$). At a wave frequency of 1 MHz, for example, the attenuation length $2c/\Gamma$ would be about 6 km. The attenuation would be quite negligible in a television feeder cable, but not in a long-distance land line.

High-frequency currents actually flow only in the outer layers of metal conductors. This is the best known example of the skin effect. (See pp. 166 and 269.) Since the skin depth is proportional to $1/\omega^{1/2}$, the value of R_0 for the cable increases as $\omega^{1/2}$, and so the attenuation length decreases as $1/\omega^{1/2}$.

In cables with solid insulators between the conductors, attenuation due to 'dielectric losses' becomes important at the highest frequencies. We saw in section 6.3 that the polarization of a dielectric will lag behind an alternating field, with consequent absorption of power from the field. The phase lag can be appreciable in a solid dielectric, even far from resonance, because the damping is greater than in a gas. In cable wave theory, dielectric losses can be allowed for by introducing a parameter representing the conductance per unit length *between* the conductors. Air spaces are usually incorporated in the insulation of 'low loss' cables.

Characteristic impedance. The characteristic impedance of a string is the force $T(\partial \psi/\partial z)$ divided by the speed $\partial \psi/\partial t$. With the help of the table we can see that the equivalent quantity for a cable is simply $|\psi_V/\psi_I|$. Assuming very light damping, we can write

$$Z_0(\omega) \approx \left(\frac{L_0}{C_0}\right)^{1/2}\left(1 - \frac{iR_0}{2\omega L_0}\right) \tag{10.28}$$

by analogy with the equivalent expression (9.45) for the string.

In most cases we can neglect R_0 entirely, and use the real expression $(L_0/C_0)^{1/2}$. Unlike the phase velocity, the value of the characteristic impedance does depend on the cable geometry. For an air-insulated coaxial cable (10.26) and (10.28) give

$$Z_0 \approx (\mu_0/4\pi^2\varepsilon_0)^{1/2} \ln(r_2/r_1)$$
$$= 60.0 \, \Omega \times \ln(r_2/r_1) \tag{10.29}$$

The gentle dependence on r_2/r_1 leaves the cable manufacturer some scope for adjusting Z_0 to a standard value such as 50 Ω, 75 Ω or 120 Ω. The use of a solid insulator will give a smaller result than the value calculated in (10.29).

Standing waves. We return to table 10.1 to discover how to interpret the standing wave spectra of fig. 9.6 for standing waves on a cable. We see a close connection between ψ_V and the slope of the string. We know that the string gradient reaches its maximum values at the nodes of a standing wave, and is zero at the antinodes. Further, a fixed end must always be a node, and a 'free' end an antinode.

In a standing cable wave, therefore, ψ_V will have its maximum amplitude at an open-circuit end ($Z_0 = \infty$) and will be zero at an end where there is a short circuit ($Z_0 = 0$). In other words, an open circuit is analogous to an anchor point, and a short circuit to a 'free' end.† Figures 9.6(a), 9.6(b) and 9.6(c) represent respectively the standing voltage waves formed on a cable with a short circuit at each end; a cable with an open circuit at each end; and a cable with a short circuit at one end ($z = 0$) and an open circuit at the other.

A length of cable makes a convenient resonant circuit of calculable properties for use at high frequencies. It is usual to make the length equal to one-quarter of the wavelength at the required frequency, and to short circuit one end. The standing wave with $n = 1$ has a frequency

$$\nu_1 = c/4L = 1/4L(L_0C_0)^{1/2}$$

where L is the length of the cable. The advantages over the simple alternative of a coil and a capacitor are the ease with which the resonant frequency can be adjusted by varying the length of the cable, and the absence of stray capacitance due to connecting wires.

Summary. Voltage (and current) fluctuations on an air-insulated cable are governed by the non-dispersive wave equation. The properties of the voltage waves can be related to those of waves on a string by the same kind of analogies as we use for resonant circuits (table 10.1).

The phase velocity with air insulation is $3 \times 10^8 \, \mathrm{m \, s}^{-1}$ for all cable geometries. The characteristic impedance can be predetermined by adjusting the cable geometry. Most air-insulated cables produce attenuation with very light damping.

† Again, if we were using ψ_I as the variable instead of ψ_V, different analogies would apply.

Problems

10.1   Which of the following travelling waves are inverted on reflection?
  (i) An acoustic pressure wave at an open end of a pipe.
  (ii) An acoustic pressure wave at a closed end of a pipe.
  (iii) A voltage wave at an open circuit.
  (iv) A voltage wave at a short circuit.

10.2   A longitudinal disturbance generated by an earthquake is observed to travel 5000 km in 15 minutes. Estimate Young's modulus for the rock through which the disturbance travels, assuming that its average density is 2700 kg m^{-3}.

10.3   For a wire carrying a longitudinal travelling wave, show that the strain is $\dot{\psi}/c$ at a point where the instantaneous longitudinal velocity is $\dot{\psi}$.

10.4   Estimate the percentage difference which the use of the isothermal compressibility instead of the adiabatic compressibility would make to the calculated value of the speed of sound.

10.5   Estimate the minimum length of an organ pipe required to produce a note of the lowest audible pitch (say 20 Hz), if a standing wave is excited with an acoustic pressure node at one end of the pipe and an antinode at the other end.

10.6   The compressibility of water is 4.9×10^{-10} m^2 N^{-1}. Estimate (a) the speed of sound in water, and (b) the characteristic impedance per unit area for water. (c) For sound waves of equal intensity in air and in water, roughly what is the ratio of the pressure amplitude of the wave in water to that of the wave in air? (d) When swimming under water, why can one not hear sounds in the air at the surface?

10.7   Acoustic plane waves are travelling in the *positive* z direction. A microphone is moving in the (positive or negative) z direction with velocity v_m. Show that the disturbance picked up by the microphone has frequency $\nu(1 - v_m/c)$ where ν and c are the wave frequency and speed respectively.
  If the transducer producing the plane waves is also moving in the z direction, with velocity v_t, show that the measured frequency is now $\nu(c - v_m)/(c - v_t)$.
  This is the *Doppler effect*.

10.8   Wishing to move his television set to another room, the owner extends the aerial feeder cable (characteristic impedance 75 Ω) by connecting it to a long piece of cable whose characteristic impedance is 120 Ω. If a 100 μV signal is received from the aerial, calculate the size of the signal transmitted across the join to the second cable.

10.9   A sinusoidal voltage wave of amplitude 25 V travels along a cable of characteristic impedance 120 Ω. Calculate the average power passing any point on the cable.

10.10   The attenuation of cables is commonly expressed in decibels (dB) per unit length. If the amplitude of a voltage wave is V_1 at point 1 and V_2 at point 2, and $V_2 < V_1$, the attenuation in bels (B) is defined as $\log_{10}(V_1^2/V_2^2)$, and 1 dB = 0.1 B.
  Calculate the attenuation of a cable with an attenuation length of 6 km.

11

Fourier theory

At this point it is worth bringing together several things we have discovered about vibrating systems in earlier chapters.

(1) If a number of harmonic driving forces act simultaneously on a linear system, the resulting steady-state vibration $\psi(t)$ is a superposition of harmonic vibrations whose frequencies are those of the driving forces: each harmonic force makes its own independent contribution to $\psi(t)$. This is an example of the principle of superposition (section 5.2).

(2) The free vibration of a non-linear system is not harmonic, but something more complicated; we found it possible, however, to express an anharmonic vibration $\psi(t)$ as a series of terms consisting of the fundamental vibration and a series of harmonics (sections 7.1 and 7.2).

(3) The principle of superposition does not hold for non-linear systems (combination frequencies appear in the vibration); but a single harmonic driving force leads to a steady-state $\psi(t)$ which contains the driving frequency and harmonics of that frequency (section 7.3).

(4) The standing waves that are possible on a *non-dispersive* string of finite length have frequencies in a sequence like $\nu_1, 2\nu_1, 3\nu_1, \ldots$; when a number of standing waves are excited simultaneously, the vibration $\psi(t)$ of any given point on the string must therefore consist of a series of superposed harmonics.

It is clear from these examples alone that the harmonic type of vibration on which we have spent so much time has a fundamental significance as the building block for more complicated motions. Obviously a means of *analyzing* a vibration $\psi(t)$ of any given form into a harmonic series would be valuable.

Harmonic analysis forms part of a mathematical system called Fourier theory after its founder. The elements of Fourier theory are introduced

here, not so much for analytical reasons as for the light that it sheds on
several aspects of vibrations and waves that we have not so far discussed.

11.1. Harmonic analysis

The cornerstone of Fourier theory is a theorem which states that almost
any periodic function can be analyzed into a series of harmonic functions
with periods $\tau, \tau/2, \tau/3, \ldots$, where τ is the period of the function under
analysis (fig. 11.1). 'Almost any periodic function' can be taken to mean

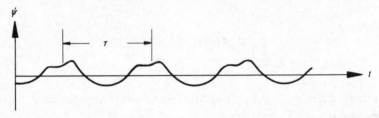

Fig. 11.1 A periodic function of time: $\psi(t) = \psi(t \pm \tau)$ for all times t, where τ is the period.

any periodic function that we are likely to come across in physics: any
function representing a free vibration, for example. (It is possible to
invent mathematical functions which cannot be analyzed.)

In form A the Fourier series for $\psi(t)$ is

$$\psi(t) = A_0 + A_1 \cos(\omega_f t + \phi_1) + A_2 \cos(2\omega_f t + \phi_2) + \ldots$$
$$+ A_n \cos(n\omega_f t + \phi_n) + \ldots \tag{11.1}$$

where

$$\omega_f \equiv 2\pi/\tau$$

is the angular frequency of the fundamental. The constant term A_0 may
be thought of as a harmonic with zero frequency. Each term in the series
has an amplitude and a phase constant; by adjusting these we can expand
the various harmonics vertically, or shift them horizontally, to make the
superposition fit the function $\psi(t)$. Harmonic analysis consists essentially
of finding A_n and ϕ_n for each value of n.

We could of course have written (11.1) in form B, form C or form D.
For analyzing real functions, form B is usually the most convenient. (See
problems 11.2 and 11.4.) But there are advantages in using form C for
developing basic ideas. In form C the series is

$$\psi(t) = A_0 + \sum_{n=1}^{\infty} [C_n \exp(in\omega_f t) + C_n^* \exp(-in\omega_f t)] \tag{11.2}$$

where the complex C_n's embody the various amplitudes and phase
constants in the usual way (1.23).

By defining new coefficients

$$c_0 \equiv A_0$$

$$c_n \equiv C_n$$

$$c_{-n} \equiv C_n^*$$

we can compress (11.2) into a form

$$\psi(t) = \sum_{n=-\infty}^{+\infty} c_n \exp(in\omega_f t) \tag{11.3}$$

which includes the constant term as the member of the series with $n = 0$. Harmonic analysis of $\psi(t)$ now becomes a matter of finding all the coefficients c_n.

How to find the coefficients. The general method for finding the c_n coefficients relies on the property that the average of any harmonic quantity over a number of complete cycles is zero. We consider two harmonic functions $\psi_1(t)$ and $\psi_2(t)$ with different periods, such as those shown in fig. 11.2(a). Over any time interval during which each function executes a number of complete cycles, the average value of the product $\psi_1\psi_2$ is zero. The average over a complete cycle of two harmonic functions with the *same* period is, however, finite, as illustrated in fig. 11.2(b).

If, therefore, we multiply our given function $\psi(t)$ (period τ) by a harmonic function of period τ/n, and integrate the product over any time interval equal to τ, the answer will be zero for all the harmonics of $\psi(t)$ except the nth.

The integral we are discussing is

$$I = \int_\tau \psi(t)[C \exp(in\omega_f t) + C^* \exp(-in\omega_f t)] \, dt$$

where the harmonic function of period τ/n has been written in form C, and τ/n has been replaced by $2\pi/n\omega_f$. It is convenient to split I into two parts

$$I = I_n + I_{-n}$$

$$I_n \equiv C \int_\tau \psi(t) \exp(in\omega_f t) \, dt \tag{11.4}$$

$$I_{-n} \equiv C^* \int_\tau \psi(t) \exp(-in\omega_f t) \, dt$$

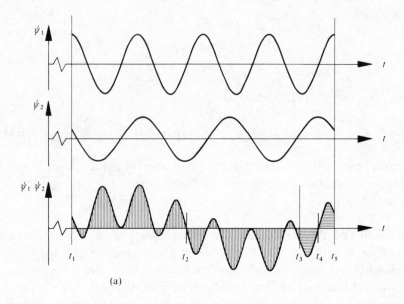

(a)

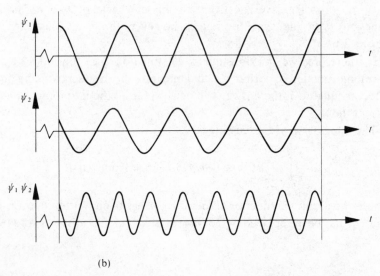

(b)

Fig. 11.2 (a) For any two harmonic functions ψ_1 and ψ_2 with different periods, the average of $\psi_1\psi_2$ over a number of complete cycles is zero. The part of the graph for $\psi_1\psi_2$ between t_1 and t_2 cancels the part between t_2 and t_3; the part between t_3 and t_4 similarly cancels the part between t_4 and t_5. (b) If ψ_1 and ψ_2 have the same period, $\psi_1\psi_2$ has a non-zero average.

When we substitute the Fourier series (11.3) for $\psi(t)$, only the nth harmonic (represented by the coefficients c_n and c_{-n}) makes any contribution to either integral in (11.4). For the second of these integrals we have

$$I_{-n} = C^* \int_\tau [c_{-n} \exp(-in\omega_f t) + c_n \exp(in\omega_f t)] \exp(-in\omega_f t)\, dt$$

$$= c_n C^* \int_\tau dt$$

$$= c_n C^* \tau$$

Thus we find

$$c_n = I_{-n}/C^* \tau$$

or, by (11.4),

$$c_n = \frac{1}{\tau} \int_\tau \psi(t) \exp(-in\omega_f t)\, dt \tag{11.5}$$

If $\psi(t)$ is known as a function of t, c_n can be found from (11.5) for all positive and negative values of n. The values of c_n, when substituted into (11.3), give the Fourier expansion for that particular $\psi(t)$. If we know all the coefficients we have complete information about $\psi(t)$.

It should be noted that the coefficient c_0 (the constant term in the series) is included in this prescription (11.5). Its value

$$c_0 = \frac{1}{\tau} \int_\tau \psi(t)\, dt$$

is simply the average of $\psi(t)$ over a complete cycle.

Example. The important predictions of Fourier theory can be illustrated by discussing the analytically simple example of a 'square' vibration, shown in fig. 11.3(a). (Although it is not a wave, this form of vibration is commonly called a 'square wave'.)

In this vibration the displacement periodically jumps from zero to some constant value H, and clicks back again to zero at some later time. We shall assume that the time spent at $\psi = H$ during each cycle is $f\tau$, where τ is the period and f is a number smaller than 1. Choosing the origin of time

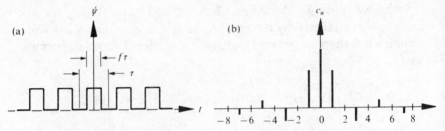

Fig. 11.3 (a) A square vibration; the example drawn has $f = \frac{1}{2}$. (b) The frequency spectrum for the same vibration.

as shown in the figure, we have

$$\psi = H \qquad (|t| < \tfrac{1}{2}f\tau)$$

$$\psi = 0 \qquad (|t| > \tfrac{1}{2}f\tau)$$

The integral (11.5) over one period may be taken between the limits $\pm\frac{1}{2}f\tau$, since ψ is zero elesewhere. Thus

$$c_n = \frac{1}{\tau} \int\limits_{-f\tau/2}^{+f\tau/2} H \exp(-in\omega_f t)\, dt$$

$$= (H/in\omega_f\tau)[\exp(inf\omega_f\tau/2) - \exp(-inf\omega_f\tau/2)]$$

Remembering that $\omega_f\tau$ is 2π, and replacing the complex exponentials by a sine, we obtain the result

$$c_n = (H/n\pi)\sin nf\pi \qquad (11.6)$$

These coefficients are all real, since $c_{-n} = c_n$ for all values of n.

We cannot get c_0 from this formula, since putting in $n = 0$ leads to $0/0$. But the general expression (11.5) gives immediately

$$c_0 = \frac{1}{\tau} \int\limits_{-f\tau/2}^{+f\tau/2} H\, dt = fH \qquad (11.7)$$

which is, of course, the average value of $\psi(t)$.

For a square vibration with $f = \frac{1}{2}$, in which the system shares its time equally between the two displacements, we find

$$c_0 = \tfrac{1}{2}H,$$

$$c_1 = H/\pi, \qquad c_2 = 0,$$

$$c_3 = -H/3\pi, \qquad c_4 = 0,$$

$$c_5 = H/5\pi, \qquad c_6 = 0,$$

$$c_7 = -H/7\pi, \qquad \ldots$$

A convenient way of displaying these results is to plot the 'frequency spectrum' shown in fig. 11.3(b).

At first sight it may appear that the negative n region of this diagram is redundant, since the spectrum is symmetric about $n = 0$. This is, however, a special circumstance resulting from our choice of time origin, which made $\psi(t)$ symmetric about $t = 0$: we say that we made the function *even*.

The nth Fourier coefficient for any even function $\psi_E(t)$ is

$$c_n = \frac{1}{\tau} \int_\tau \psi_E(t) \exp\left(-in\omega_f t\right) dt$$

We may replace t by $-t$ in this formula to find

$$c_n = \frac{1}{\tau} \int_\tau \psi_E(-t) \exp\left(in\omega_f t\right) dt$$

$$= \frac{1}{\tau} \int_\tau \psi_E(t) \exp\left(in\omega_f t\right) dt$$

$$= c_{-n}$$

where we have used the fact that

$$\psi_E(-t) = \psi_E(t)$$

The condition $c_n = c_{-n}$ means $C_n^* = C_n$ (making C_n real) in the form C representation of the nth harmonic. A real C_n corresponds to a phase constant ϕ_n of zero (if C_n is positive) or π (if C_n is negative). This is merely another case of arranging convenient phase constants by judicious choice of the time origin.

Some functions, such as the one shown in fig. 11.4, can be made *odd*; for any odd function $\psi_O(t)$ we have

$$\psi_O(-t) = -\psi_O(t)$$

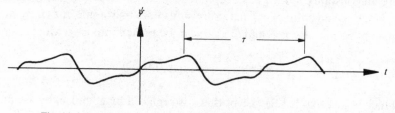

Fig. 11.4 A periodic function of t which, with the origin shown, is odd.

By reasoning similar to that used for even functions, we can show that $c_{-n} = -c_n$ for all the Fourier coefficients of an odd function.

A function like the one in fig. 11.1, which cannot be made either even or odd, has $|c_{-n}| \neq |c_n|$ in general. Its Fourier coefficients are therefore complex, and could not be plotted in a simple diagram like fig. 11.3(b). It is obviously worth making our function even or odd at the outset if we possibly can.

Harmonic analysis in space. From a mathematical point of view $\psi(t)$ is just a function of some variable t. There is no reason to restrict the use of the theory to functions of time only.

We saw in section 9.3 that the standing waves on a stretched string of finite length could all be written in the form

$$\psi_n(z, t) = A_n u_n(z) \cos(\omega_n t + \phi_n) \tag{11.8}$$

where the eigenfunctions $u_n(z)$ are sinusoidal functions of the space variable z. The arguments of these sinusoidal functions contain terms $k_n z$; for a string fixed at both ends (or free at both ends) the k_n values belong to the sequence $k_1, 2k_1, 3k_1, \ldots$.

In general, all of these standing waves will be present, in varying strengths. The vibration of the string will then be described by

$$\psi(z, t) = \sum_{n=1}^{\infty} \psi_n = \sum_{n=1}^{\infty} A_n u_n(z) \cos(\omega_n t + \phi_n) \tag{11.9}$$

If we photograph the string at some instant during this vibration, its appearance in the photograph can be described by the series

$$\psi(z) = \sum_{n=1}^{\infty} A'_n u_n(z) \tag{11.10}$$

whose coefficients

$$A'_n \equiv A_n \cos(\omega_n t + \phi_n)$$

are all constants, since t has been fixed by the camera.

Since the eigenfunctions $u_n(z)$ are sinusoidal (harmonic in z) they can be expressed in form C and (11.10) can be turned into a Fourier series

$$\psi(z) = \sum_{n=-\infty}^{\infty} c_n \exp(ink_1 z)$$

which is just like (11.3) except that t is replaced by z, and ω_f by k_1, the smallest wavevector consistent with the boundary conditions in force.

What is valuable is of course the ability to analyze a given $\psi(z)$ into its eigenfunctions, just as we can analyze any $\psi(t)$ into its harmonics. The coefficients are

$$c_n = \frac{k_1}{2\pi} \int_{2\pi/k_1} \psi(z) \exp(-ink_1 z) \, dz$$

This formula is just (11.5) with k_1 again appearing instead of ω_f, and τ now replaced by the space 'period', namely the maximum wavelength $2\pi/k_1$.

We can, for example, calculate the future motion of a string of length L, fixed at both ends and set into vibration by being held at rest in some non-equilibrium shape and then released. We first invent an odd periodic function of z which matches the string over half a period (fig. 11.5). The

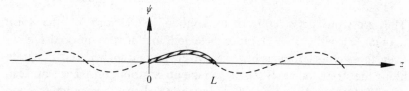

Fig. 11.5 The shape of a vibrating string of length L, at some fixed instant. For the purposes of Fourier analysis we can imagine the string to be part of a periodic function of z; the wavelength is then taken as $2L$, to construct the Fourier series for the string shape.

initial shape can now be analyzed into eigenfunctions by taking $2L$ as the wavelength. These eigenfunctions are used to construct the complete vibration with the aid of a series like (11.9). Naturally these eigenfunctions will stretch to infinity in both directions, but it is enough for our purpose that they should correctly reproduce the physical conditions *at the string*.

Recognizing the eigenfunctions as building blocks of the vibration allows us to see that the standing waves possess one of the attributes of modes: the property of making independent contributions to the total energy (section 8.1).

The instantaneous energy density $w(z, t)$ at any point on the string is given by (9.32). The instantaneous *total* energy can be found by integrating $w(z, t)$ along the length of the string: from $z = 0$ to $z = L$, say.

If the vibration consisted of mode n (11.8) excited on its own, its total energy would be

$$W_n(t) = \frac{ZA_n^2}{2c} \int_0^L [\omega_n^2 u_n^2(z) \sin^2(\omega_n t + \phi_n) + c^2 u_n'^2(z) \cos^2(\omega_n t + \phi_n)] \, dz$$

where

$$u'_n(z) \equiv \frac{du_n(z)}{dz}$$

In general the motion is a superposition of all the modes (11.9). In that case the energy density is

$$w(z, t) = \frac{Z}{2c}\{[\sum \omega_n A_n u_n(z)]^2 \sin^2(\omega_n t + \phi_n)$$

$$+ [\sum c A_n u'_n(z)]^2 \cos^2(\omega_n t + \phi_n)\}$$

the summations being taken from $n = 1$ to $n = \infty$.

The first term on the right contains the factor

$$[\sum \omega_n A_n u_n(z)]^2 = \sum \omega_n^2 A_n^2 u_n^2(z) + \text{cross-products}$$

The cross-products referred to contain expressions of the form $u_n(z)u_m(z)$, with $n \neq m$. Because the eigenfunctions are sinusoidal, however, these cross-products have zero averages and so, when we integrate along the string to find $W(t)$, they make no contribution.† The same can be said of the products $u'_n(z)u'_m(z)$ appearing when we expand the second term, since the $u'_n(z)$ functions are sinusoidal also. Thus the total energy may be written

$$W(t) = \sum \frac{Z}{2c}\int_0^L A_n^2[\omega_n^2 u_n^2(z) \sin^2(\omega_n t + \phi_n) + c^2 u'^2_n(z) \cos^2(\omega_n t + \phi_n)]\,dz$$

$$= \sum W_n(t)$$

Since the total energy is just the sum of the energies obtained for the separate standing waves, we can confidently identify the standing waves with the modes of the string.

Summary. A periodic function $\psi(t)$ can be written as the series

$$\psi(t) = \sum_{n=-\infty}^{+\infty} c_n \exp(in\omega_f t)$$

† Figure 11.2 suggests that it is necessary to integrate over complete wavelengths of both $u_n(z)$ and $u_m(z)$ to obtain this result, whereas the length L comprises only integral numbers of half wavelengths, or even of quarter wavelengths (fig. 9.6). The integrals from $z = 0$ to $z = L$ are zero (and therefore the corresponding averages are zero) in this case because the boundary conditions are the same for all standing waves on a given string. This corresponds to choosing the same phase constant for ψ_1 and ψ_2 in fig. 11.2.

in which the coefficients are given by

$$c_n = \frac{1}{\tau} \int_\tau \psi(t) \exp(-in\omega_f t) \, dt$$

These coefficients will be complex unless the origin of t can be adjusted to make $\psi(t)$ either even or odd. In these equations τ is the period of the function being analyzed and ω_f is $2\pi/\tau$.

A function $\psi(z)$ can be expanded as an analogous series of sinusoidal functions. The shape of a vibrating string at any instant can, for example, be written as a series of eigenfunctions (the shapes of its standing waves). Because the eigenfunctions are sinusoidal, the various standing waves make independent contributions to the energy of the vibration.

11.2. Modulation

When information in the form of time signals, speech, music or television pictures is broadcast, the current in the transmitting aerial, instead of being made to follow the signal itself, oscillates harmonically at some standard high frequency, and this high-frequency oscillation is *modulated* in accordance with the signal. The modulated 'carrier wave' arriving at the receiving aerial produces a forced oscillation which is modulated in the same way as the transmitter current. This oscillation is selectively amplified as outlined in section 6.1, and the signal is recovered from it by means of suitable circuitry.

The beating vibration illustrated in fig. 5.10 is an elementary example of *amplitude modulation*. The same example suggests a role for Fourier theory in the discussion of modulation, since we recall that it was the result of superposing two harmonic driving forces of slightly different frequencies. The Fourier spectrum of this particular modulated vibration consists of just these two frequencies.

In this section we consider the general problem of how to find the frequency spectrum for an amplitude-modulated vibration. This is of practical importance, since a radio receiver, for example, must be designed to accept the entire range of frequencies present in the modulated wave.

As an alternative to amplitude modulation it is possible to modulate the frequency of the vibration (as in stereo radio and television sound transmissions) or even the phase 'constant'. We shall not pursue these possibilities here.

Fourier series for an amplitude-modulated vibration. Any amplitude-modulated vibration may be represented by writing

$$\psi(t) = M(t)[C\,e^{iN\omega t} + C^*\,e^{-iN\omega t}] \tag{11.11}$$

Here $M(t)$ is the modulation envelope, which at present we assume to be periodic (fig. 11.6). The basic harmonic vibration (the 'carrier') has been given the angular frequency $N\omega$ where $2\pi/\omega$ is the *modulation* period τ, and N is a number large enough to ensure that the harmonic vibration is much faster than the amplitude changes resulting from the modulation.

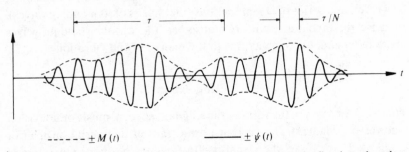

Fig. 11.6 An amplitude-modulated vibration. The almost-harmonic vibration takes place within an envelope defined by the periodic function $M(t)$. Compare fig. 5.10, in which $M(t)$ is itself harmonic.

We can always fix our origin of time so that the harmonic vibration is even; this allows us to simplify (11.11) by putting $C = C^*$. By further choosing $C = \tfrac{1}{2}$ we give the carrier unit amplitude, so that the over-all amplitude of $\psi(t)$ is governed by $M(t)$ only. The function to be analyzed is now

$$\psi(t) = \tfrac{1}{2}M(t)[e^{iN\omega t} + e^{-iN\omega t}]$$

The Fourier coefficient of the nth harmonic (11.5) is

$$c_n = \frac{1}{2\tau}\int_\tau M(t)[e^{iN\omega t} + e^{-iN\omega t}]\,e^{-in\omega t}\,dt$$

$$= \frac{1}{2\tau}\int_\tau M(t)[e^{-i(n-N)\omega t} + e^{-i(n+N)\omega t}]\,dt$$

This has two parts. The first is the same (apart from the factor $\tfrac{1}{2}$) as the $(n-N)^{\text{th}}$ Fourier component of the modulation function $M(t)$, and the second is the same as the $(n+N)^{\text{th}}$ component. In other words, the frequency spectrum can be found simply by adding together two spectra which are identical in shape to the spectrum for $M(t)$, one of these two being centred on $n = -N$ instead of $n = 0$, and the other one being centred on $n = +N$.

Example. A simple example will help to clarify this result. We take as $M(t)$ the square vibration with $f = \frac{1}{2}$ whose frequency spectrum we calculated previously (fig. 11.3). We use this to modulate a carrier whose period is $\tau/19$, so that the harmonic vibration is alternately switched on for $9\frac{1}{2}$ cycles and switched off for $9\frac{1}{2}$ cycles, as shown in fig. 11.7(a). (For simplicity we have taken an odd value for N, and made both the carrier vibration and the modulation function even.)

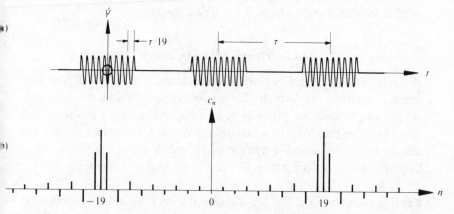

Fig. 11.7 (a) Harmonic vibration modulated by the function illustrated in fig. 11.3(a). (b) The frequency spectrum. In this example the carrier frequency is 19 times the modulation frequency.

The frequency spectrum consists of the spectrum of fig. 11.3(b) drawn twice: once with its centre at $n = -19$ and once at $n = +19$. Each spectrum is half the height of the parent spectrum. To obtain a given c_n we should, strictly, add the relevant members of these two spectra: the coefficient c_{20}, for example, has a small contribution from 'c_{39}' in the left-hand spectrum as well as a large one from 'c_1' in the right-hand spectrum. But in practice this is not usually necessary since one spectrum generally dominates the other, except in the region near $n = 0$ where both sets of coefficients are in any case small.

The important thing to realise is that changing the carrier frequency has no effect at all on the shapes of the two half-spectra, but only changes their separation. Thus the spread of frequencies present (the *bandwidth*) is entirely dependent on how the carrier is modulated. A circuit which has to amplify an unmodulated harmonic oscillation can be designed on the assumption that it need only work properly at that frequency; but a circuit to amplify the bursts in fig. 11.7(a) must be satisfactory over a range of frequencies extending above and below the frequency of the vibration that occurs during a burst.

We shall investigate the relationship between the bandwidth and the form of the modulation in the following section.

Summary. To find the frequency spectrum for a harmonic vibration of period τ/N, modulated by a function of period τ, we first calculate the frequency spectrum for the modulation function. This spectrum (with all its coefficients halved) is plotted at $n = N$ and also at $n = -N$. The required spectrum is the sum of these two spectra.

11.3. Pulses and wave groups

We return to the simple square vibration shown in fig. 11.3. The frequency spectrum plotted in that figure corresponds to $f = \frac{1}{2}$; if f is progressively reduced, $\psi(t)$ will begin to look less like a vibration and more like a sequence of separate *pulses*. We wish to explore in turn the effects on the frequency spectrum of (1) increasing the length of time between two successive pulses, and (2) shortening the pulses themselves.

Varying the pulse rate. We can increase the intervals between pulses (so decreasing the pulse rate) by increasing τ while we decrease f, in such a way that the pulse width maintains some constant value Δt. In fig. 11.8 are plotted the spectra obtained from (11.6) with $\tau/\Delta t$ equal to 2 (corresponding to $f = \frac{1}{2}$, which we have already seen), 4 and 8.

What the figure shows is that the pulse rate has no effect on the envelope within which the frequency spectrum is formed. The *range* of frequencies required to reproduce $\psi(t)$ does not alter, although the fact that ω_f is getting smaller means that an increasing *number* of harmonics will be involved.

It is easy to understand why more harmonics should be needed when the pulses are more widely spaced. To make $\psi(t)$ zero, the superposed harmonics must cancel each other completely; complete cancellation for a longer time obviously requires more components, with more closely spaced frequencies.

There appears to be no reason why we should not let the spacing become so great that the next pulse in the train never appears at all. The theory should, in fact, be applicable to single pulses as well as to a regular

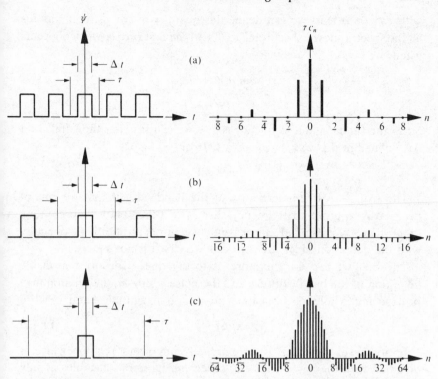

Fig. 11.8 How the frequency spectrum changes as the pulse rate is decreased. The pulse width Δt is kept constant while the vibration period τ is increased. To make the spectra comparable with each other τc_n is plotted rather than c_n, and the scales of n are adjusted to make them equal in units of frequency. (a) $\tau = 2 \Delta t$; (b) $\tau = 4 \Delta t$; (c) $\tau = 8 \Delta t$.

succession of pulses. At the end of the chapter we shall show very briefly how this is done mathematically. At present it is sufficient to realise that these ideas can be carried to such a conclusion.

Varying the pulse length: bandwidth theorem. We have seen that the range of frequencies necessary to reproduce the pulse train is the same for all values of the pulse rate; how does this frequency range (the bandwidth) vary with the pulse width?

No precise answer to this question will be possible unless we can stipulate exactly how the bandwidth is defined. The total frequency spread in any of the spectra shown is actually infinite, although the spectra themselves indicate that the lowest frequencies are by far the most significant.

A reasonable rough measure of the bandwidths of these particular spectra is the frequency spread between zero and the lowest frequency with $c_n = 0$; at least we can define this quantity in a consistent way for all the spectra shown. According to (11.6) the first zero coefficient occurs when

$$\sin nf\pi = \pi$$

$$n = 1/f$$

Since the fundamental frequency is $1/\tau$ we can say that the bandwidth $\Delta \nu$, defined in this way, is equal to $1/f\tau$ or

$$\Delta\nu\Delta t = 1$$

The trouble with this criterion of the bandwidth is that we cannot necessarily apply it to frequency spectra of other kinds, representing other shapes of pulse. Indeed, we would encounter an additional difficulty with the definition of Δt itself for a pulse which is not square.

The best we can do, therefore, is to say that, with any reasonable definition of the bandwidth $\Delta \nu$ and the pulse width Δt, the relation may be expected to hold to within about an order of magnitude,† and we write

$$\Delta\nu\Delta t \sim 1 \qquad\qquad (11.12)$$

Imprecise as it is, this is an extremely important relation which is dignified with the name *bandwidth theorem*. Its importance lies mainly in the fact that the product $\Delta\nu\Delta t$ has a lower limit, not equal to zero, though we do not know exactly how big that limit is. As we have already seen, the pulse rate has no effect on $\Delta\nu$; the only thing that matters is the pulse width Δt. The narrower the pulses are, the wider the spread of frequencies in their Fourier series. In the extreme case, to produce a train of absolutely short pulses ($\Delta t = 0$) would require a spectrum extending upwards to infinite frequencies.

Modulation pulses. The bandwidth theorem should also hold for a harmonic vibration which is modulated by a train of pulses (or, in the last resort, by an isolated pulse). As we know, the effect of changing from a given vibration $\psi(t)$ to a harmonic vibration amplitude-modulated by the same $\psi(t)$ is to split the frequency spectrum into two identical parts; these two parts are relocated at $n = \pm N$ where N is the ratio of the modulation period to the harmonic vibration period. The shapes, and in particular the widths, of these spectra are the same as those of the parent spectrum.

---

† Even that statement is optimistic. Merely superposing a set of harmonics with the correct frequencies and amplitudes will not give the desired result if their phase constants are not correctly arranged also: the superposition must be constructive during the pulses and destructive during the spaces. A more general statement is thus $\Delta\nu\Delta t \gtrsim 1$.

Thus the frequencies required to synthesize the modulated vibration lie in the range $\nu \pm \Delta\nu$, where ν is the vibration frequency and $\Delta\nu$ is given by (11.12). Again, increasing the interval between successive bursts of vibration does not change the shape of the frequency spectrum, but brings the harmonics closer together in frequency and so increases their number. In the limiting case of a single burst we expect the harmonics to merge into a continuous smear of frequencies; but the width depends only on how long the burst lasts.

We used the bandwidth theorem to demonstrate the impossibility of absolutely sharp pulses. At the opposite extreme, we can now see that a very long burst of harmonic vibration ($\Delta t \to \infty$) can be produced with a very narrow spread of frequencies ($\Delta\nu \to 0$), centred on the vibration frequency: in fact, by a very rudimentary spectrum consisting of the vibration frequency on its own. Of course, an infinitely long burst simply means an unmodulated harmonic vibration, and it is no surprise to find that it can synthesize itself.

Travelling wave groups. Most of this chapter has been about vibrations. There are important implications for wave motion, however.

A train of pulses applied to the end of a stretched string will cause a train of disturbances to be propagated along the string and, from what we learnt in chapter 9, we can anticipate that the shape of the wave on the string at any given moment will resemble the graph of ψ against t. Figure 11.9 shows the effect of applying regular bursts of harmonic motion, thereby generating separate trains of sinusoidal travelling waves. We shall refer to such isolated travelling disturbances as *wave groups*.

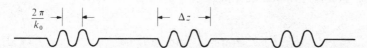

Fig. 11.9 Train of wave groups produced by applying regular bursts of vibration (harmonic in this example) to the end of the string. The diagram shows the shape of the string at a fixed time.

Like that of a standing wave, a 'photograph' of a train of wave groups on a string is amenable to Fourier analysis. From our earlier discussion we know that k takes the place of ω in Fourier theory whenever t is changed to z, and so we can immediately write down the bandwidth theorem for wave groups,

$$\Delta k \Delta z \sim 2\pi \qquad (11.13)$$

This shows that the spread of superposed wavevectors Δk required to reproduce a train of wave groups is inversely proportional to the group length Δz. If the wave groups are modulated sinusoidal waves of wavevector k_0 (fig. 11.9) the range of wavevectors required is $k_0 \pm \Delta k$. As with an isolated pulse, the wavevector spectrum representing an isolated wave group will be continuous.

There is a vital difference to be noticed between the analysis of a travelling wave group and that of a standing wave. In the latter case (fig. 11.5) we may invent a periodic function of z, part of which coincides with the non-periodic string, and analyze that function. The results of the mathematical analysis have no significance beyond the ends of the string.

In the present case, however, the analysis must include all the points where the string is momentarily undisplaced. It obviously takes many more sinusoidal components to produce cancellation in the gaps than it would take merely to reproduce the string shape within a group.

Fourier transform. We end with a brief look at the problem of adapting the Fourier series to deal with single pulses rather than pulse trains.

The series form is inappropriate because, as we have seen, making the pulse rate infinitely slow makes ω_f infinitely small; the frequencies $n\omega_f$ thus merge into a continuum. To handle this situation we need to replace $n\omega_f$ by a continuous variable ω, and to integrate over ω instead of summing over n.

We must also substitute a smooth function of ω for the discrete coefficients c_n. We shall introduce the function $c(\omega)$ and use as the 'coefficient' in front of exp $(i\omega t)$ the quantity $c(\omega) \, d\omega$, where the angular frequency lies somewhere in the range between ω and $\omega + d\omega$. The expansion (11.3) then becomes

$$\psi(t) = \int_{-\infty}^{+\infty} c(\omega) \, e^{i\omega t} \, d\omega \qquad (11.14)$$

The range of integration here includes 'negative frequencies', which are physically absurd. But this is merely a mathematical device whereby both halves of form C can be written in the same way; a 'negative frequency' really means an ordinary positive frequency with a minus sign in front of it.

If we know $c(\omega)$ we can generate $\psi(t)$ with the aid of (11.14); but how do we find the $c(\omega)$ for a given $\psi(t)$? The formula (11.5) that provided the c_n's for a periodic $\psi(t)$ must be turned into a formula that will give $c(\omega)$ when $\psi(t)$ is aperiodic.

Once more we replace $n\omega_f$ by ω. The frequency 'step' ω_f is replaced by $d\omega$, and so the factor $1/\tau$ in front of the integral in (11.5) becomes $d\omega/2\pi$. The result of these substitutions is

$$c(\omega) = \frac{1}{2\pi} \int_{-\infty}^{+\infty} \psi(t) \, e^{-i\omega t} \, dt \tag{11.15}$$

This function gives us the (continuous) frequency spectrum for the (aperiodic) function $\psi(t)$; it is known as the *Fourier transform* of $\psi(t)$.

To analyze non-repeating functions in space we would use, instead of (11.14) and (11.15), the corresponding relations

$$\psi(z) = \int_{-\infty}^{+\infty} c(k) \, e^{ikz} \, dk$$

$$c(k) = \frac{1}{2\pi} \int_{-\infty}^{+\infty} \psi(z) \, e^{-ikz} \, dz \tag{11.16}$$

Taking our stock example, we can find the Fourier transform for a rectangular pulse of height H and width Δt. The calculation is very similar to the evaluation of c_n for a square vibration in section 11.1, and leads to the result

$$c(\omega) = \frac{H}{\pi} \left(\frac{\sin \frac{1}{2}\omega \, \Delta t}{\omega} \right)$$

This function is plotted in fig. 11.10; it is clearly of the correct form to act as a template for the frequency spectra in fig. 11.8.

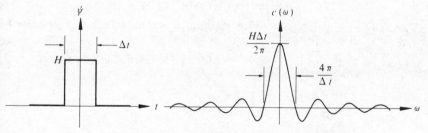

Fig. 11.10 A single rectangular pulse and its Fourier transform.

Summary. The width of the frequency spectrum representing a train of pulses is independent of the pulse rate but inversely proportional to the pulse width. For all pulse shapes, the frequency spread $\Delta\nu$ and the pulse width Δt are related by the bandwidth theorem

$$\Delta \nu \Delta t \sim 1$$

If the pulses are bursts of almost harmonic vibration the bandwidth theorem indicates the necessary spread of frequencies about the vibration frequency.

The wavevector spread Δk and the length Δz of a travelling wave group obey a similar relation

$$\Delta k \Delta z \sim 2\pi$$

Problems

11.1 Sketch the frequency spectrum representing a harmonic vibration of frequency 10ν, modulated harmonically at a frequency ν. (This is the motion illustrated in fig. 5.10.)

11.2 (a) Show that the Fourier series for an even function may be written

$$\psi(t) = b_0 + \sum_{n=1}^{\infty} b_n \cos n\omega_f t$$

where

$$b_0 = \frac{1}{\tau} \int_{\tau} \psi(t)\, dt$$

$$b_n = \frac{2}{\tau} \int_{\tau} \psi(t) \cos n\omega_f t\, dt \quad (n \neq 0)$$

(b) Show that $b_0 = c_0$ and $b_n = 2c_n$ $(n \neq 0)$ where c_0 and c_n have the same meanings as in the text.

11.3 Sketch the frequency spectrum representing the modulated carrier

$$\psi(t) = (A + B \cos \omega t) \cos N\omega t$$

where N is a large integer. Confirm that your answer to problem 11.1 is obtained when $A = 0$, $N = 10$.

11.4 (a) Show that the Fourier series for an odd function may be written

$$\psi(t) = b_0' + \sum_{n=1}^{\infty} b_n' \sin n\omega_f t$$

where

$$b_0' = \frac{1}{\tau} \int_{\tau} \psi(t)\, dt$$

$$b_n' = \frac{2}{\tau} \int_{\tau} \psi(t) \sin n\omega_f t\, dt \qquad (n \neq 0)$$

(b) Show that $b_0' = c_0$ and $b_n' = \mathrm{i}2c_n$ $(n \neq 0)$ where c_0 and c_n have the same meanings as in the text.

11.5 A single period of a 'sawtooth' vibration may be written $\psi(t) = At/\tau$, where A is the distance separating the maxima of $|\psi|$, and τ is the period. (a) Sketch the graph of ψ against t over several periods. (b) Choosing an origin that makes ψ odd, find the c_n coefficients for the Fourier expansion of $\psi(t)$. (The analysis is easiest in form B: see problem 11.4.)

11.6 A string is stretched between two anchor points separated by a distance L. It is then pulled aside a distance d at its centre. (a) Using the axes and origin indicated in fig. 11.11, write down the periodic function $\psi(z)$ drawn (broken line)

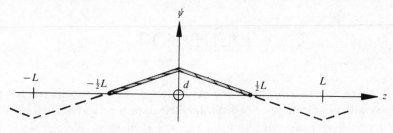

Fig. 11.11.

between $z = -L$ and $z = +L$. (b) Find the c_n coefficients for the Fourier expansion of $\psi(z)$. (The analysis is easiest in form B: see problem 11.2.) (c) Sketch the wavevector spectrum. (d) If the centre of the string is released at time $t = 0$, obtain a series representing the shape of the string $\psi(z, t)$ at all subsequent times.

11.7 A note played by a musical instrument will have a frequency spread determined by the bandwidth theorem. If the note lasts for a very short time, the spread may be too wide for the listener to be able to identify the pitch correctly.

Estimate the duration of a note which will just allow the listener to place its pitch to the nearest semitone (i.e. within 6 per cent in frequency) for (a) a high note on a piccolo ($\nu \approx 3.7$ kHz), and (b) the lowest note on a contra-bassoon ($\nu \approx 30$ Hz).

11.8 A stone dropped into a pond generates a travelling wave group approximately 1 m long; the waves within the group have a wavelength of about 0.1 m. Estimate the range of the wavevector spectrum representing the group.

12

Dispersion

The wave equation we discussed in chapter 9 has the special property that it allows a disturbance of arbitrary form to be propagated indefinitely as a travelling wave, without having its shape changed. We met several examples of such non-dispersive waves in chapter 10.

Non-dispersive waves are exceptional. In this chapter we examine possible sources of dispersion in a stretched string.

12.1. Stiff strings

The stretched string of chapter 9 was assumed to be perfectly flexible, so that there were no transverse return forces other than those due to the tension. Real strings, such as violin and piano strings, are 'stiff' and tend to straighten out even when unstretched. The extra return forces due to this lateral stiffness make the string dispersive.

These return forces come from the stresses within curved parts of the string. The stress forces at any cross-section of the string will have components acting along the string direction, and components acting in the plane perpendicular to the string. Each set of components can be replaced by a single force and a single torque: for the components parallel to the string we have the force of *tension*, and a *bending moment*, while the perpendicular components give a *shear force* tending to break the string across, and a *twisting moment*. Since we already know the return force due to the tension, and since the string is presumably not twisted, we need consider only the other two. These are shown in fig. 12.1(a): here $f(z)$ is the shear force acting at position z along the string, and $\tau(z)$ is the corresponding bending moment.

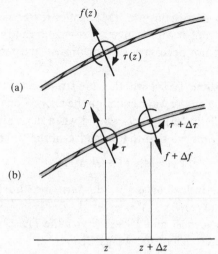

Fig. 12.1 (a) Shear force $f(z)$ and bending moment $\tau(z)$ acting at position z along a stiff string. These stresses are assumed to act on the material immediately to the right of z; equal and opposite stresses act on the material to the left of z. (b) Forces and torques acting on the two ends of the segment between z and $z+\Delta z$.

What we are interested in is the sideways force on a segment of the string. Figure 12.1(b) shows the external forces and torques acting on such a segment. As well as the resultant force Δf there is a resultant torque $(f\Delta z-\Delta\tau)$: but on our usual 'small displacement' assumption any rotation of the segment will be much less important than its bodily transverse movement, and so we may set the torque to zero. For a vanishingly small segment this means

$$f \approx \frac{\partial\tau}{\partial z}$$

If we consider only long-wavelength disturbances (and small displacements), Δf will act almost vertically, and we may take it as the return force. We have

$$\Delta f = \left(\frac{\partial f}{\partial z}\right)\Delta z \approx \left(\frac{\partial^2\tau}{\partial z^2}\right)\Delta z$$

Now the value of the bending moment at any point will be proportional to the curvature of the string there, and when ψ is small the curvature is just $\partial^2\psi/\partial z^2$. Thus Δf is proportional to $\partial^4\psi/\partial z^4$ and we can replace the wave equation (9.3) by

$$\frac{\partial^2\psi}{\partial t^2} = c^2\left[\frac{\partial^2\psi}{\partial z^2} - \alpha\left(\frac{\partial^4\psi}{\partial z^4}\right)\right]$$

where α is some positive constant. (Mechanics textbooks show how α can be found for a string or wire of given cross-section and given elasticity, but such details are not necessary for our present purposes.)

The dispersion relation. To see whether the string can still support wave motion, we substitute in the wave equation the expression for a sinusoidal travelling wave (9.7) in form D, as we did when discussing attenuation (section 9.5). This time we find that ω and k in the wave must satisfy

$$\omega = \pm ck(1 + \alpha k^2)^{1/2} \tag{12.1}$$

If the string is perfectly flexible ($\alpha = 0$), this turns into the simpler relation (9.6) which we found earlier.

The relation between ω and k determines the phase velocity of the wave (9.8). The wave speed that we get from (12.1)

$$|v_\phi| = c(1 + \alpha k^2)^{1/2}$$

depends on the magnitude of k (fig. 12.2), and therefore *depends on the wavelength*. At very long wavelengths (small k) the string is only gently curved, and its stiffness has a correspondingly small effect; $|v_\phi|$ then has nearly the same value as a flexible string, namely c. As the wavelength is decreased (and k becomes greater) the influence of the stiffness is increasingly felt, and $|v_\phi|$ rises.

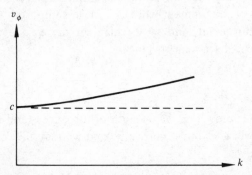

Fig. 12.2 Phase velocity of a sinusoidal travelling wave on a stiff string, plotted against the wavevector k. For a flexible string the phase velocity is c for all values of k (broken line).

The variation of phase velocity with wavelength (or frequency) is the hallmark of dispersion; wave systems showing this feature are said to be dispersive. The all-important equation connecting ω and k is the *dispersion relation* for the system.

In a dispersive system, arbitrary (non-sinusoidal) disturbances cannot be propagated as travelling waves without change of shape. We cannot argue, as we did in section 9.1, that any function of $ct \mp z$ will satisfy the wave equation. Thus the simple travelling wave equation (9.12) is inapplicable, since it relied on the old dispersion relation (9.6).

Slightly dispersive systems. If the string is only slightly stiff, and the wavelength is not too short, the condition $\alpha k^2 \ll 1$ may be satisfied. This will allow us to use the binomial expansion to write (12.1) in the approximate form

$$\omega \approx \pm ck(1 + \tfrac{1}{2}\alpha k^2)$$

We take the plus (minus) sign for waves travelling to the right (left), thereby keeping ω positive and making the dispersion relation an even function of k. A similar procedure will be possible for other systems, and we shall find it useful to adopt

$$\omega = ck - dk^3 \tag{12.2}$$

as the *standard form of dispersion relation for a slightly dispersive system*. We have dropped the $\pm$ sign for convenience: that just means that we are deciding to discuss waves travelling in the positive z-direction from now on. The choice of a minus sign on the right is conventional: it makes the constant d negative for a stiff string, but in other cases it can be positive.

This equation, unlike (12.1), expresses ω as a simple series of powers of k. One advantage of such a form is that it allows us to go back to a differential equation in ψ with the help of the form D expression for a sinusoidal travelling wave (9.7). The relevant derivatives in this case are

$$\frac{\partial \psi}{\partial t} = i\omega\psi$$

$$\frac{\partial \psi}{\partial z} = -ik\psi$$

$$\frac{\partial^3 \psi}{\partial z^3} = ik^3\psi$$

With their aid we can turn (12.2) into

$$\frac{\partial \psi}{\partial t} + c\left(\frac{\partial \psi}{\partial z}\right) + d\left(\frac{\partial^3 \psi}{\partial z^3}\right) = 0 \tag{12.3}$$

which is the travelling wave equation for a slightly dispersive system.

Wave groups. In the case of an amplitude-modulated disturbance such as a travelling wave group, we can get a very clear understanding of why the shape is not maintained, from Fourier theory. We saw in section 11.3 that any isolated wave group may be viewed as a superposition of many sinusoidal waves. These component waves have a continuous spread of k values, the range of k covered being inversely proportional to the length of the group (11.13).

In a dispersive system these component waves travel at different speeds. To produce a wave group they must have not only the correct amplitudes and frequencies but must also be suitably aligned along the z direction; it is this spatial arrangement that is spoilt with the passage of time. What will usually happen is that the group will become smeared out as it progresses, and may eventually disappear entirely.

If the dispersion is slight, however, the degradation will take place slowly, and the group may survive long enough to convey useful information. The question will then arise of how fast the group as a whole travels.

We consider a group, such as the one shown in fig. 12.3, in which $|\psi|$ has a fairly well-defined maximum whose progress we can observe. At time t that maximum will be found at that position z where the components are constructively superposed. To be constructively superposed they must all have the same value of the phase angle $\omega t - kz + \phi$, although the individual values of ω, k and ϕ will be different. Thus the group maximum will always be found at a position z and a time t for which $\omega t - kz + \phi$ is the same for all wave frequencies.

The quantity $\omega t - kz + \phi$ will be independent of frequency if

$$\left[\frac{\mathrm{d}}{\mathrm{d}\omega}(\omega t - kz + \phi) \right]_{\bar{\omega}} = 0$$

The suffix $\bar{\omega}$ indicates that we should evaluate the derivative when ω is equal to the angular frequency of the almost-sinusoidal wave within the group. Carrying out the differentiation, we obtain

$$t - \left(\frac{\mathrm{d}k}{\mathrm{d}\omega} \right)_{\bar{\omega}} z = 0$$

This is satisfied by all points z and times t whose quotient z/t is equal to

$$v_g(\bar{\omega}) \equiv \left(\frac{\mathrm{d}\omega}{\mathrm{d}k} \right)_{\bar{\omega}} \qquad (12.4)$$

This is the *group velocity* at angular frequency $\bar{\omega}$.

The phase velocity v_ϕ that we discussed before is the velocity of points of constant phase angle. The group velocity is the velocity of points of

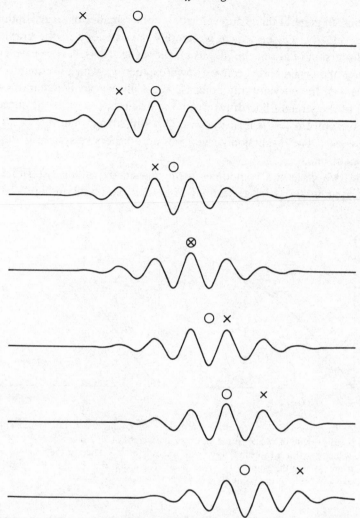

Fig. 12.3 A travelling wave group. The diagrams show the group at times one period apart. In this example the dispersion is 'normal', with the group velocity equal to half the phase velocity. A specimen wave crest (marked ×) enters the group at the trailing edge; its amplitude increases to a maximum and then decreases again as it passes forwards through the group, to disappear again at the leading edge of the group. The position of maximum disturbance in the group is marked ○ in each picture.

constant amplitude. Both v_ϕ and v_g can be found from the dispersion relation. For the flexible string we have, from (9.6),

$$v_g = \pm c = v_\phi$$

for all values of $\bar{\omega}$: the group travels at the same speed as all the individual components. This is, however, a special property of non-dispersive systems. In general the group velocity is either smaller or larger than the phase velocity. To see which, we merely have to look at the graph of the dispersion relation: the *dispersion curve* (fig. 12.4). If the curve goes through the origin and is convex upwards, then v_g, which is equal to the gradient of the curve at any point (k, ω), is always smaller than v_ϕ, the slope of the straight line drawn from the origin to (k, ω). If the curve is concave upwards, as it is in the case of a stiff string, then $v_g > v_\phi$ at all frequencies. The dispersion curve for a non-dispersive system is simply a straight line.

The two cases are sometimes distinguished by calling the situation illustrated in fig. 12.4(a) *normal dispersion*, and that in fig. 12.4(b)

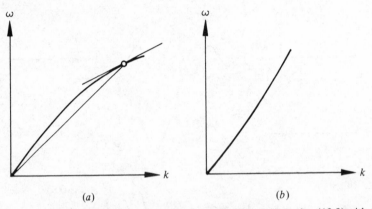

Fig. 12.4 Dispersion curves. The curves plotted are the graphs of equation (12.2) with the same positive value of c, and equal and opposite values of d: (a) $d > 0$, (b) $d < 0$. The phase velocity v_ϕ for a travelling wave with wavevector k is equal to the gradient of the straight line joining the points $(0, 0)$ and (k, ω). The group velocity v_g is the gradient of the dispersion curve at the point (k, ω).

anomalous dispersion. The nomenclature comes from optics, where 'normal' dispersion is more common. (See p. 266.) In a system with 'normal' dispersion sinusoidal waves of low frequency travel faster than those of high frequency, and the group velocity is smaller than the phase velocity at all frequencies. According to this convention a stiff string exhibits 'anomalous' dispersion.

Because we deal with non-sinusoidal disturbances in a dispersive system by treating them as Fourier superpositions of sinusoidal components, the dispersion relation tends to be more useful than the wave equation in such cases. As we shall see, it can often be found without having to discover the wave equation first. When the dispersion relation turns out to be a power series in k, we can recover a wave equation from it merely by replacing the powers of k by appropriate space derivatives. In subsequent chapters we shall nearly always look for the dispersion relation for a system rather than its wave equation.

Standing waves. Another important consequence of dispersion is its effect on the eigenfrequencies of the standing waves in a closed system. For illustration we consider a string fixed at both ends, and assume that the dispersion is slight enough to be described by (12.2).

The dispersion relation can be used for standing waves, because a standing wave has the same wavelength $2\pi/|k|$ as the incident and reflected travelling waves (with wavevectors k and $-k$) from which it is formed. The allowed values of k are the same as before (9.26) since the boundary conditions are the same. The eigenfunctions will be identical to those plotted in fig. 9.6(a); but the eigenfrequencies are now given by

$$\omega_n = ck_n - dk_n^3$$

which leads to

$$\nu_n = nc/2L - n^3 d\pi^2/2L^3$$

instead of (9.31).

Some of the new eigenfrequencies are compared with the old in fig. 12.5, showing how the differences arise from the different dispersion relations (9.6) and (12.2) used for the two systems. The eigenfrequencies of the stiff string do not form a series of harmonics: their spacing becomes greater with increasing frequency.

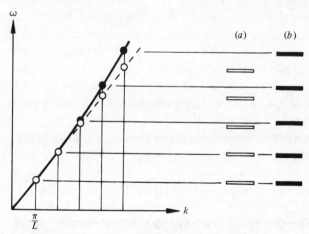

Fig. 12.5 Frequencies of the first five standing waves on a string of length L, fixed at both ends, when the string is (a) flexible, and (b) stiff. In both cases the allowed values of k are the same, and equally spaced, but the different dispersion relations lead to different eigenfrequencies.

A practical consequence of this departure from the harmonic series occurs in the tuning of pianos. For two strings an octave apart, the piano tuner will adjust the tension of one string until he can detect no beats between the second mode (eigenfrequency ν_2) of the low-frequency string and the fundamental mode (eigenfrequency ν_1') of the high-frequency string: that is, he makes $\nu_1' = \nu_2$. Since $\nu_2 > 2\nu_1$ he will end up with $\nu_1' > 2\nu_1$, which means that the frequencies of the 'tuned' strings are actually more than a true octave apart. This does not matter, of course, because the result sounds in tune.

Summary. A dispersive system can propagate a sinusoidal travelling wave, but its phase velocity ω/k varies with the frequency. A non-sinusoidal disturbance cannot be propagated without change of shape.

The speed of a wave group is found, for any system, by taking the derivative $d\omega/dk$ at the average or dominant frequency of the waves within the group. The phase velocity and the group velocity are equal only if the system is non-dispersive.

Stiff strings of finite length have the same eigenfunctions (equally spaced in k) as the corresponding flexible string, but the eigenfrequencies do not form a series of harmonics.

12.2. Lumpy strings

Wave motion is essentially a feature of continuous systems. There is, strictly, no such thing as a truly continuous system; but in most cases the ultimate graininess due to the atomic structure of matter is not obvious, since most ordinary waves have wavelengths very much greater than the typical spacings involved. There are exceptions, however, one of the most important being the propagation of high-frequency elastic waves through a crystal. In such cases certain characteristics due to the lumpiness of the medium, and its regularity, become apparent in the wave motion.

We can discover what these characteristics are by considering a perfectly elastic string (of the chapter 9 type) to which have been attached a number of identical, equally spaced beads. The mass of each bead is m, and we neglect the mass of the string itself. In equilibrium (fig. 12.6)

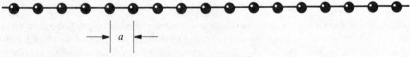

Fig. 12.6 A beaded string.

the string is stretched to a tension T, and the distance between adjacent beads is a.

In fig. 12.7 we see a short section of the string during transverse wave motion. The picture shows a specimen bead, identified by the label r, and its immediate neighbours bead $r-1$ and bead $r+1$. Each bead has some instantaneous transverse displacement. These displacements are

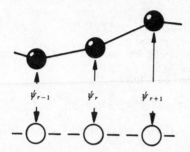

Fig. 12.7 Bead r and its neighbours during transverse wave motion.

exaggerated in the figure: we shall actually assume that they are small enough for us still to take the tension in each segment of string to be approximately T, and the length of each segment as approximately a.

The approximate transverse return force acting on bead r can then be found simply by resolving the tensions of each side (fig. 12.8). In this way we obtain the equation

$$
\begin{aligned}
m\ddot{\psi}_r &\approx -T(\psi_r - \psi_{r-1})/a + T(\psi_{r+1} - \psi_r)/a \\
&= (T/a)(\psi_{r-1} - 2\psi_r + \psi_{r+1})
\end{aligned}
\tag{12.5}
$$

For all possible transverse motions of the string, every bead and its neighbours must obey an equation like this one.

Now we specify that the motion is wave-like, and that the shape of the string at any given moment is roughly sinusoidal. It can never be

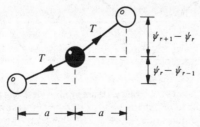

Fig. 12.8 The transverse return force acting on bead r is due to unequal transverse components of the tensions on each side of the bead.

exactly sinusoidal, because the segments of massless string between the beads are always straight; but the beads themselves always lie on an exact sine curve.

If we measure distance along the string from a bead which we choose as 'bead zero', we can put $z = ra$ for bead r, and similarly for all the other beads. Then the various displacements appearing in (12.5) are, in form A,

$$\psi_{r-1} = A \cos \left[\omega t - k(r-1)a \right]$$

$$= A[\cos (\omega t - kra) \cos ka - \sin (\omega t - kra) \sin ka]$$

$$\psi_r = A \cos (\omega t - kra)$$

$$\psi_{r+1} = A \cos \left[\omega t - k(r+1)a \right]$$

$$= A[\cos (\omega t - kra) \cos ka + \sin (\omega t - kra) \sin ka]$$

and we also know that

$$\ddot{\psi}_r = -\omega^2 \psi_r$$

Substituting these into (12.5) leads quickly to the dispersion relation

$$\omega = \omega_c \sin \tfrac{1}{2}ka \tag{12.6}$$

where

$$\omega_c \equiv 2(T/ma)^{1/2}$$

Travelling waves. The dispersion curve (fig. 12.9) shows a new feature: it repeats itself endlessly along the k axis, so that there is an infinite number of values of k for every value of ω. There is no longer a unique wavelength for each wave frequency.

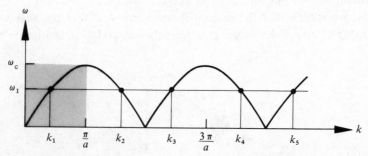

Fig. 12.9 The dispersion curve for a string carrying beads with spacing a. Any wave angular frequency ω_1 corresponds to an infinite number of different wavevectors $k_1, k_2, k_3, \ldots$, but all possible motions of the string can be described in terms of wavevectors within the shaded region. The value of ω_c depends on the tension, the mass of each bead; and the bead spacing.

It is not hard to understand why this is so. The first maximum in the curve occurs when k has the value π/a, corresponding to a wavelength $2a$. In a travelling wave of this wavelength, the displacement of bead r is

$$\psi_r = A \cos(\omega t - r\pi)$$

and the displacements of its neighbours are

$$\psi_{r-1} = A \cos[\omega t - (r-1)\pi] = -\psi_r$$
$$\psi_{r+1} = A \cos[\omega t - (r+1)\pi] = -\psi_r$$

Thus in this particular wave adjacent beads are vibrating in antiphase with each other (fig. 12.10).

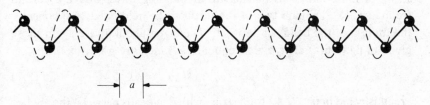

Fig. 12.10 In a travelling wave of wavelength $2a$, adjacent beads vibrate in antiphase with each other.

Values of k larger than π/a correspond to waves with wavelengths smaller than $2a$, but the structure of the string is too coarse to allow it to register such detail. The physical motion accompanying any wave with $\lambda < 2a$ can in fact be described equally well by another wave with $\lambda > 2a$, as fig. 12.11 demonstrates. Thus *the first half loop of the dispersion curve contains all the physical information.*[†]

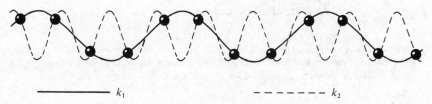

Fig. 12.11 Two travelling waves in which the motion of the string is identical. The waves shown have the wavevectors marked k_1 and k_2 in fig. 12.9, and the corresponding phase velocities. Only in the longer-wavelength example does the shape of the string resemble the sinusoidal wave profile. (For clarity the connecting strings are not shown.)

[†] The range $|k| \leq \pi/a$ is known as the *first Brillouin zone.*

At the lowest frequencies, for which the wavelength is much larger than the bead spacing a, we have

$$ka \ll 1$$

$$\sin \tfrac{1}{2}ka \approx \tfrac{1}{2}ka$$

$$\omega \approx (Ta/m)^{1/2}k$$

If m/a is replaced by μ, the last equation turns into the dispersion relation for a flexible string (9.6): thus the lumpiness of the string has no appreciable effect at these long wavelengths.

Standing waves. There is a corresponding limit to the ability of a lumpy string of finite length to participate in standing waves above a certain frequency. We imagine two of the beads to be held fixed, and consider the motion of the beads in between. The eigenfunctions are again those given in fig. 9.6(a). The eigenfrequencies, which are given by

$$\omega_n = \omega_c \sin \tfrac{1}{2}k_n a \qquad (12.7)$$

are illustrated in fig. 12.12 for a string with five beads between the anchor points. The striking new thing is that there are only five eigenfrequencies.

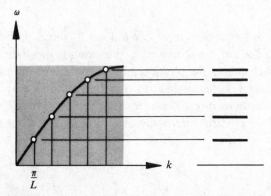

Fig. 12.12 The eigenfrequencies of a string with five beads, fixed at both ends. The dispersion curve used is the first half-loop of fig. 12.9. In contrast to the examples shown in fig. 12.5, higher allowed values of the wavevector do not lead to new modes of the system.

Here we are reminded of the two-coordinate system, which had two mode frequencies. The result that the number of modes is equal to the number of coordinates is quite general, as we shall shortly see. A corollary is that a truly continuous closed system has an infinite number of possible standing waves: all the examples we have looked at so far showed this feature.

In the case of the beaded string the eigenfunction merely acts as a kind of template for determining the relative displacements of the beads when the system is vibrating in a mode. With a large number of beads the motion will appear wave-like, and the eigenfunction gives us a useful means of describing the string profile in a given mode, but it has an exact physical significance only at those points where there are beads.

A string with N beads between the anchor points has a length $(N+1)a$, and so the wavevectors (9.26) to be used in (12.7) are

$$k_n = n\pi/(N+1)a \qquad (n = 1, 2, 3, \ldots)$$

The eigenfunction describing 'standing wave $N+1$' (the one with $\omega = \omega_c$) is

$$u_{N+1}(z) = \sin k_{N+1}z = \sin(\pi z/a)$$

But *at the beads* we have

$$u_{N+1}(ra) = \sin r\pi = 0$$

since beads are to be found only where r is integral. Thus this 'standing wave' consists of no motion at all.

Considering next 'standing wave $N+2$', we first write

$$n = N+2 = 2(N+1) - N$$

Since, when r is integral, we have

$$\sin\left(2r\pi - \frac{Nr\pi}{N+1}\right) = -\sin\left(\frac{Nr\pi}{N+1}\right) \tag{12.8}$$

then

$$u_{N+2}(ra) = -\sin\left(\frac{Nr\pi}{N+1}\right) = -u_N(ra)$$

Thus there is no observable difference between 'standing wave $N+1$' and standing wave N if they are excited equally, apart from a formal phase difference of π which will make no difference to the actual motion when definite initial conditions are allowed for. Moreover, since (12.8) is satisfied by $r = \frac{1}{2}$ as well as by integral values, we have

$$\omega_{N+2} = \omega_N$$

by (12.7). In spite of its unique eigenfunction, 'standing wave $N+2$' describes motion which is identical to that described by standing wave N.

The replication of previously recognized modes continues as we increase the index n: 'standing wave $N+3$' describes mode $N-1$, and 'standing wave $2N+1$' describes mode 1. The frequency is a falling function of n in this region.

We can similarly see that 'standing wave $2N+3$' describes mode 1, and the whole pattern begins to repeat itself. The frequency rises again until n equals $3N+2$ and $3N+4$, and then falls until n equals $4N+3$. All this time the shape of the string is becoming less and less like the corresponding eigenfunction.

We conclude that *there are only N physically distinct modes*,† and that all possible free vibrations of the system can be represented by standing waves of wavelength $2a$ and larger.

In fig. 12.13 are illustrated the eigenfunctions and string profiles (for equal amplitudes and at identical times) for a string with five beads, for values of n from 1 to 18. It is indeed hard to envisage any new kind of vibration after mode 5, in which adjacent beads are in antiphase with each other.

Summary. The beaded string can support sinusoidal travelling waves of wavelength down to twice the bead spacing. At low wave frequencies the system is approximately non-dispersive. A wave whose wavelength is less than twice the bead spacing produces physical motion of the beads which is indistinguishable from that observed under another wave of longer wavelength.

The modes of a beaded string of finite length can be described in terms of standing waves. The number of transverse modes is equal to the number of beads on the string, and they can be completely catalogued by eigenfunctions of wavelength not less than twice the bead spacing.

12.3. Evanescent waves

The dispersion curve in fig. 12.9 shows another feature which we have not commented on so far: a range of frequencies for which there are apparently no corresponding wavevectors. There are no points on the curve with $\omega > \omega_c$. Some other systems have dispersion curves with *lower* frequency limits. It is presumably possible to shake the end of the string at any frequency we please; what kind of motion is induced if we choose a frequency beyond the limiting value for the system?

Low-frequency cut-off. As a model system with a lower frequency limit we consider the artificial but instructive case of an *anchored string* (fig.

---

† If we included the longitudinal modes and the independent transverse modes in the plane at right angles to the plane of fig. 12.7, the total number of modes would be $3N$.

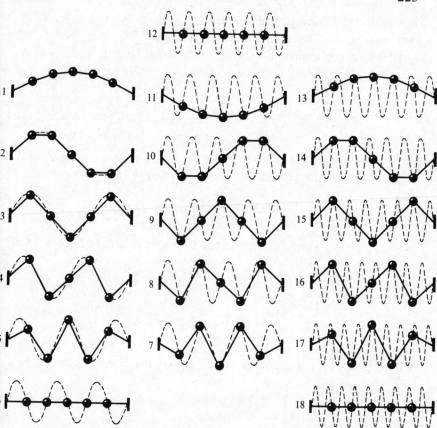

Fig. 12.13 Modes of a string with five beads. The eigenfunctions (broken curves) numbered 1 to 5 cover all possible modes. Eigenfunctions 6, 12, 18,... have zeros at the positions of the beads, and therefore describe no motion. Other eigenfunctions describe modes already covered by eigenfunctions 1 to 5.

12.14). Here the string itself is assumed to be perfectly flexible, but some lateral stiffness is applied by means of a number of springs connecting the string to a rigid anchoring post parallel to its equilibrium direction. These springs are identical, linear, uniformly spaced and close together. The essential feature of the resulting system is that the extra return force

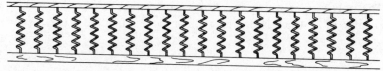

Fig. 12.14 An anchored string in equilibrium. Transverse disturbances are produced in the plane of the springs.

contributed by the springs does not require the string to be curved, but acts equally at all wavelengths, however large.

If the springs contribute stiffness σ per unit length, they give the segment of string at z an extra outward force $-(\sigma\Delta z)\psi(z, t)$. The equation of motion

$$\frac{\partial^2 \psi}{\partial t^2} \approx \frac{T}{\mu}\left(\frac{\partial^2 \psi}{\partial z^2}\right) - \left(\frac{\sigma}{\mu}\right)\psi$$

is obtained by adding the new force, divided by $\mu\Delta z$, to the right of (9.2). We write this equation in a standard form†

$$\frac{\partial^2 \psi}{\partial t^2} = c^2 \left(\frac{\partial^2 \psi}{\partial z^2}\right) - \omega_{c'}^2 \psi$$

where c has the same meaning as previously (9.4) and we have defined

$$\omega_{c'}^2 \equiv \sigma/\mu$$

The dispersion relation

$$\omega^2 = c^2 k^2 + \omega_{c'}^2 \tag{12.9}$$

is easily found. The dispersion curve (fig. 12.15) shows low-frequency cut-off at $\omega = \omega_{c'}$. Of course the dispersion relation turns into (9.6) when

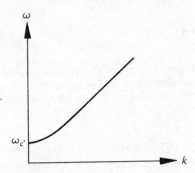

Fig. 12.15 The dispersion curve for an anchored string.

the springs are removed ($\sigma = 0$, making $\omega_{c'} = 0$). It also does so approximately, at frequencies well above cut-off ($\omega \gg \omega_{c'}$).

The only way in which the dispersion relation can be satisfied by a real value of ω which is smaller than $\omega_{c'}$ is to allow the wavevector

$$k = \frac{(\omega^2 - \omega_{c'}^2)^{1/2}}{c}$$

† This particular wave equation is known as the *Klein–Gordon equation*.

to become imaginary. We can see what an imaginary wavevector means physically by putting $K = 0$ in (9.40), which was the result of making k *complex*. We find

$$\psi = \mathrm{Re}\{D\exp[\mathrm{i}(\omega t + \mathrm{i}\kappa z)]\} = \exp(-\kappa z)\,\mathrm{Re}[D\exp(\mathrm{i}\omega t)] \quad (12.10)$$

There are two things to notice here.

(1) The phase angle does not now contain z; thus all points on the string vibrate in phase with each other, a characteristic we associate with a standing wave.

(2) The shape of the string at a given time is not sinusoidal, but exponential. The amplitude of the in-phase vibration falls off exponentially as we go along the string from the driven end. The string profiles at times half a period apart are shown in fig. 12.16.

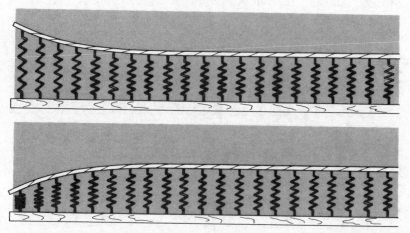

Fig. 12.16 An evanescent wave on an anchored string. The figure shows the shape of the string at times separated by half a period.

We shall call this kind of disturbance an *evanescent wave*. Whereas a travelling wave affects all points identically (though at different times), in an evanescent wave points much farther along the string than $1/\kappa$ will hardly move at all. The anchored string acts as a *high-pass filter*: if it is correctly terminated by a suitable damper much farther than $1/\kappa$ from the driven end, the damper will receive only those disturbances whose frequencies satisfy $\omega > \omega_{c'}$. As ω decreases below $\omega_{c'}$ the value of κ becomes larger, and so the cut-off becomes increasingly severe.

Reflection and tunnelling. Because an evanescent wave has the phase characteristics of a standing wave, it cannot carry power along the string.

(Figure 9.8 is relevant.) This allows the possibility of reflection of a travelling wave arriving at a discontinuity in the string, if the wave is cut off beyond the discontinuity.

If a sudden increase in the value of σ (or a sudden decrease in the value of μ, or both) occurs at $z = z_0$, the string beyond z_0 will have a higher cut-off frequency than the string in front of z_0. A sinusoidal travelling wave of angular frequency ω coming up to z_0 will be evanescent beyond z_0 if ω is smaller than the cut-off frequency on that side.

Because no power can be carried past z_0 and none is dissipated by damping forces, the incident wave must be totally reflected. The disturbance on the left-hand string in fig. 12.17(a) is therefore a standing wave. The point $z = z_0$ is not necessarily a node or an antinode; in principle we can position the nearest node or antinode wherever we wish, by correctly adjusting the values of σ and μ on the right.

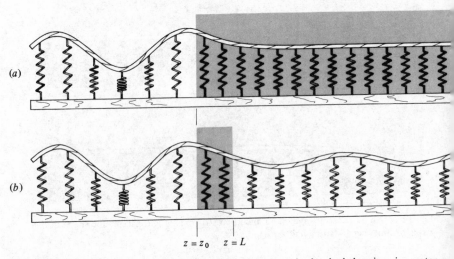

$$z = z_0 \qquad z = L$$

Fig. 12.17 Cut-off and tunnelling. The cut-off frequency in the shaded regions is greater than the wave frequency. (a) Reflection of a travelling wave arriving at $z = z_0$ from the left gives a standing wave on the left. Note that there is neither a node nor an antinode at $z = z_0$. (b) If the cut-off region is short enough, a travelling wave can appear beyond it.

If a second discontinuity occurs a certain distance L to the right of the first, and if the wave frequency, although below cut-off for the middle string, is above cut-off for $z > z_0 + L$, we may have a travelling wave in the latter region, as in fig. 12.17(b). For the amplitude on the third string to be appreciable, L should clearly be about $1/\kappa$ or less.

This phenomenon, whereby a travelling wave can appear beyond a short region which is 'forbidden', in the sense that travelling wave propagation is impossible in it, is known as *tunnelling*. It turns up in optics, in acoustics and, most spectacularly, in quantum mechanics. To analyze the effect completely we have to apply the correct boundary conditions at both discontinuities. The resulting algebra, though not difficult, is cumbersome, and we shall not pursue it. We can see, however, that one effect of the second boundary must be to reintroduce some z-dependence into the phase angle of the wave in the second region, so enabling it to carry energy through to the third region. The wave in the 'forbidden' region ceases to be truly evanescent, in fact.

High-frequency cut-off. Putting $\omega > \omega_c$ in the dispersion relation for the beaded string (12.7) leads to values of $\sin \frac{1}{2}ka$ larger than 1; this suggests that the argument $\frac{1}{2}ka$, and therefore k itself, becomes complex (not merely imaginary) above cut-off. We write

$$k = K - i\kappa$$

as before (9.39).

In terms of K and κ our present dispersion relation may be written

$$\omega/\omega_c = \sin\left(\tfrac{1}{2}Ka - i\tfrac{1}{2}\kappa a\right)$$
$$= \sin \tfrac{1}{2}Ka \cos\left(i\tfrac{1}{2}\kappa a\right) - \cos \tfrac{1}{2}Ka \sin\left(i\tfrac{1}{2}\kappa a\right) \tag{12.11}$$

The factor $\sin\left(i\tfrac{1}{2}\kappa a\right)$ on the right is imaginary, whereas the other three factors are all real. (The easiest way to see why is to write out the four expressions in terms of exponentials.) The ratio ω/ω_c must always be real, however; we can ensure this by restricting K so that

$$\cos \tfrac{1}{2}Ka = 0$$

$$Ka = \pi, 3\pi, 5\pi, \ldots$$

If we take the option $Ka = \pi$ we shall also make k pass the cut-off frequency smoothly, since we know from our earlier discussion that k is equal to π/a when $\omega = \omega_c$.

We conclude that, for the beaded string above cut-off, we have

$$k = (\pi/a) - i\kappa$$

where κ may be found as a function of frequency by putting $K = \pi/a$ into (12.11). This is left as an exercise (problem 12.8).

Our interest here is to discover the nature of the wave motion above cut-off. We put $K = \pi/a$ and $z = ra$ in (9.40) to find the displacement of bead r,

$$\psi(ra) = \exp(-\kappa ra) \, \mathrm{Re}\{D \exp[i(\omega t - r\pi)]\}$$

Apart from the change from z to ra, this result differs from that for low-frequency cut-off (12.10) in one important respect: there is a phase difference of 180° between any bead and its neighbours, expressed by the factor $\exp(-ir\pi)$ which reverses the sign of ψ every time r is increased or decreased by 1. Thus the phase relations between the beads that we saw at $\omega = \omega_c$ (fig. 12.10) are maintained above the cut-off frequency, but their amplitudes now decrease exponentially along the line, becoming smaller by a factor e after a distance $1/\kappa$. Thus the beaded string can be regarded as a *low-pass filter*. (An anchored lumpy string would act as a *band-pass filter*.)

The evanescent wave illustrated in fig. 12.18 should be compared with the corresponding behaviour with low-frequency cut-off in fig. 12.16.

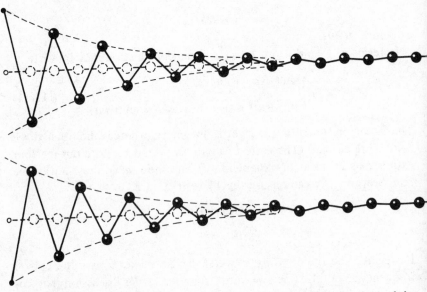

Fig. 12.18 An evanescent wave on a beaded string. The figure shows the shape of the string at times separated by half a period.

Summary. The anchored string can exhibit low-frequency cut-off. Its motion at frequencies below cut-off is an evanescent wave in which all parts of the string move in phase, with amplitudes which fall off exponentially with increasing distance along the string.

Evanescent waves carry no energy along the string. A travelling wave meeting a discontinuity beyond which it is evanescent will be totally reflected. It can, however, 'tunnel' through a short length of string on which it is evanescent.

The beaded string can exhibit high-frequency cut-off. The characteristic feature of the evanescent wave in that case is that adjacent parts of the system vibrate in antiphase with each other.

Problems

12.1 The end of a string is given a transverse displacement $\psi = \cos \omega_1 t + \cos \omega_2 t$ where the two frequencies are almost equal, and $\omega_1 > \omega_2$. Show that the resultant motion is a travelling wave of angular frequency $(\omega_1 + \omega_2)/2$, modulated by an envelope which is itself a travelling wave of angular frequency $(\omega_1 - \omega_2)/2$.

Show that the speed of the envelope (the 'group') is $(\omega_1 - \omega_2)/(k_1 - k_2)$ where k_1 and k_2 are the wavevectors which the component waves would have separately. (This result exemplifies the general result that the group velocity is $d\omega/dk$.)

If the individual component waves have the same phase velocity, show that the envelope also has this velocity.

12.2 For a system with the dispersion relation

$$\omega = ak^r$$

where a and r are constants, show that

$$v_g = rv_\phi$$

at all wave frequencies.

12.3 A wave group is formed by the superposition of harmonic waves of average wavelength λ. If $v_\phi(\lambda)$ is the phase velocity of a wave whose wavelength is λ, show that the group velocity is

$$v_g = v_\phi - \lambda(dv_\phi/d\lambda)$$

12.4 A piano string approximately obeys the dispersion relation (12.2) with $d/c = -1.0 \times 10^{-4} \, \text{m}^2$. Estimate the percentage increase in the tenth eigenfrequency of a piano string 1.0 m long, relative to that of a perfectly flexible string of the same length.

12.5 A sinusoidal travelling wave of frequency 5.0 Hz and amplitude 100 mm is sent along a system of three anchored strings like the one in fig. 12.17(b). The length of the string in the middle is 580 mm. The phase velocity is $3.0 \, \text{m s}^{-1}$ for all three strings; the cut-off frequencies are 4.0 Hz for the outer strings and 6.0 Hz for the middle string.

(a) Find the ratio of $1/\kappa$ for the middle string to the wavelength on the outer strings.

(b) Estimate the amplitude of the wave on the third string.

12.6 If the string carrying the travelling wave group shown in fig. 12.3 obeys dispersion relation (12.9), show that the wave frequency is $2^{1/2}$ times the cut-off frequency.

12.7 The motion of a string is given by

$$\psi(z, t) = e^{-\kappa z} \, \text{Re}\,(D_1 \, e^{i\omega t}) + e^{+\kappa z} \, \text{Re}\,(D_2 \, e^{i\omega t})$$

where D_1 and D_2 are complex, and $D_1 \neq D_2$. Does power flow along the string?

12.8   For a beaded string above cut-off, show that the dependence of κ on frequency is given by

$$\omega = \omega_c \cosh \tfrac{1}{2}\kappa a$$

Sketch the graph of ω against κ.

12.9   Calculate the upper cut-off frequency for a flexible beaded string with beads of mass 0.010 kg spaced at intervals of 0.010 m, when the tension is 10 N.

12.10   The beaded string (fig. 12.6) may be taken to represent a line of atoms in a one-dimensional 'crystal'. To make the model better, we should allow each atom to be attracted by *all* the other atoms, and write

$$m\ddot{\psi}_r = \sum_{n=1}^{\infty} S_n(\psi_{r-n} - 2\psi_r + \psi_{r+n})$$

in place of (12.5). Here S_n is the stiffness due to each of the forces acting between atom r and atoms $r \pm n$.

(a) Show that the dispersion relation for the 'crystal' is

$$m\omega^2 = 2 \sum_{n=1}^{\infty} S_n(1 - \cos nka)$$

(b) Verify that this turns into (12.6) when only nearest-neighbour atoms are presumed to interact with each other ($n = 1$ only) and $S_1 = T/a$.

(c) The new dispersion relation may be written

$$-\tfrac{1}{2}m\omega^2 = S_0 + \sum_{n=1}^{\infty} S_n \cos nka$$

where

$$S_0 \equiv -\sum_{n=1}^{\infty} S_n$$

This is a Fourier series, in k, of the even function $\omega^2(k)$. (Compare problem 11.2.) Show that

$$S_n = -\frac{ma}{2\pi} \int_{-\pi/a}^{+\pi/a} \omega^2 \cos nak \, dk$$

(This illustrates how measured dispersion relations can be used to investigate the forces between atoms in solids.)

(d) Verify that the result $S_1 = T/a$ is obtained when the ω^2 given by (12.6) is used in the above formula for S_n.

13

Water waves

The next three chapters are about waves which are, except in special circumstances, dispersive. For each type of wave we shall find the dispersion relation. When we have found it we shall quarry in it for physics, exploring particularly those extreme forms which can be so instructive.

To most people the word 'wave' suggests a stormy sea or ripples on the surface of a pond. Our discussion of these, the most familiar of all waves, falls into three main parts. In the first section we consider the purely *geometrical* problem of how the water must move in order to make a sinusoidal travelling wave on the surface. Then we find the dispersion relation, which tells us what sinusoidal waves are actually possible: this is a *physical* problem. Finally we conduct our discussion of the dispersion relation under various interesting extreme conditions.

Most of the physical results are to be found in the third section. If you wish to skip the preliminaries, you should first study fig. 13.5 and its caption, and the summary at the end of section 13.1, and then proceed directly to the dispersion relation (13.19).

13.1. The nature of the wave motion

We study the water in a long, rectangular canal of depth h (fig. 13.1). In equilibrium the water surface is flat and horizontal. With luck, a very special kind of breeze might blow along the length of the canal, exciting a sinusoidal travelling wave. All the water will be involved to some extent in the wave motion, but all an observer sees is a transverse sinusoidal disturbance of the *surface*.

It is sufficient to consider the two-dimensional motion of the water in a vertical plane parallel to the canal banks. We choose the intersection of

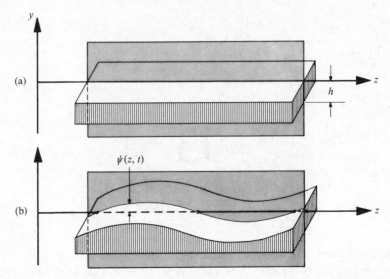

Fig. 13.1 A stretch of canal water; for clarity the canal banks are not shown. (a) The water in equilibrium. (b) The water with a sinusoidal travelling wave on its surface.

this plane with the equilibrium water surface as the z axis; instantaneous vertical displacements of the surface from this axis are specified as usual by the function $\psi(z, t)$.

The properties of water. We shall be making a number of guesses about the motion of the water, on the basis of our previous experience of wave systems. These guesses can be tested to see whether they are consistent with the known properties of water. First we must decide what these properties are.

From the point of view of wave motion, the most important characteristics of water are:

(1) it is hard to compress;

(2) it flows easily.

Numerically, the compressibility of water is about $5 \times 10^{-10} \, \mathrm{m}^2 \, \mathrm{N}^{-1}$. Thus if water is piled up to a height of, say, 1 m in the crest of a wave, the resulting pressure increase under the crest ($\sim 10^4 \, \mathrm{N} \, \mathrm{m}^{-2}$) will increase the density of the water there by only 0.05 per cent or so. Clearly it is a very good approximation to treat the water as completely incompressible, and we do so from now on.

The ease with which a liquid flows depends on its viscosity. Viscosity leads to damping, and we can expect water waves to be attenuated as a result. For simplicity, however, we shall not specifically allow for viscosity in our analysis.

Incompressibility places restrictions on the way in which water can move when a wave is present on the surface. If we imagine a downward displacement ($\psi < 0$) of some point on the surface, we can see immediately that some of the water beneath that point must have flowed away sideways to a region where the surface displacement is upwards ($\psi > 0$). The water cannot simply move up and down like a string carrying a transverse wave; every 'water particle' must move in two dimensions. What we have to do is to discover the details of this two-dimensional motion.

How the water moves: general considerations. We first introduce a new coordinate y which specifies the vertical position, measured from the z axis, of a given water particle when the water is in equilibrium. Water particles on the undisplaced surface have $y = 0$.

When the wave is present, the particle labelled (y, z) will, at time t, be found at a new position which we can specify by means of vertical and horizontal displacements $\psi_y(y, z, t)$ and $\psi_z(y, z, t)$. These displacements (fig. 13.2(a)) are measured from (y, z), that is, from the equilibrium position of the particle in question. To know everything about the motion of the water, we need to know how ψ_y and ψ_z vary with time, for all values of y and z.

For the particular case of a sinusoidal surface wave we can make a number of plausible statements about $\psi_y(y, z, t)$ and $\psi_z(y, z, t)$. We shall not attempt to justify these statements rigorously.

(1) The water is in a steady state of motion, in the sense in which that term has been used in chapter 5 and elsewhere. Therefore, *both ψ_y and ψ_z vary harmonically with the angular frequency ω of the wave*. The motion of each water particle is a superposition of two harmonic vibrations at right angles to each other.

(2) We expect the phase angles of both ψ_y and ψ_z to contain a term $-kz$, because the motion belongs to a travelling wave. We have no such preconceptions about how the phase angles depend on y, if at all. We shall therefore make the simplest possible assumption, namely that *the phase angles of ψ_y and ψ_z are independent of y*: particles which in equilibrium are stacked vertically, move in step with each other, both up-and-down and sideways, when there is a wave. If this assumption is wrong, we shall be made aware of the fact when we try to incorporate it in our picture of the motion responsible for a wave.

(3) This leaves open the question of the phase difference between the y and z motions. If we think about the water half way between a crest and a trough, where ψ_y is instantaneously zero, we might guess that ψ_z should

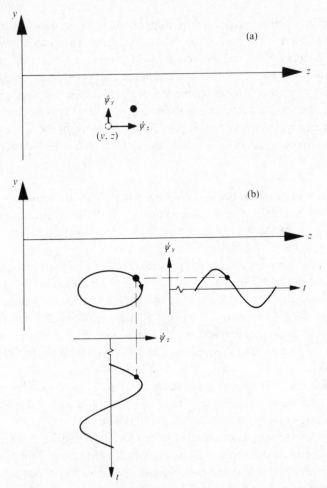

Fig. 13.2 (a) The coordinates ψ_y and ψ_z measure the vertical and horizontal displacements of a water particle from its equilibrium position (y, z). (b) For small displacements, the horizontal and vertical components of its motion (plotted in the inset graphs) are harmonic vibrations of the same frequency, in quadrature with each other. In the resultant motion the water particle moves round its equilibrium position in an elliptical orbit.

have its maximum value there since that is where the slope of the surface has its greatest value and where the maximum horizontal rearrangement of water must have occurred. Such a state of affairs will come about if *the vertical and horizontal components of the vibration are in quadrature with each other*.

(4) We expect the movement in the water to become increasingly gentle as the depth beneath the surface increases. In an infinitely long canal, however, there should be no dependence on distance along the canal.

Therefore we assume that *the vertical and horizontal amplitudes depend on y only.*

We can summarize these four statements by saying that we might expect ψ_y and ψ_z to be of the form

$$\psi_y(y, z, t) = A_y(y) \cos(\omega t - kz)$$
$$\psi_z(y, z, t) = A_z(y) \sin(\omega t - kz)$$

$$(13.1)$$

This is the simplest answer we can think of. We shall show that it can be made to reproduce the properties of water and to satisfy boundary conditions: we shall then be confident that we have hit upon the correct answer.

For any values of y, z and t the displacements (13.1) satisfy

$$(\psi_y/A_y)^2 + (\psi_z/A_z)^2 = 1$$

This is the equation of an ellipse whose principal axes lie along the ψ_y and ψ_z axes; these are vertical and horizontal respectively, and meet at the point (y, z). Thus *the path of each water particle is an upright ellipse, centred on the particle's equilibrium position* (fig. 13.2(b)).

The size and shape of a given ellipse – whether it is flattened or elongated in the vertical direction, and by how much – depend on the values of A_y and A_z, and therefore on the depth. We do not yet know how A_y and A_z vary with y.

As with acoustic waves, the displacement is not always the most useful variable to use. Fluid motion is usually described in terms of the velocity of the fluid at a given point (a measurable quantity) rather than the positions of water particles. At this stage we make a small-amplitude approximation: we differentiate (13.1) and call the result the velocity at (y, z), although it is strictly the velocity at $(y + \psi_y, z + \psi_z)$. Specifically, we write

$$v_y(y, z, t) = -\omega A_y(y) \sin(\omega t - kz)$$
$$v_z(y, z, t) = \omega A_z(y) \cos(\omega t - kz)$$

$$(13.2)$$

and all results from now on will apply to small-amplitude waves only.

Equations (13.1) and (13.2) are, so far, based only on inspired guess-work. We must show that it is actually possible for the particles to move like this in water. Can they move in elliptical paths without crowding together at some places and spreading apart at others, in a way that would be incompatible with the assumed incompressibility of the water? What particular elliptical shapes are actually possible?

To find the answers to such questions, we now impose two conditions: one which reflects the property of incompressibility, and another reflecting the absence of viscosity. These conditions will restrict the ways in which the amplitudes A_y and A_z vary with y. Together with boundary conditions, the water properties fix the shapes of the various elliptical paths.

The incompressibility condition. We imagine any closed surface lying entirely within some incompressible water. The mass of water inside this surface must remain constant, whatever the water as a whole is doing. This is the basis of the mathematical formulation of the incompressibility property.

We consider the surface enclosing a small rectangular volume $\Delta x \Delta y \Delta z$ at (y, z), where Δx is measured across the canal as shown in fig. 13.3. We wish to calculate the net outward flow of mass per unit time, in terms of $v_y(y, z, t)$ and $v_z(y, z, t)$. By setting the answer to zero we shall end up with a condition that must be satisfied by v_y and v_z in incompressible water.

The mass of water crossing any face per unit time is the density ρ multiplied by the area of the face and by the normal component of the velocity at that face. The two faces parallel to the canal banks contribute nothing, since there is never any velocity component across the canal.

For the top and bottom faces, the relevant velocity component is v_y. For an infinitely small volume, v_y at the top face exceeds v_y at the bottom face by an amount $(\partial v_y / \partial y) \Delta y$. Thus the net mass leaving the volume across these horizontal faces is $\Delta x \Delta z (\partial v_y / \partial y) \Delta y$.

In a similar way we can show that the net outward flow across the vertical faces is $\Delta x \Delta y (\partial v_z / \partial z) \Delta z$. The sum of these two quantities must be zero in incompressible water, and so the required condition† is

$$\frac{\partial v_y}{\partial y} + \frac{\partial v_z}{\partial z} = 0 \qquad (13.3)$$

This equation must be satisfied at all points in the water, and at all times.

The no-viscosity condition. We assumed that the water in the canal was originally at rest, and that the wind generated a sinusoidal travelling wave. Because the water has no viscosity, it is possible to make a general statement about the kind of motion it has been given: we can say that the water cannot have gained any angular momentum in the form of eddies or

† In the notation of vector calculus, this condition is the two-dimensional version of div $v = 0$.

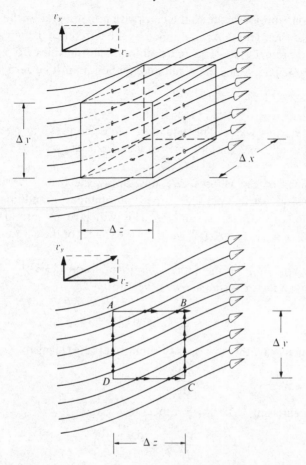

Fig. 13.3 The properties of water. The net flow of incompressible water out of the small volume $\Delta x \, \Delta y \, \Delta z$ must be zero: the mass entering and the mass leaving must cancel. The circulation of non-viscous water round the small loop *ABCD* must be zero: the vector components of the velocity clockwise round the loop must cancel.

whirlpools. Rotation could only be induced in this case through the agency of shear forces, which cannot exist in a non-viscous fluid.

The formal statement of this restriction is that the *circulation* of any loop of water must remain zero at all times, whatever may be happening to its shape. Circulation means the line integral of the fluid velocity round the loop.

We consider the chain of water particles lying round the small loop *ABCD* in fig. 13.3. By setting the circulation round this loop to zero we can find the way in which v_y and v_z are restricted by the absence of viscosity.

The contribution made to the clockwise line integral by the sides AB and CD is their length Δz multiplied by the difference in their v_z values, namely $\Delta z\,(\partial v_z/\partial y)\Delta y$. The contribution from the sides BC and DA is $-\Delta y(\partial v_y/\partial z)\,\Delta z$. Since the resultant circulation must be zero, we have

$$\frac{\partial v_y}{\partial z}-\frac{\partial v_z}{\partial y}=0 \tag{13.4}$$

for non-viscous water. This equation,† like the previous one, is to be satisfied at all times for all values of y and z in the water.

The shapes of the ellipses. We now apply the incompressibility and no-viscosity restrictions, and a pair of boundary conditions, to our expressions (13.2) for v_y and v_z. This will give us formulas for the amplitudes $A_y(y)$ and $A_z(y)$, which fix the shapes of the ellipses at a given depth.

After applying the incompressibility condition (13.3) to (13.2), and cancelling a factor $\omega \sin(\omega t - kz)$, we find

$$\frac{dA_y}{dy}-kA_z(y)=0 \tag{13.5}$$

In a similar way, the no-viscosity condition (13.4) leads to

$$\frac{dA_z}{dy}-kA_y(y)=0 \tag{13.6}$$

Differentiating (13.5) with respect to y gives

$$\frac{d^2A_y}{dy^2}-k\left(\frac{dA_z}{dy}\right)=0$$

Now we can use (13.6) to eliminate dA_z/dy, thereby obtaining

$$\frac{d^2A_y}{dy^2}-k^2A_y=0$$

This is a standard differential equation with the general solution

$$A_y(y)=A'\,e^{ky}+B'\,e^{-ky} \tag{13.7}$$

where A' and B' are arbitrary constants which will, as usual, be fixed by boundary conditions.

We need two boundary conditions, and we find them at the top and bottom surfaces of the water. At the top $(y=0)$ we have a sinusoidal travelling wave

$$\psi_y(0, z, t)=\psi(z, t)=A\cos(\omega t - kz)$$

---

† It is the two-dimensional version of **curl** $v=0$. Fluid flow with **curl** $v=0$ everywhere is said to be *irrotational*.

Substituting (13.7) into (13.1) and then applying this boundary condition gives

$$A_y(0) = A' + B' = A \tag{13.8}$$

At the bed of the canal $(y = -h)$ there can be no vertical movement: water on the bottom must remain on the bottom, and so

$$A_y(-h) = A' e^{-kh} + B' e^{kh} = 0 \tag{13.9}$$

Now we can find A' and B' by solving (13.8) and (13.9). The results are

$$A' = \frac{A e^{kh}}{e^{kh} - e^{-kh}}$$

$$B' = -\frac{A e^{-kh}}{e^{kh} - e^{-kh}}$$

and so (13.7) becomes

$$A_y(y) = \frac{A[e^{k(h+y)} - e^{-k(h+y)}]}{e^{kh} - e^{-kh}} \tag{13.10a}$$

There is no objection to leaving $A_y(y)$ in this exponential form, but a convenient short-hand notation is available in the form of hyperbolic functions. In that notation we have, as an alternative version of the last equation,

$$A_y(y) = \frac{A \sinh [k(h+y)]}{\sinh kh} \tag{13.10b}$$

The corresponding expression for $A_z(y)$ can be found by substituting the $A_y(y)$ we have just found into (13.5). The result is

$$A_z(y) = \frac{A[e^{k(h+y)} + e^{-k(h+y)}]}{e^{kh} - e^{-kh}} \tag{13.11a}$$

or, in the hyperbolic notation,

$$A_z(y) = \frac{A \cosh [k(h+y)]}{\sinh kh} \tag{13.11b}$$

From these two expressions we can find the shape of any ellipse, measured by the ratio of its height to its length. That ratio is equal to $\tanh [k(h+y)]$ at depth y, and can never be greater than 1 (fig. 13.4). Thus the orbital ellipses all have their longer axes lying horizontally. The flattening of the orbits becomes more pronounced as the depth increases. The orbits at the bottom of the canal, where there is no vertical motion, are completely flattened into horizontal lines: the water merely rubs backwards and forwards against the bottom.†

† In real (viscous) water the fluid in contact with the bottom is stationary, and transition to backwards-and-forwards motion takes place across a 'boundary layer'. The boundary layer contains horizontal vortices whose roller action is responsible for the familiar ridges on a sandy shore.

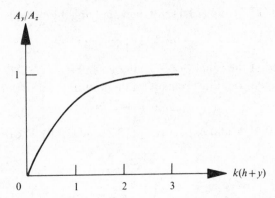

Fig. 13.4 The ratio of the height of an elliptical orbit to its length is tanh $[k(h+y)]$, which behaves as shown. On the canal bottom ($h+y = 0$) the orbits are flattened into horizontal lines. In deep water ($kh \gg 1$) they are nearly circular.

The horizontal movement also decreases with increasing depth, but more slowly, since cosh $[k(h+y)]$ falls off more slowly than sinh $[k(h+y)]$ as y becomes more negative.

Clockwise or anticlockwise? We can also work out whether the water particles move clockwise or anticlockwise round their elliptical paths.

For a wave travelling from left to right ($k > 0$), ψ_z *lags* ψ_y by 90°. This is because $A_y(y)$ and $A_z(y)$ in (13.1) are both positive, as we can see by looking at (13.10) and (13.11) while remembering that $h+y$ never becomes negative. The sense of the particle motion round its orbit is

---

Fig. 13.5 How the water moves when there is a sinusoidal travelling wave on the surface. Adjacent pictures are separated in time by one-eighth of a period, with time increasing from top to bottom. Each picture shows the position of some sample 'water particles' at that instant, calculated from equations (13.1), (13.10) and (13.11) with $kh = 1$. When the water was in equilibrium the particles were uniformly spaced, horizontally and vertically.

The diagram illustrates the following points:

(1) Each particle traces out a clockwise elliptical path. (The ellipses are drawn for the stack of particles at the extreme left.) These ellipses are all longer than they are high, and are flattened into horizontal lines on the canal bottom. Particles whose equilibrium positions were in a vertical line keep in phase with each other as they move round their ellipses.

(2) The water is never compressed or expanded. The sample of water between the 11th and 16th columns of particles is distorted in various ways as the wave passes, but its volume (proportional to the area of the section shown) never changes.

(3) There is no circulation in the water. The distortions of the same water sample never involve a net clockwise or anticlockwise rotation of the particles on the perimeter.

The formulas used are valid for small-amplitude waves, but for the sake of clarity a large amplitude has been adopted in the figure. It is interesting to note that this has sharpened the peaks and flattened out the troughs of the surface wave, making it look more like a real, large-amplitude water wave. (Real water waves rarely look sinusoidal.)

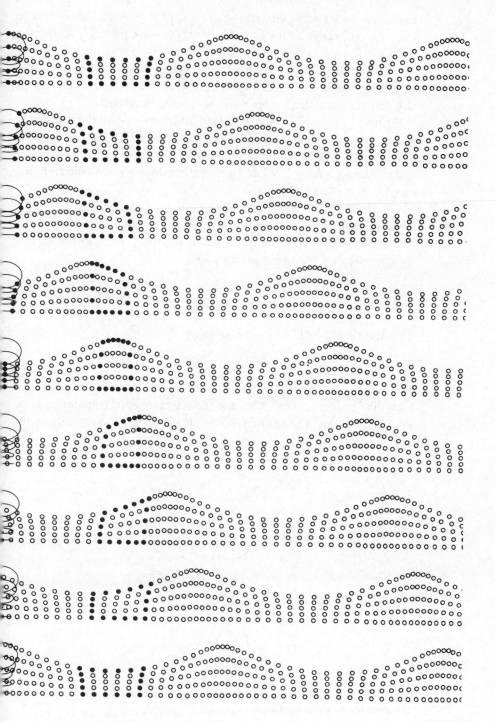

therefore clockwise as viewed in the positive x direction. (It is easy to show that the sense is anticlockwise when the wave travels from right to left.)

It follows that any water particle vertically beneath a wave crest must be moving directly forwards, i.e. in the direction of the wave. Conversely, any particle beneath a trough must be moving backwards.

Figure 13.5 shows how the combined orbital motions of the particles of water produce a sinusoidal travelling wave on the surface, while maintaining water-like behaviour. In this example the wavelength was chosen to be $2\pi h$, giving $kh = 1$. When kh is very large or very small the motion is simpler, as we shall now see.

The motion in deep water. If the water is many wavelengths deep ($kh \gg 1$) and we are only interested in the water relatively close to the surface ($|y| \ll h$), we may use the approximations

$$\sinh kh \approx \tfrac{1}{2} e^{kh}$$

$$\sinh [k(h+y)] \approx \cosh [k(h+y)] \approx \tfrac{1}{2} e^{k(h+y)}$$

The water motion (13.1) is then given by

$$\psi_y(y, z, t) \approx A\, e^{ky} \cos (\omega t - kz)$$

$$\psi_z(y, z, t) \approx A\, e^{ky} \sin (\omega t - kz)$$

Now the flattening is negligible, and the orbits are approximately circular at all depths (fig. 13.6). Their diameters decrease exponentially as the

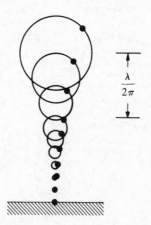

Fig. 13.6 Particle orbits in deep water. The orbits are circular, and their diameters decrease exponentially with increasing depth.

depth $-y$ increases, the motion being negligible at depths much greater than $1/|k|$. *Deep-water waves do not appreciably disturb the water more than a wavelength or so below the surface*, a property which is used by the designers of floating structures such as ocean oil rigs (fig. 13.7).

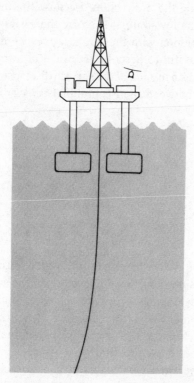

Fig. 13.7 A semi-submersible rig of the type used for exploratory drilling in deep water. If the buoyant parts are at depth h, the structure will not be appreciably affected by waves of wavelength $\lambda \lesssim h$.

The motion in shallow water. At the other extreme $(kh \ll 1)$ we have

$$\sinh kh \approx kh$$

$$\sinh [k(h+y)] \approx k(h+y)$$

$$\cosh [k(h+y)] \approx 1$$

In this case (13.1) becomes

$$\psi_y(y, z, t) \approx A(1+y/h) \cos (\omega t - kz)$$

$$\psi_z(y, z, t) \approx A(1/kh) \sin (\omega t - kz)$$

$$(13.12)$$

Now the horizontal amplitude is approximately the same at all depths. Its value A/kh is, moreover, very large compared with the surface wave amplitude A. The vertical amplitude falls off almost linearly from its surface value A to zero on the bottom. Thus *shallow-water waves are essentially longitudinal waves*: the water at any point just sloshes bodily back and forth along the z-direction. Because the motion is essentially one-dimensional, shallow-water waves are the easiest kind to deal with, having much in common with the simple waves we met in chapter 10.

It is sometimes useful to describe the wave in terms of this relatively violent sloshing motion instead of the surface displacement. A particularly simple connection between the longitudinal *velocity* $\dot{\psi}_z$ and ψ

$$\dot{\psi}_z(z, t) \approx (\omega A/kh) \sin(\omega t - kz)$$

$$= (\omega/kh)\psi_y(0, z, t) \qquad (13.13)$$

$$= (\omega/kh)\psi(z, t)$$

is easily obtained from (13.12).

Summary. We have worked out how the water must move to produce a small-amplitude sinusoidal travelling wave on the surface. The motion has to allow for the assumed properties of water (incompressibility and lack of viscosity) and must satisfy boundary conditions at the top and bottom surfaces of the water. The result is a combination of transverse and longitudinal motions in quadrature.

In deep water, the water particles move in almost circular orbits whose diameters decrease exponentially with increasing depth, with a decay distance of less than a wavelength. In shallow water the motion is nearly all longitudinal.

13.2. The dispersion relation

We have discovered how to describe the motion of a stretch of canal water with a sinusoidal travelling wave on its surface, but we still do not know what kinds of sinusoidal surface wave are possible. In order to find the dispersion relation, we must involve Newton's second law in some way.

In fluid mechanics the counterpart of Newton's second law is *Bernoulli's theorem*. This theorem applies, however, to steady-state situations only: that is, to problems in which the fluid velocity at every point is independent of time.† Our velocity components (13.2) do not meet this requirement.

We can surmount this difficulty by means of a simple trick of coordinate transformation. We imagine a new set of coordinate axes that travel in the z direction at the same speed as the wave, namely ω/k. Referred to these moving axes, a point whose distance along the canal was z in the fixed system has the new coordinate

$$z' = z - \omega t/k$$

and the velocity components (13.2) become

$$v_y(y, z') = \omega A_y(y) \sin kz'$$

$$v_{z'}(y, z') = \omega A_z(y) \cos kz' - \omega/k$$

These are *independent of time*. The surface of the water appears stationary (and sinusoidal) in the new coordinate system; each water particle seems to flow generally backwards, rising and falling as it does so. This new way of looking at the motion is illustrated in fig. 13.8.

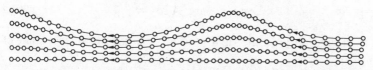

Fig. 13.8 Motion of the water particles viewed in a coordinate system which moves to the right with the wave. The particles appear to move to the left along the streamlines shown. Their speeds are greatest where they are most widely spaced out on the streamline.

Bernoulli's theorem states that the total energy per unit mass

$$W = p/\rho + \tfrac{1}{2}v^2 + V \tag{13.14}$$

has the same value at every point lying on a given *streamline*; a streamline is a path followed by a fluid particle taking part in steady-state flow. In (13.14) p is the pressure, v is the fluid speed and V is the potential energy per unit mass. We shall apply the theorem to a streamline lying along the surface of the water.‡

† This is a more restrictive use of the term 'steady state' than we have met previously. As applied to vibrations and waves (section 5.1), the phrase embraces repetitive motion.
‡ When the motion is irrotational, as it is here, the theorem states that W is the same at *all* points in the fluid. The more elementary statement is adequate for our present purpose, however.

There are two contributions to the pressure at the surface. One is the atmospheric pressure p_a, which is the same at every point. The other is a term due to surface tension, which gives rise to a pressure difference across any curved boundary. Curvature in this case occurs in the z direction only; the radius of curvature is $1/(\partial^2\psi/\partial z^2)$, and the total pressure is therefore

$$p = p_a - \sigma(\partial^2\psi/\partial z^2) \tag{13.15}$$

where σ is the surface tension (0.073 N m^{-1} for a boundary between air and water at 20°C). The minus sign appears on the right of (13.15) because a positive value of $\partial^2\psi/\partial z^2$ means that the surface is concave upwards, giving a pressure *reduction*. For a sinusoidal surface wave we have, in the moving coordinate system,

$$\psi = A \cos kz'$$

$$\frac{\partial^2\psi}{\partial z^2} = -k^2 A \cos kz'$$

and so (13.15) becomes

$$p = p_a + \sigma k^2 A \cos kz' \tag{13.16}$$

The second term on the right of (13.14) is, for the surface, half of

$$v^2 = v_y^2(0, z') + v_{z'}^2(0, z')$$

$$= \omega^2 A^2(\sin^2 kz' + \coth^2 kh \cos^2 kz') + (\omega/k)^2 \tag{13.17}$$

$$-2\omega^2(A/k) \coth kh \cos kz'$$

Finally, we have for the potential energy term simply

$$V = g\psi = gA \cos kz' \tag{13.18}$$

since only gravity is involved. Putting (13.16), (13.17) and (13.18) into (13.14), and neglecting the first term on the right of (13.17) since $kA \ll 1$ for small-amplitude waves, we get

$$W = p_a/\rho + (\sigma k^2/\rho)A \cos kz' + \tfrac{1}{2}(\omega/k)^2$$

$$-(\omega^2/k) \coth kh\, A \cos kz' + gA \cos kz'$$

which, according to Bernoulli's theorem, must be independent of z'. This will be so only if the sum of the coefficients of $\cos kz'$ vanishes, and thus the mechanical constraint placed on the surface waves is

$$\sigma k^2 / \rho - (\omega^2 / k) \coth kh + g = 0$$

This is in fact the dispersion relation; with a little rearrangement it may be written

$$\omega^2 = (gk + \sigma k^3 / \rho) \tanh kh \qquad (13.19)$$

Summary. The results of section 13.1 were essentially geometrical. By adding physics, in the shape of Bernoulli's theorem, we have found the dispersion relation. This contains a term which depends on the surface tension only, and a term which depends on gravity only.

13.3. Examples of water waves

In this section we discuss extreme forms of the dispersion relation (13.19) and examples of waves which these forms describe.

Deep water and shallow water. The water depth h enters the dispersion relation only through the factor $\tanh kh$. (As a reminder of how a hyperbolic tangent behaves, see fig. 13.4.) For deep water ($kh \gg 1$) we may put

$$\tanh kh \approx 1$$

and the dispersion relation becomes independent of the depth.

For smaller values of kh, there is a useful series expansion

$$\tanh kh = kh - \tfrac{1}{3}(kh)^3 + \tfrac{2}{15}(kh)^5 - \ldots \qquad (13.20)$$

which is valid when $kh < \tfrac{1}{2}\pi$. In the case of shallow water ($kh \ll 1$) we may drop terms beyond the second.

In the following discussion we shall always assume that the water is either deep or shallow (relative to the prevailing wavelength) and use one of these approximations for $\tanh kh$.

Ripples. The dispersion relation has two terms on the right, representing the two kinds of return force acting on the displaced surface. The first term depends on g but not on σ, and represents the tendency of water which is piled up in the wave crests to fall under gravity. The second term represents the effect of surface tension, which tends to straighten out the curves on the surface. We can show that the surface tension term is significant only for the short waves usually called *ripples*.

The two terms are equal when

$$k^2 = \rho g / \sigma$$

which corresponds to a wavelength

$$\lambda = 2\pi(\sigma/\rho g)^{1/2}$$

This critical value of the wavelength is 17 mm for water at 20°C (σ = 0.073 N m^{-1}, ρ = 1000 kg m^{-3}). Waves much shorter than this, making the surface more highly curved, will be dominated by surface tension.

If the gravity term is neglected and we assume that $kh \gg 1$ (which will be true in all but the shallowest puddles), the dispersion relation becomes

$$\omega \approx (\sigma k^3/\rho)^{1/2} \qquad (13.21)$$

Since a $k^{3/2}$ graph is concave upwards, we have here a case of 'anomalous' dispersion (p. 216).

The phase velocity of the ripples is

$$v_\phi = \omega/k \approx (\sigma k/\rho)^{1/2}$$
$$= (2\pi\sigma/\rho\lambda)^{1/2}$$

and the shortest waves travel fastest. (For a wavelength of 1 mm in water at 20°C, v_ϕ has the value 0.68 m s^{-1}.) The group velocity is

$$v_g = d\omega/dk \approx \tfrac{3}{2}(\sigma k/\rho)^{1/2} = \tfrac{3}{2}v_\phi$$

Whatever the wavelength, the ratio of the group and phase velocities is always 3:2. (See problem 12.2.) Since $v_g > v_\phi$, individual ripples appear to travel backwards through the parent group as it propagates.

Gravity waves in deep water. For any wave whose wavelength is much larger than 17 mm, the effect of surface tension is quite negligible. Therefore most water waves are 'gravity waves'. We shall first assume that $kh \gg 1$ is still true (i.e. the water is many wavelengths deep). This is the kind of wave in which all the water particles move in circles.

The dispersion relation for deep-water gravity waves is simply†

$$\omega \approx (gk)^{1/2} \qquad (13.22)$$

and indicates 'normal' dispersion. The phase and group velocities are

$$v_\phi \approx (g/k)^{1/2} = (g\lambda/2\pi)^{1/2}$$
$$v_g \approx \tfrac{1}{2}(g/k)^{1/2} \approx \tfrac{1}{2}v_\phi \qquad (13.23)$$

---

† It is interesting to notice that the density does not appear in (13.22), which therefore holds for *any* incompressible, non-viscous liquid.

(See problem 12.2 again.) A group with $v_g = \frac{1}{2}v_\phi$ was shown in fig. 12.3.

The prime example of a deep-water wave is ocean swell. A storm raging far out to sea will make itself felt initially by the arrival of waves of long wavelength, since these travel fastest. Wavelengths up to about 100 m are typical; equation (13.23) gives the order of magnitude of their phase velocity as 10 m s^{-1}, or 36 km h^{-1}. But the energy put into wave motion by the storm travels at the group velocity, which is only half this value.

It is possible in principle to estimate how far away the storm is by observing the rate of decrease of the wavelength at the shore. In practice it is more convenient to time the arrival of successive wave crests, which are separated by a time interval

$$\tau = \lambda / v_\phi \approx (2\pi\lambda/g)^{1/2} \qquad (13.24)$$

If the storm struck at time t_0, and at a distance L out to sea, waves whose wavelength is λ will reach the coast at time

$$t = t_0 + L/v_\phi = t_0 + L\tau/\lambda \qquad (13.25)$$

Elimination of λ from (13.24) and (13.25) gives

$$\tau \approx 2\pi L/g(t - t_0)$$

This period decreases at a rate

$$-\frac{d\tau}{dt} \approx 2\pi L/g(t-t_0)^2 = g\tau^2/2\pi L \qquad (13.26)$$

Thus L may be found by measuring τ and $-d\tau/dt$.

To find how rapidly τ decreases in a typical case, we can consider a storm 1000 km out to sea. When the period of the waves reaching the coast is 10 s, the value of $-d\tau/dt$ is, from (13.26), approximately 1.6×10^{-4} (about 0.6 s per hour). Clearly the method works best when the measurements can be made over many hours, if not days. It has been used to help to establish that swell can sometimes be traced to very distant storms indeed: almost half way round the earth in some cases.

Gravity waves in shallow water. Waves in shallow water are, as we have seen, approximately longitudinal. When $kh \ll 1$ we may use the first two terms of (13.20), and the dispersion relation becomes

$$\omega^2 \approx ghk^2(1 - \frac{1}{3}h^2k^2)$$

Because the second term in the bracket is much smaller than 1, the dispersion is slight and we may turn the dispersion relation into the standard form (12.2)

$$\omega \approx ck - dk^3$$

with

$$c \equiv (gh)^{1/2}$$
$$d \equiv \tfrac{1}{6}ch^2 \qquad\qquad (13.27)$$

The dispersion is 'normal' since $d > 0$. Often we can neglect it completely, and then

$$v_\phi = v_g \approx c = (gh)^{1/2} \qquad\qquad (13.28)$$

When the dispersion is negligible, the longitudinal water velocity is not only in phase with the surface displacement (13.13), but their ratio is the same at all frequencies since

$$\dot{\psi}_z \approx (c/h)\psi \approx (g/h)^{1/2}\psi \qquad\qquad (13.29)$$

In fact $(g/h)^{1/2}$ is just the characteristic impedance of the system (apart from a factor with the dimensions of mass), since we may look on ψ (the amount by which the water surface is raised) as a measure of the hydrostatic pressure increase underneath.

The most familiar shallow-water waves are those seen at the sea shore. If the beach shelves smoothly, the approaching waves will slow down as the value of h gradually decreases. This makes their amplitude increase, since the power they carry must be constant. Our analysis does not hold for large-amplitude waves; their ultimate fate is to 'break', and the outcome is turbulent motion of the water known as surf.

Essentially similar but more spectacular behaviour is found in the *tsunami* ('tidal wave') produced by a submarine earthquake. Here we see the characteristics of shallow-water waves displayed on a gigantic scale.

A massive but slow upheaval of the sea bed generates a wave of immense wavelength, long enough to make even the Pacific Ocean seem 'shallow' in comparison. The amplitude at the site of the earthquake is usually quite small (possibly less than 1 m), and so the wave may be undetectable there. In an open ocean of depth 4 km, say, (13.27) gives its speed as about 200 m s^{-1}; but it is slowed down to a small fraction of this value as it approaches the coast, while its amplitude increases correspondingly. Since there is very little dispersion the energy is concentrated into a few mighty wave crests, with devastating consequences.

Summary. Surface tension is the controlling influence on a water wave when the wavelength is much smaller than 17 mm. At all wavelengths of such size the group velocity of the surface tension waves is 50 per cent higher than their phase velocity.

The group velocity of deep-water gravity waves is half their phase velocity, at all wavelengths. Shallow-water gravity waves are almost non-dispersive, and their speed increases as the square root of the depth.

Problems

13.1 For a canal 5.0 m deep, estimate the phase velocity and the group velocity for waves of wavelength (a) 10 mm, (b) 1.0 m, and (c) 100 m.

13.2 Calculate the phase velocities for waves of wavelength (a) 10 m, (b) 20 m, and (c) 50 m, on water of depth 1.6 m.

13.3 If a fairly large stone of diameter D is dropped into deep water, most of the energy will go into gravity waves whose wavelengths are roughly $2D$. Show that the ring of disturbed water has an approximate diameter $(gDt^2/\pi)^{1/2}$ where t is the time interval after the stone was dropped in.

13.4 Write down a travelling wave equation for shallow-water gravity waves.

13.5 Show that the phase velocity v_ϕ for deep-water waves influenced by both gravity and surface tension is given by

$$v_\phi^2 \approx \frac{g}{k} + \frac{\sigma k}{\rho}$$

Show that v_ϕ is a minimum when $k^2 = \rho g/\sigma$ (i.e. $\lambda = 17$ mm), and calculate this minimum value.

13.6 For deep-water waves influenced by both gravity and surface tension, show that

$$v_g v_\phi \approx \frac{g}{2k} + \frac{3\sigma k}{2\rho}$$

Hence show that

$$v_g/v_\phi \approx \tfrac{1}{2}(1 + 3\sigma k^2/\rho g)/(1 + \sigma k^2/\rho g)$$

when $k^2 = \rho g/\sigma$ (i.e. $\lambda = 17$ mm).

13.7 It was discovered in the nineteenth century that resistance to the motion of a shallow-draught, horse-drawn canal boat could be substantially reduced by jerking the boat forwards so that it travelled on its own bow wave.

If horses can tow at a speed of 4.0 m s^{-1}, estimate the maximum depth of canal in which this trick will work. (Treat the bow wave as a two-dimensional shallow-water wave.)

13.8   (a) Show that shallow-water waves influenced by both gravity and surface tension will be almost non-dispersive if $\rho g h^2 = 3\sigma$.

(b) Simple wave behaviour may be conveniently demonstrated with the aid of water waves in a 'ripple tank'. How deep must the water be for the waves to be non-dispersive?

13.9   A small obstacle such as a twig in a fast-flowing stream of deep water sometimes produces a stationary pattern consisting of waves travelling upstream with speed (relative to the water) equal to the water speed. Waves of other speeds do not contribute to the stationary pattern.

(a) For a stream flowing at 0.30 m s^{-1}, calculate the wavelengths of contributing gravity waves and surface tension waves.

(b) What are the corresponding group velocities? Why do the longer waves appear downstream and the shorter waves upstream of the obstacle?

14

Electromagnetic waves

As travelling voltage and current waves move along a cable, they are accompanied by a moving pattern of electric and magnetic fields. The particular pattern belonging to a coaxial cable is illustrated in fig. 14.1. The electric field E is radial and perpendicular to the cable axis; its magnitude oscillates in step with the voltage ψ_V across the cable. The magnetic field B is circumferential; its magnitude oscillates in step with the current ψ_I in the conductors. At every point inside the cable E and B are at right angles to each other.

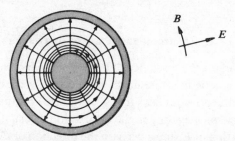

Fig. 14.1 A transverse electromagnetic wave inside an air-insulated coaxial cable with no resistance. The figure shows the instantaneous electric and magnetic fields in a plane at right angles to the cable, for a wave travelling out of the page.

Clearly both E and B are propagated along the inside of the cable as transverse waves. Like the voltage and current waves, they are interdependent. In a cable without resistance they are perpetually in phase (or in antiphase) with each other, since ψ_V and ψ_I are in phase (or in antiphase) with each other.

We can show that the electric and magnetic fields can be propagated as a 'free' wave without the guiding influence of the cable. We concentrate on electromagnetic *plane waves* in which, at any given moment, E (or B) has the same value all over any plane perpendicular to the propagation direction. Figure 14.2 shows the instantaneous situation in such a plane. The wave inside a coaxial cable is not a plane wave, since both fields get weaker as the distance from the axis of the cable increases (fig. 14.1).

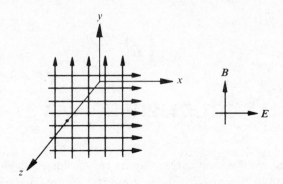

Fig. 14.2 An electromagnetic plane wave. The figure shows the instantaneous electric and magnetic fields in a plane at right angles to the direction of propagation, for a wave travelling out of the page. The pattern extends indefinitely over the plane. Coordinate axes are chosen in the directions shown.

Somewhere there must be a suitable system of charges and currents generating the wave, but here we are interested in its propagation rather than its production.

14.1. Electromagnetic waves in a vacuum

The basic reason for the existence of a wave is the fact that time-varying electric and magnetic fields influence each other. We can find the wave equation from the physical laws that govern these mutual influences.

With the wave propagating in the z direction as usual, we set up an x axis in the direction of E, and a y axis in the direction of B, as shown in fig. 14.2. We know that a voltage is induced round a circuit if there is a magnetic field with a component perpendicular to the circuit, and if that field varies with time. Electromagnetic theory explains this phenomenon by saying that the time-varying magnetic field gives rise to an electric field, and that this will happen even when there is no actual circuit present. The effect is governed by Faraday's law of induction; we shall apply this law to an element of area $\Delta x \, \Delta z$ lying perpendicular to B at position z, as in fig. 14.3(a).

If the magnitude of the field at z is E, and that at $z + \Delta z$ is $E + \Delta E$, the voltage that would be induced in an imaginary circuit placed round the

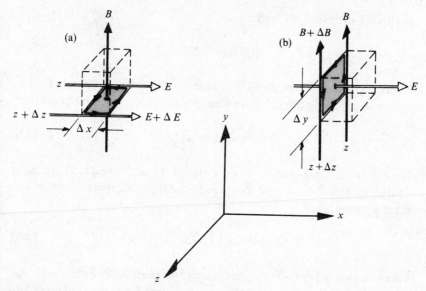

Fig. 14.3 (a) The line integral of E round the loop $\Delta x\,\Delta z$ is $\Delta E\,\Delta x$. (b) The line integral of B round the loop $\Delta y\,\Delta z$ is $-\Delta B\,\Delta y$.

perimeter of $\Delta x\,\Delta z$ is

$$\oint E\,\mathrm{d}l = \Delta E\,\Delta x$$

where the integral is taken clockwise round the circuit as viewed in the positive y direction. The electric fields at z and $z + \Delta z$ tend to cancel, and there are no contributions from the other two edges since these are perpendicular to the field.

Faraday's law of induction states that the induced voltage is equal to the rate of *decrease* of the magnetic flux $B\,\Delta x\,\Delta z$ through the imaginary circuit. Thus

$$\Delta E\,\Delta x = -\left(\frac{\partial B}{\partial t}\right)\Delta x\,\Delta z$$

or, if we make the circuit vanishingly thin in the z direction,

$$\frac{\partial E}{\partial z} = -\frac{\partial B}{\partial t} \tag{14.1}$$

Time-varying electric fields give rise to magnetic fields in a parallel if less familiar way. We consider the element $\Delta y\,\Delta z$ perpendicular to E at z, as shown in fig. 14.3(b). Going round this area clockwise as viewed in the positive x direction, we find

$$\oint B \, dl = -\Delta B \, \Delta y \tag{14.2}$$

where the magnitude of the field increases from B to $B + \Delta B$ when we go from z to $z + \Delta z$. In this situation we apply Ampère's law, which says that

$$\oint B \, dl = \mu_0 I \tag{14.3}$$

Here I is the total current flowing across the area $\Delta y \, \Delta z$. There is, of course, no actual flow of charge through the loop, but there is a 'displacement current'

$$I_D = \varepsilon_0 \left(\frac{\partial E}{\partial t} \right) \Delta y \, \Delta z \tag{14.4}$$

Maxwell showed that displacement currents must be included with true currents when time-varying fields are involved; the concept allows us to think of currents 'passing through' capacitors which are being charged or discharged.

Putting $I = I_D$ in (14.3), and using (14.2), we obtain

$$-\Delta B \, \Delta y = \mu_0 \varepsilon_0 \left(\frac{\partial E}{\partial t} \right) \Delta y \, \Delta z$$

For a vanishingly thin area this reads

$$\frac{\partial B}{\partial z} = -\mu_0 \varepsilon_0 \left(\frac{\partial E}{\partial t} \right) \tag{14.5}$$

Equations (14.1) and (14.5) must be satisfied simultaneously; they are the basis of the wave equation. If we differentiate (14.1) with respect to z, differentiate (14.5) with respect to t, and equate the resulting expressions for $\partial^2 B / \partial z \, \partial t$ and $\partial^2 B / \partial t \, \partial z$, we get

$$\frac{\partial^2 E}{\partial t^2} = \frac{1}{\mu_0 \varepsilon_0} \left(\frac{\partial^2 E}{\partial z^2} \right)$$

Alternatively, if we differentiate (14.1) with respect to t, differentiate (14.5) with respect to z, and equate $\partial^2 E / \partial z \, \partial t$ and $\partial^2 E / \partial t \, \partial z$, we get

$$\frac{\partial^2 B}{\partial t^2} = \frac{1}{\mu_0 \varepsilon_0} \left(\frac{\partial^2 B}{\partial z^2} \right)$$

We have found that the electric and magnetic fields satisfy the same wave equation, which is non-dispersive. The phase velocity, which is the same for both fields, and which we have already evaluated (10.27), is exactly equal to the speed of light in empty space. This was one of the most exciting results of nineteenth-century theoretical physics. Years before the discovery of radio waves, it led Maxwell to suggest that light consists of electromagnetic waves. Since then, a very wide variety of other electromagnetic waves have become familiar.

Travelling waves. Since these are non-dispersive waves, the travelling wave equation (9.12) is obeyed. The electric field magnitude, for example, satisfies

$$\frac{\partial E}{\partial t} \pm c \left(\frac{\partial E}{\partial z} \right) = 0$$

It is also tied to the magnetic field magnitude through (14.1) and (14.5); using either of these equations, we find

$$\frac{\partial E}{\partial t} = \pm c \left(\frac{\partial B}{\partial t} \right)$$

which can be integrated to yield the very simple relationship

$$E = \pm cB \tag{14.6}$$

We have not included arbitrary constants here, because these would merely add to any steady fields already present; as usual we are ignoring steady values.

We saw at the outset that E would always be proportional to B in a travelling wave inside a cable, since the voltage and the current vary in proportion to each other. We have now shown that the same is true in the free wave, the constant factor linking the two fields being simply the phase velocity c.

If the travelling wave is sinusoidal, the proportionality of E and B implies that they vary in phase or in antiphase with each other. To find out which it is, we write

$$E(z, t) = E_0 \cos (\omega t - kz)$$

$$B(z, t) = B_0 \cos (\omega t - kz + \phi)$$

where E_0 and B_0 are the field amplitudes; we then have

$$\frac{\partial E}{\partial z} = kE_0 \sin (\omega t - kz)$$

$$-\frac{\partial B}{\partial t} = \omega B_0 \sin (\omega t - kz + \phi)$$

These expressions must be equal, everywhere and always (14.1). If the wave is travelling in the positive z direction, k will be positive and so ϕ must be zero; if the wave is travelling in the negative z direction, k will be negative and so ϕ must be equal to π.

We can summarize these phase relationships in another way, by saying that *the direction of travel of the wave is always the direction of the vector product* $E \times B$ (fig. 14.4).

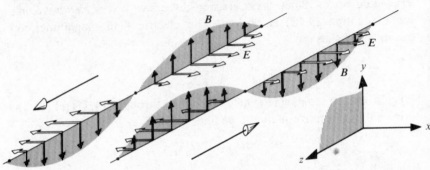

Fig. 14.4 If a sinusoidal wave is travelling in the positive z direction, E and B oscillate in phase with each other. If the wave is travelling in the negative z direction, the fields oscillate in antiphase. In each case the direction of travel is the direction of the vector $E \times B$.

We see that E and B in a travelling wave are very closely linked to each other. In fact, we know all about $B(z, t)$ if we know all about $E(z, t)$, and *vice versa*. Either E or B could be used as the basic variable; it is conventional, however, to choose E for this purpose, and we shall see why shortly.

Intensity of a travelling wave. The instantaneous power *per unit area* carried by a travelling electromagnetic plane wave will be $\pm cw(z, t)$, where $w(z, t)$ is the energy density *per unit volume*. This follows from an easy adaptation of (9.35) to suit plane waves.

Electromagnetic theory gives the energy density at a point where there are fields of magnitude E and B as

$$w(z, t) = \tfrac{1}{2}\varepsilon_0 E^2 + \tfrac{1}{2}B^2/\mu_0$$

When E and B belong to a travelling plane wave, we can eliminate B from the expression on the right with the aid of (14.6); this leads to the result

$$w(z, t) = \varepsilon_0 E^2$$

and so the power per unit area is $c\varepsilon_0 E^2$.

If the travelling wave is sinusoidal, with amplitude E_0, its intensity will be

$$I(z, t) = \langle c\varepsilon_0 E^2 \rangle = \tfrac{1}{2} c\varepsilon_0 E_0^2 \tag{14.7}$$

The intensity is the easiest property of the wave to measure; if we know the intensity we can deduce E_0 from (14.7), and thus B_0 from (14.6). As a specific and familiar example, we think about the radiation coming from the sun. The intensity of bright sunlight at the surface of the earth is found to be of the order of $1\,\text{kW m}^{-2}$. This gives a value in the region of $1\,\text{kV m}^{-1}$ for E_0. An electric field of this size can be rated as modest by the standards of modern technology: a field whose amplitude is about 100 times larger exists between the conductors in an ordinary AC mains cable.

The corresponding magnetic field amplitude in the wave is E_0/c, which is about $3\,\mu\text{T}$. For most purposes such a field can be described as weak: even the earth's magnetic field is about 10 times larger. Although the wave cannot exist unless both E and B are present, from the point of view of the effects produced when the wave passes through matter E is much more important than B. As we shall see, we can usually ignore the effects due to B.

Characteristic impedance. We found in section 10.2 that the intensity of a sinusoidal acoustic plane wave is $A_p^2/2\mathscr{Z}_0$, where A_p is the amplitude of the basic variable (the acoustic pressure) and $\mathscr{Z}_0$ is the characteristic impedance per unit area (10.19). For an electromagnetic plane wave the basic variable is E, and it is therefore convenient to define $\mathscr{Z}_0$ so that the intensity can be written $E_0^2/2\mathscr{Z}_0$. Comparison of this expression with the one on the right of (14.7) suggests that we use

$$\mathscr{Z}_0 \equiv 1/c\varepsilon_0 = \mu_0 c = (\mu_0/\varepsilon_0)^{1/2} \tag{14.8}$$

as the characteristic impedance (per unit area) of an electromagnetic wave in a vacuum. This is a real quantity with the numerical value $376.7\,\Omega$ at all wave frequencies.[†] It is the value of E/H, where H is $(1/\mu_0)B$, at any point and any time, for any travelling electromagnetic wave in a vacuum.

Summary. The oscillating electric and magnetic fields in a travelling electromagnetic wave in a vacuum are:

(1) transverse (i.e. perpendicular to the propagation direction);

[†] Strictly, $\mathscr{Z}_0$ is a surface resistivity, not a resistance; but the units are the same.

(2) perpendicular to each other;

(3) proportional to each other (and therefore in phase or in antiphase with each other in a sinusoidal travelling wave). They obey the same wave equation, and that equation is non-dispersive. The phase velocity $(3 \times 10^8 \, \text{m s}^{-1})$ and the characteristic impedance per unit area $(377 \, \Omega)$ are fixed by the values of ε_0 and μ_0, and by these alone.

14.2. Electromagnetic waves in a dielectric

Every material contains electrons. When an electromagnetic wave travels through a material, the oscillating electric field in the wave will set some of these electrons into forced vibration, and the vibrating electrons will generate new waves of their own. In section 6.2 we discussed the process known as scattering, in which the electrons are driven incoherently. For the forced vibrations to be incoherent, the sites of individual scatters must be far apart compared with the wavelength of the radiation involved.

If the participating electrons are sufficiently close together they will be driven coherently, with quite different results. In this case the scattered waves can be superposed with the 'direct' wave, giving rise to a new disturbance which will be 'the wave in the material'.

In an electrically insulating, non-magnetic material, the coherent problem can be analyzed by treating the material as a uniform medium whose response to any electric field (such as that due to a wave) can be described in terms of its complex electric susceptibility $\chi_e(\omega)$. As we saw in section 6.3, it is possible to relate the bulk property $\chi_e(\omega)$ to properties of the atoms or molecules of which the material is composed.

Dispersion relation for a dielectric. From the point of view of electromagnetic wave propagation, the essential difference between a dielectric and a vacuum is that real currents will be produced in the dielectric as charges are moved backwards and forwards by the oscillating field; in a vacuum there are 'displacement currents' only. When we are dealing with a dielectric, the quantity I which features in Ampère's law (14.3) must include both kinds of current.

The effect of a steady electric field on a dielectric is to polarize it. The polarization field $\mathscr{P}$ (section 6.3) measures the dipole moment produced by the field, per unit volume. To produce a dipole moment $|\mathscr{P}| \, \Delta x \, \Delta y \, \Delta z$ in the slab $\Delta x \, \Delta y \, \Delta z$ (fig. 14.3) requires the transport of charges $\pm |\mathscr{P}| \, \Delta y \, \Delta z$ through the slab to its opposite faces.

In a time-varying field the polarization also changes with time. At any given moment $\mathscr{P}$ will be increasing at a rate

$$\frac{\partial \mathscr{P}}{\partial t} = \chi_e(\omega)\varepsilon_0\left(\frac{\partial E}{\partial t}\right)$$

which will lead to a current

$$I_P = \left(\frac{\partial \mathscr{P}}{\partial t}\right)\Delta y\,\Delta z = \chi_e(\omega)\varepsilon_0\left(\frac{\partial E}{\partial t}\right)\Delta y\,\Delta z$$

It is easy to see that adding this current to the displacement current (14.4) will introduce a factor $1 + \chi_e(\omega)$ into all subsequent expressions containing $\partial E/\partial t$. This factor will be carried through to the wave equation, which takes the new form

$$[1 + \chi_e(\omega)]\left(\frac{\partial^2 E}{\partial t^2}\right) = \frac{1}{\mu_0\varepsilon_0}\left(\frac{\partial^2 E}{\partial z^2}\right)$$

Obvious changes will be made to expressions we found in the vacuum case. For example, the phase velocity in the dielectric is

$$v_\phi = \frac{c}{[1 + \chi_e(\omega)]^{1/2}}$$

where c is the phase velocity in the vacuum. (From now on we shall give c this meaning only.) The characteristic impedance, and the ratio E/B, are likewise equal to their vacuum values divided by $[1 + \chi_e(\omega)]^{1/2}$. The new dispersion relation is

$$\omega^2 = \frac{c^2 k^2}{1 + \chi_e(\omega)} \tag{14.9}$$

These changes are, of course, purely formal. They will tell us about the physics of wave propagation in the dielectric only if we know how $\chi_e(\omega)$ behaves at different frequencies; but once we do know enough about $\chi_e(\omega)$ we can solve the dispersion relation (14.9) for k in the usual way.

Refractive index. In optics, the dispersion is more commonly analyzed in terms of a quantity

$$n \equiv c|k|/\omega \tag{14.10}$$

known as the *refractive index* of the medium. The refractive index of a non-dispersive medium is obviously 1 at all wave frequencies, dispersion being indicated by values of n either smaller or larger than 1 (fig. 14.5). The most straightforward methods of measuring n (and thus the phase

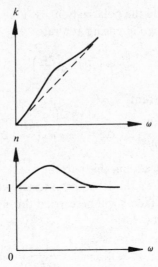

Fig. 14.5 Departures from the dispersion curve $\omega = ck$ correspond to departures of the refractive index from $n = 1$. For comparison, the dispersion curve is plotted with the frequency axis horizontal in this figure.

velocity) in non-dispersive media involve refraction (section 17.1). If there is attenuation, n, like k, will be complex.

In terms of the refractive index, our dispersion relation (14.9) takes on the simpler form

$$n^2 = 1 + \chi_e(\omega) \tag{14.11}$$

At very low frequencies the quantity on the right will approach the value $1 + \chi_e$ (the 'dielectric constant') where χ_e is the steady-field susceptibility. If the susceptibility is small enough (as in the case of a dielectric gas, for example) we can write (14.11) even more simply as

$$n \approx 1 + \tfrac{1}{2}\chi_e(\omega) \tag{14.12}$$

Thus the refractive index gives direct information about the susceptibility at 'optical' frequencies.

In section 6.3 we related the bulk susceptibility of a gas to molecular properties: the resonance angular frequency ω_0 and the damping width γ for the driven electrons (6.8). We now have, in (14.12), a relationship by which we can connect these molecular properties with something measurable, the refractive index. Combining (6.8) and (14.12) gives

$$n \approx 1 + \frac{\tfrac{1}{2}\omega_p^2}{\omega_0^2 - \omega^2 + i\gamma\omega} \tag{14.13}$$

where ω_p^2 measures the number density of electrons which can scatter

(6.7). We can use (14.13) to predict the behaviour of electromagnetic waves in the gas at any chosen frequency.

Normal dispersion. In chapter 6 we saw that forced vibration of an electron attached to an atom or a molecule will usually be well below the resonance frequency, if visible light is used to provide the driving force. For visible light in a dielectric gas we can therefore use an off-resonance approximation to (14.13),

$$n \approx 1 + \frac{\frac{1}{2}\omega_p^2}{\omega_0^2 - \omega^2} \qquad (14.14)$$

In this frequency region n is essentially real (and therefore k is real also), showing that the light is propagated with negligible attenuation: the gas is transparent.

The value of n is always slightly greater than 1, and it increases as ω increases: red light travels faster than blue. By (14.10) the speed of the wave is c/n: the refractive index is merely the factor by which the speed and the wavelength are reduced. In this case the coherently scattered waves combined with the direct wave to produce a new sinusoidal travelling wave: one that travels more slowly, and therefore has a shorter wavelength, than a wave of the same frequency in a vacuum.

Dense transparent materials. All this strictly refers to gases only. The connection (6.8) between $\chi_e(\omega)$ and the molecular electrons was derived on the assumption that the driven electron feels only the field in the wave, and not any extra fields produced by its polarized neighbours.

In a dense medium such as water or glass, the field driving the electron will usually be *proportional to E*. The susceptibility will then exceed the calculated value (6.8) by an unknown but constant factor f, and the refractive index (14.14) will become†

$$n \approx 1 + \frac{\frac{1}{2}f\omega_p^2}{\omega_0^2 - \omega^2}$$

The unknown factor f can be eliminated by comparing values of the refractive index at two different frequencies. We can, for example, write

$$\frac{n_b - 1}{n_r - 1} \approx \frac{\omega_0^2 - \omega_r^2}{\omega_0^2 - \omega_b^2} \qquad (14.15)$$

† For a dense material (14.12) may not be an adequate approximation, and it may be necessary to use expressions derived from the exact relationship (14.11) if accurate calculations have to be done. Here we are merely estimating orders of magnitude.

where the suffixes b and r mean 'blue' and 'red' respectively. This gives us a means of estimating ω_0 for any dielectric which is transparent to visible light.

For water at 20°C the relevant values are

$$n_b = 1.340 \quad (\text{at } \omega_b = 4.35 \times 10^{15} \text{ s}^{-1})$$

and

$$n_r = 1.328 \quad (\text{at } \omega_r = 2.46 \times 10^{15} \text{ s}^{-1})$$

The value of ω_0 predicted by (14.15) is $2 \times 10^{16} \text{ s}^{-1}$, corresponding to a resonance frequency of 3×10^{15} Hz for an electron on a water molecule.

Resonance and anomalous dispersion. In the resonance region the imaginary part of n will become large, indicating attenuation of the wave due to absorption of light by the material. We write

$$n = n_1 - in_2$$

where n_1 and n_2 are related to the real and imaginary parts of k by

$$n_1 = c|K|/\omega$$

$$n_2 = c|\kappa|/\omega$$

We can appreciate the general behaviour of n_1 and n_2, if ω_p^2 is not too large, by using approximation (14.12). In terms of the molecular compliance, we have

$$n_1 \approx 1 + \tfrac{1}{2}m\omega_p^2 \operatorname{Re} K(\omega)$$

$$n_2 \approx -\tfrac{1}{2}m\omega_p^2 \operatorname{Im} K(\omega)$$

The real and imaginary parts of $K(\omega)$ have the kind of frequency variation shown in fig. 5.8, and so n_1 and n_2 are as in fig. 14.6: as resonance is passed, n_1 swings upwards and downwards before settling to the value 1, while n_2 has the familiar resonance shape. The depth of the swing, and the height of the resonance, increase with increasing ω_p^2 (electron density).

The curve for the real part n_1 shows a region, immediately around the resonance frequency, where the phase velocity c/n_1 *increases* with increasing frequency. This departure from the more usual behaviour was the original example of 'anomalous' dispersion.

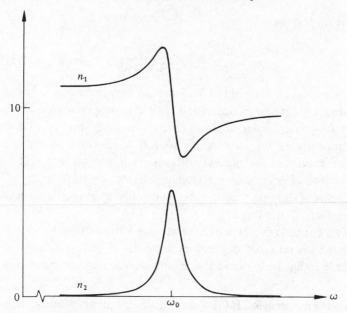

Fig. 14.6 Real part n_1 and negative imaginary part n_2 of the refractive index, in the region of the resonance frequency. In this example $Q = 5$ and $\omega_p = \frac{1}{2}\omega_0$.

Summary. Results derived for electromagnetic waves in a vacuum can be carried over to a dielectric with complex susceptibility $\chi_e(\omega)$, by replacing c wherever it appears by $c/[1 + \chi_e(\omega)]^{1/2}$. We deduced the properties of $\chi_e(\omega)$, and thus the behaviour of the waves, from the picture of electrons in coherent forced vibration.

At frequencies far below resonance, the wave propagation shows dispersion (v_ϕ decreases with increasing ω) but negligible attenuation. In the resonance region there is severe attenuation; the dispersion is also pronounced, and may be 'anomalous'.

14.3. Electromagnetic waves in a plasma

In a plasma, the only charges we usually need to consider are the free electrons. If we think of each of these as 'belonging' to one of the positive ions, we can write

$$\chi_e(\omega) = \frac{\omega_p^2}{-\omega^2 + i\gamma\omega}$$

for the susceptibility of the plasma: we have merely put $\omega_0 = 0$ in (6.8), since the electron has no stiffness binding it to the ion.

The dispersion relation

$$\frac{c^2 k^2}{\omega^2} = 1 + \frac{\omega_p^2}{-\omega^2 + i\gamma\omega} \qquad (14.16)$$

follows immediately from (14.9). Since ω_p now measures the density of *free* electrons, it can be identified with the plasma angular frequency (2.10). It is clearly going to play a crucial role here, since there is now no resonance frequency. It remains merely a measure of the electron density, however: the plasma stiffness itself has no influence on the propagation of *transverse* electromagnetic waves, since layers of electrons merely slide past each other with only their inertia to resist the driving force of the wave.†

As we examine the dispersion relation in various frequency ranges, for plasmas with different degrees of damping, it will be helpful to refer regularly to fig. 14.7, where the results are summarized pictorially.

High wave frequencies. If the wave angular frequency is much larger than the damping width $(\omega \gg \gamma)$ the dispersion relation (14.16) may be written

$$\omega^2 \approx c^2 k^2 + \omega_p^2$$

This is just like the dispersion relation for an anchored string (12.10).

If ω is also much larger than ω_p the plasma is transparent and non-dispersive, and the phase velocity of the waves is $\pm c$. These are just the properties of a vacuum. The motion of the electrons is mass controlled at all driving frequencies, since $\omega_0 = 0$; at the highest frequencies they simply cannot follow the rapid fluctuations of the field (5.10) and cease to have any effect on the wave.

The approximate dispersion relation will remain valid for frequencies around the plasma frequency, provided the plasma is very lightly damped $(\gamma \ll \omega_p)$. Cut-off will be observed at the plasma frequency: the plasma is transparent but dispersive if $\omega > \omega_p$, but waves of lower frequency are evanescent, and the plasma totally reflects any such travelling waves falling on it.

Low wave frequencies. At the other extreme $(\omega \ll \gamma)$ the dispersion relation becomes

$$\omega^2 - i(\omega_p^2/\gamma)\omega \approx c^2 k^2$$

---

† The plasma stiffness is involved when longitudinal waves are propagated in a plasma, since these cause electron density fluctuations.

This is like the dispersion relation for a damped string (9.38) with

$$\Gamma = \omega_p^2/\gamma \qquad (14.17)$$

and indicates attenuation but no dispersion.

In section 9.5 we examined the behaviour when Γ/ω was very small and very large. When $\Gamma/\omega \gg 1$ we had the skin effect: very severe attenuation, with an attenuation length which was inversely proportional to $\omega^{1/2}$. The plasma will have this effect on a wave if the frequency is low enough. 'Low enough' means $\omega \ll \omega_p^2/\gamma$; this is fulfilled automatically (because we have already assumed $\omega \ll \gamma$) except in the case of a very heavily damped plasma ($\gamma \gg \omega_p$).

With $\Gamma/\omega \ll 1$ the attenuation length was the same at all frequencies, and absorption was efficient. The plasma will have these properties if $\omega \gg \omega_p^2/\gamma$; this will be true if $\gamma \gg \omega_p$ since we have also assumed that $\gamma \gg \omega$. Thus a very heavily damped plasma efficiently absorbs waves with $\omega_p \ll \omega \ll \gamma$.

Intermediate wave frequencies. We have now examined the dispersion relation (14.16) in the extremes of very high and very low wave frequencies. In the former case k is either purely real or purely imaginary: the wave is either sinusoidal (with dispersion but without attenuation) or evanescent. In the latter, k is complex and its real part K is independent of frequency: the wave is attenuated, but there is no dispersion. The frequency regions where these results apply are shown schematically in fig. 14.7 for very lightly damped, moderately damped and very heavily damped plasmas.

When the wave frequency is neither very high nor very low we must use the full formula (14.16). We shall not solve it here, but qualitatively the results are obvious enough. In general we expect that k will be complex, and that both K and κ will vary with frequency: the wave will be attenuated, and its phase velocity will depend on the frequency.

We now return to the examples of plasmas that we discussed in chapter 2 and chapter 4. Using our previous estimates of the plasma frequencies and collision widths, we can draw some general conclusions about wave propagation in these plasmas with the aid of fig. 14.7.

The ionosphere. As shown in table 4.1, the F_2 layer of the ionosphere is a very lightly damped plasma ($\gamma \ll \omega_p$). We estimated its plasma frequency to be ~ 10 MHz, corresponding to a wavelength of about 30 m. This result, combined with fig. 14.7(a), shows that all radio waves with wavelengths longer than about 30 m are cut off in the F_2 layer, and will

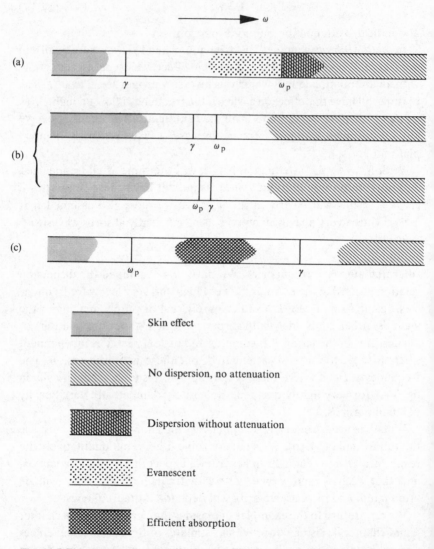

Fig. 14.7 Electromagnetic waves in plasmas. The diagrams indicate the regions where approximations discussed in the text are applicable. Wave angular frequency ω increases logarithmically from left to right.

(a) A very lightly damped plasma ($\gamma \ll \omega_p$).

(b) Moderately damped plasmas ($\gamma \sim \omega_p$).

(c) A very heavily damped plasma ($\gamma \gg \omega_p$).

In unmarked regions the waves are both dispersive and attenuated.

therefore be reflected back to the ground, enabling them to be received at long distances from the transmitter in spite of the curvature of the earth. (For the same reason, frequencies below 10 MHz are obviously useless for communicating with space ships or planets.)

In order to reach the reflecting F_2 layer, the waves must first penetrate the intervening D layer. The D layer is moderately damped ($\gamma \sim \omega_p$) and has a plasma frequency of about 300 kHz, which corresponds to a wavelength of 1 km. Figure 14.7(b) is relevant in this case; it shows that the D layer is transparent to Short Waves, which have wavelengths between 10 m and 80 m ($\omega \gg \omega_p$) but not to Medium Waves, whose wavelengths are between 195 m and 600 m. Thus long-distance radio communication requires Short Waves, which can penetrate the D layer and are reflected by the F_2 layer (fig. 14.8).

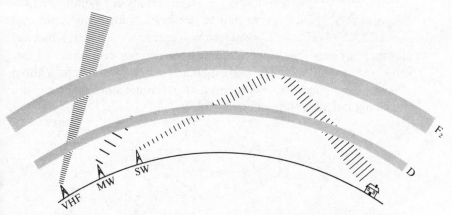

Fig. 14.8 Short Wave radio broadcasts can be received from distant stations in spite of the curvature of the earth; the Short Waves penetrate the D layer of the ionosphere and are reflected by the F_2 layer. Medium Waves cannot penetrate the D layer during daylight. Very High Frequency waves penetrate both layers.

The frequent collisions which give the D layer its large damping width also lead to rapid recombination of the free electrons and the positive ions, causing the D layer to disappear quickly after sunset. Medium Wave reception from distant stations thus becomes possible at night.

Metals. Adopting copper as a typical metal, we found $\omega_p \approx 2 \times 10^{16}\,\mathrm{s}^{-1}$ and $\gamma \approx 4 \times 10^{13}\,\mathrm{s}^{-1}$. Radio waves will have $\omega \ll \gamma$, and will therefore undergo the skin effect. Any radio waves falling on the metal are almost totally reflected, with a phase change of nearly 180°. Highly conducting metals will be the most efficient reflectors, since Γ is inversely proportional to γ (14.17) and γ is inversely proportional to the conductivity (4.14).

The above estimates for ω_p and γ in copper give $\Gamma \approx 1 \times 10^{19}\,\text{s}^{-1}$; for 30 mm waves ($\omega \approx 6 \times 10^{10}\,\text{s}^{-1}$) we have a skin depth $(2c^2/\Gamma\omega)^{1/2}$ of about 0.5 μm. The value will be smaller for a better conductor, and the smaller the skin depth is, the smaller are the 'losses' due to the forced vibration of the damped plasma. Therefore resonant cavities and wave-guides should have clean, highly conducting inner surfaces.

The wavelength of a wave with $\omega = \omega_p$ is about 100 nm for copper. Visible light has wavelengths in the approximate range 400 nm to 750 nm, and is therefore cut off; this is why metals appear shiny. A few are, however, transparent to ultraviolet light. These are the so-called alkali metals which include lithium, sodium and potassium. In these metals the atoms are relatively far apart, and the conduction electron densities are correspondingly low.

If we calculate the conduction electron density for potassium ($\rho = 860\,\text{kg m}^{-3}$, $A = 39.1$) as we did in section 2.3 for copper, we find $N \approx 1.32 \times 10^{28}\,\text{m}^{-3}$. This is only 16 per cent of the copper value; the plasma frequency should therefore be about 40 per cent of the copper value, or $6.6 \times 10^{15}\,\text{s}^{-1}$. The corresponding cut-off wavelength is about 290 nm; experimentally, it is found that potassium stops reflecting at a wavelength of about this value (fig. 14.9).

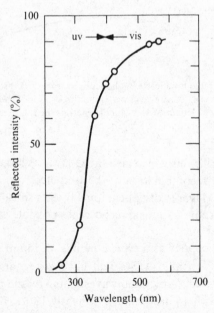

Fig. 14.9 Reflecting power of a thin film of potassium. (After H. E. Ives and H. B. Briggs, *J. Opt. Soc. Amer.* **26** (1936) 238.)

Since the opacity of a metal to visible light is a cut-off effect, it should be possible to see through a film of metal by tunnelling, if its thickness L is no greater than about $1/|\kappa|$. We require

$$L \lesssim c/(\omega_c^2 - \omega^2)$$

where ω_c is the cut-off angular frequency. For visible light ($\omega \approx 3 \times 10^{15}\,\text{s}^{-1}$) in copper ($\omega_c = \omega_p \approx 2 \times 10^{16}\,\text{s}^{-1}$) we find a limiting thickness of about 15 nm.

Infrared waves in copper have $\omega \ll \omega_p$, but ω is now comparable with γ. (A wave with $\omega = \gamma$ would have a wavelength of about 50 μm.) This brings us into the region where a more exact treatment of the dispersion relation is necessary. We can, however, say that infrared waves will certainly be attenuated to some extent in the metal. In fact, it is possible for the same metal film to be reasonably transparent to visible light, while strongly absorbing infrared. The gold film on the visor of an astronaut's space helmet is just one application.

Summary. The most compact summary of this section is fig. 14.7, which says the following:

(1) from the wave point of view a plasma behaves like a vacuum if $\omega_p \ll \omega$ and $\gamma \ll \omega$;

(2) the skin effect occurs at low frequencies in any plasma;

(3) in a very lightly damped plasma ($\gamma \ll \omega_p$) the plasma frequency acts as a lower cut-off frequency;

(4) a very heavily damped plasma ($\gamma \gg \omega_p$) will almost totally absorb incident radiation for which $\omega_p \ll \omega \ll \gamma$.

Problems

14.1 Show that the instantaneous magnitude and direction of the power per unit area carried by an electromagnetic plane wave in a vacuum are those of the vector $\boldsymbol{E} \times \boldsymbol{H}$, where $\boldsymbol{H} \equiv (1/\mu_0)\boldsymbol{B}$. (This is known as the *Poynting vector*.)

14.2 (a) Calculate the characteristic impedance for visible light in polystyrene ($n = 1.58$).

(b) If a sinusoidal travelling wave in polystyrene has $E_0 = 100\,\text{V m}^{-1}$, what is B_0?

14.3 An electromagnetic plane wave travelling in a vacuum is incident normally on the plane surface of a material whose refractive index is n. Show that the reflection coefficient is

$$R = \frac{n-1}{n+1}$$

Calculate the fraction of the *power* reflected if the material is glass with $n = 1.50$.

14.4   Show that, when an electromagnetic plane wave travelling in glass is incident normally on a plane surface separating glass and air, the reflected wave is inverted with respect to the incident wave.

14.5   The dielectric constant of water is about 80, giving $\chi_e \approx 79$ for its steady-field susceptibility. The refractive index of water is about 1.33.

Estimate the susceptibility of water at optical frequencies. Compare the result with χ_e and explain why the difference is so great.

14.6   Estimate the skin depth for 10 km radio waves in the D layer of the ionosphere. (Assume that $\gamma = \omega_p$ in the D layer.)

14.7   Show that the group velocity in a very lightly damped plasma is approximately nc, where n is the refractive index of the plasma and c is the phase velocity in a vacuum.

14.8   For a very lightly damped plasma with $\omega_p \ll \omega$, show that the refractive index is approximately $1 - \omega_p^2/2\omega^2$.

The phase velocity of transverse electromagnetic waves of frequency 14 MHz in a certain plasma is $1.01c$. Estimate the free electron density.

15

De Broglie waves

During the 1920's it became apparent that the mathematics of wave motion could be successfully applied to problems involving electrons and other sub-atomic particles, for which classical laws were known to be deficient. *Wave mechanics* is the name given to a semi-quantum theory which has been constructed on this basic idea. It works for particles moving at non-relativistic speeds, and acted on by simple forces.

Underlying wave mechanics is the hypothesis, due to de Broglie, that every particle has associated with it a wavelength which is inversely proportional to the magnitude of its momentum p. The constant of proportionality is the Planck constant h (6.63×10^{-34} J s). We write

$$\lambda = h/|p|$$

or, in terms of the wavevector k,

$$p = \hbar k \tag{15.1}$$

where the symbol $\hbar$ denotes $h/2\pi$. Experiments confirm that this relationship holds for particles of all kinds, under conditions which make their wave-like behaviour observable.

In most cases the possession of a wavelength in fact makes very little difference to the behaviour of a particle. We study particles by means of experiments which enable us to observe and modify their movements in space. Waves travelling in space undergo diffraction (chapter 18) when they meet obstacles, but the effects of diffraction are hard to observe unless the waves have a wavelength which is comparable in size with the obstacles. The same is true of particles which behave like waves: differences between 'wave' behaviour and 'particle' behaviour will be observed only with those particles which have the largest de Broglie wavelengths.

To have a large de Broglie wavelength we require a particle of low mass, moving slowly. An electron (mass 9.1×10^{-31} kg) accelerated through 1 volt has a kinetic energy of 1.6×10^{-19} J, and so its momentum is 5.4×10^{-25} kg m s^{-1}. Its de Broglie wavelength is thus only 1.2 nm. As a working rule we can say that a 1 eV electron has a de Broglie wavelength of about 1 nm.

The behaviour of the electrons in a colour television tube is therefore particle-like, because the smallest objects in their path (the holes in the mask behind the screen) are much larger than 1 nm across, and the electrons have energies much larger than 1 eV. The behaviour of electrons in atoms, on the other hand, can be expected to be wave-like.

15.1. Wave functions

By the time of de Broglie's wavelength–momentum hypothesis for particles, it had become clear that light and other electromagnetic radiation could be absorbed or emitted by matter only in energy quanta ('photons') of size

$$W = \hbar\omega \qquad (15.2)$$

By applying the same relationship to particles other than photons we can find a dispersion relation for de Broglie waves.

A particle whose potential energy is V has total energy

$$W = T + V = \frac{p^2}{2m} + V$$

Using the connections between total energy and frequency (15.2) and between momentum and wavevector (15.1) we find the dispersion relation

$$\omega = \left(\frac{\hbar}{2m}\right)k^2 + \frac{V}{\hbar} \qquad (15.3)$$

This is plotted in fig. 15.1 for a 'free' particle: one for which V has the same value everywhere, indicating that no force is acting.

The dispersion relation allows us to find the group velocity (12.4). For a free particle we obtain

$$v_g = \hbar k/m$$

which is simply p/m, the classical velocity of the particle itself. Thus we can form the simple mental picture of a wave group travelling along instead of the particle, and we can think of the particle at any given

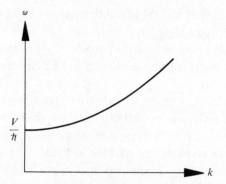

Fig. 15.1 The dispersion curve for de Broglie waves representing free particles. The absolute vertical position of the curve depends on the zero from which the potential energy V is measured, but shifting the graph up or down has no effect on the group velocity.

moment as being 'where the wave group is'. (We shall make this statement more quantitative shortly.)

The position of a wave group is, of course, a more diffuse concept than the position of a classical particle; this lack of definition is an intrinsic feature of quantum physics.

Interpretation of the wave function. We have talked about a wave, but we have not said what is 'waving'. What quantity should we use for ψ in a de Broglie wave?

Wave mechanics has no straightforward answer to this question. It treats ψ as a kind of dummy variable, not directly related to any measurable quantity. There are, however, several rules by which the values of measurable things can be predicted once the *wave function* ψ has been found. We examine one of these rules below.

Wave functions are not, in general, real but complex. This need not worry us, since we are not using ψ to measure any simple physical entity such as a displacement. A complex ψ can carry more information about the physics of a problem because its value is always twofold: it has a real part and an imaginary part.

In other respects the wave function behaves like all other ψ's we have met elsewhere. The only new features are interpretive: once the classical picture of a particle has been replaced by a wave picture, the solution of a problem proceeds by the methods used for any other wave problem. This is important; however greatly the predictions of wave mechanics may differ from those of classical physics in any given situation, the behaviour of the waves themselves is not different from that of any other waves. If 'strange' results are predicted, they are a consequence of accepting a wave picture of the problem in the first place.

The rules for turning the results of the wave calculation into physical predictions give probabilities, not certainties. We shall state only one of these rules, which tells us 'where the particle is' at any given moment. For simplicity we consider a particle moving in one dimension, along the z axis.

The rule says that if, at time t, we carry out a rapid experiment designed to detect the particle, the probability that we shall find it somewhere between z and $z + \Delta z$ is proportional to the quantity $\psi^*(z, t)\psi(z, t) \Delta z$. It is quite helpful to visualize the particle as being 'smeared' along the z axis, over those values of z where $\psi^*\psi$ is appreciable. This is a useful pictorial way of thinking about the problem, but we must remember that what is actually smeared is not the particle but the probability. A detection experiment will never find a fraction of the particle: it either detects the particle or not, and wave mechanics predicts the success rate when the experiment is repeated many times.

The rule will give unique answers only if the function ψ has no breaks in it. When looking for the solution to a problem we therefore accept only wave functions which are continuous in space.

We can illustrate these ideas by considering a simple travelling wave function of the form

$$\psi(z, t) = D \exp\left[i(\omega t - kz)\right] \tag{15.4}$$

This looks much like a sinusoidal travelling wave (9.7) in form D, except that we have not taken the real part.†

The product of this $\psi(z, t)$ and its complex conjugate $\psi^*(z, t)$ is simply D^*D, a real constant. Thus the probability of finding a particle between z and $z + \Delta z$ is the same for all values of z, and independent of time. In fact this wave function describes a uniform steady *beam* of particles all having the same energy $\hbar\omega$. In a case like this the probability rule can be thought of as giving the number density of particles as a function of position and time.

Only a complex wave function will give a detection probability (or particle density) which is constant and independent of position. If we had taken the real part of ψ in (15.4) our expression for $\psi^*\psi$ (or ψ^2 in that case) would have included both z and t.

The probability interpretation of the wave function means that no physical significance can be attached to the absolute value of the phase angle. Adding any ϕ to the phase angle in (15.4), for example, is

† Quantum mechanics textbooks usually reverse the terms in the phase angle and write

$$\psi = D \exp\left[i(kz - \omega t)\right]$$

Thus our wave (15.4) would be called ψ^*.

equivalent to multiplying ψ by exp $(i\phi)$; but the product $\psi^*\psi$ is multiplied by exp $(-i\phi + i\phi)$, which is simply 1, and so the predictive properties of ψ are unaltered.

Since the phase of ψ does not affect the physics, we might expect that the phase velocity should be equally insignificant. In fact the absolute value of ω is arbitrary, since it contains a term in the potential energy which can be measured from any zero we wish; thus the phase velocity ω/k is also arbitrary. The significant quantity is the group velocity which, as we have seen, has the classical value p/m for a free particle.

Summary. In wave mechanics, physical information about particles is embodied in complex wave functions from which probability predictions can be obtained. Effects which differ from those predicted by classical mechanics can be observed when the de Broglie wavelength $h/|p|$ is significant.

15.2. Physical implications

In this section we consider briefly several important consequences that follow when we adopt a wave picture of particle behaviour.

The uncertainty principle. The probability interpretation of the wave function allows us to quantify the intuitive idea that a free particle is instantaneously located 'where the de Broglie wave group is'. The detection probability is proportional to $\psi^*\psi$ and is therefore large where ψ is large: namely, within the group.

Precise location of the particle is, however, possible only if the wave group is short. We know from the bandwidth theorem for wave groups (11.13) that a short wave group has a poorly defined wavelength, and a poorly defined de Broglie wavelength means a poorly defined particle momentum.

The wave picture thus points to a fundamental restriction, enshrined in the *Heisenberg uncertainty principle*, on the amount of information we can obtain about a particle at any given moment. The momentum spread Δp (in the z direction) implied by a wavevector spread Δk is

$$\Delta p = \hbar\,\Delta k$$

and is therefore related to the position spread Δz by

$$\Delta p\,\Delta z \sim h$$

Thus the precision with which we can measure the momentum must always be inversely proportional to the precision with which we can *simultaneously* measure the position. Even under ideal experimental conditions, the product $\Delta p \, \Delta z$ can never be zero.

Energy levels. The wave picture also allows us to understand why an electron in an atom can exist only in a series of allowed energy levels. The essential property of an electron attached to an atom is its localization within a small region of space. We can represent this in a crude way by thinking of an electron contained in a one-dimensional 'box' of length L. The electron will never be found outside the box, and so its wave function must be zero everywhere outside. Inside the box it is quite free, and therefore obeys the dispersion relation (15.3) with a constant V.

We have seen that a wave function must be continuous in space, in order to give unique answers. The wave function representing the electron in the box must therefore come down to zero at the edges of the box, because its value is zero just outside the box. This is precisely the condition that must be satisfied by a standing wave on a string fixed at both ends, and we can say immediately that the wave function must be one of the standing waves† in fig. 9.6(a). Only certain values of k are possible (9.26) and therefore only certain eigenfrequencies. For a de Broglie wave that means only certain allowed energies.

An electron in an atom is not quite like a particle in a box, of course. Its potential energy, and thus its wavelength, vary with its position in the atom, and so the standing waves involved have *non-sinusoidal* eigenfunctions, like the standing waves on a *non-uniform* string. Advanced wave mechanics consists largely of methods for finding the eigenfunctions in cases where the equations cannot be solved analytically.

Cut-off and tunnelling. The dispersion relation (15.3) shows a lower cut-off frequency equal to V/h. Below cut-off, the wave function for a beam of particles (15.4) will take the form

$$\psi(z, t) = D \exp\left[i(\omega t + i\kappa z)\right]$$

where we have as usual written the imaginary wavevector as $-i\kappa$. The particle detection probability given by this evanescent de Broglie wave is proportional to

$$\psi^*(z, t)\psi(z, t) = \exp\left(-2\kappa z\right)D^*D$$

This is real, though its value is different at different points along the beam.

---

† The fact that the wave functions are real in this case is related to the absence of any electron *current*: in this problem the electron is trapped.

Apparently it is quite possible to find particles in a region where their de Broglie waves are evanescent.

This result is in direct conflict with classical physics. A de Broglie wave with an imaginary wavevector corresponds to a particle with imaginary momentum (15.1) or negative kinetic energy: a classical absurdity. It is as though our beam of particles were to come to a point on the z axis where there was a very strong force acting in the negative z direction, and some of them were nevertheless to be found downstream of that point. The force at any point is always equal to the negative potential energy gradient at that point, and so we can represent a large negative force by a steeply rising potential energy, as shown in fig. 15.2(a). If the potential energy step is large enough, particles on the right of the step may have $V > W$, making their kinetic energy $W - V$ negative.

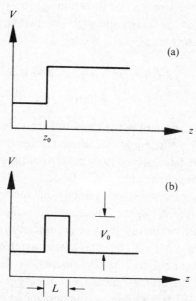

Fig. 15.2 (a) A sudden increase in the potential energy at z_0 indicates a strong force acting (from right to left) at z_0. For a sufficiently large potential step, a travelling wave in the region $z < z_0$ may become evanescent in the region $z > z_0$. (b) A potential barrier is formed by positive and negative potential steps at points separated by a distance L, and indicates forces of opposite sign, acting at these points. In this example the two potential steps are equal.

The behaviour of the de Broglie waves *as waves* is quite normal, and corresponds to the situation on the anchored strings shown in fig. 12.17(a). It is, as always, the interpretation of the waves in terms of particles which leads to non-classical behaviour.

It is hard to think of any experimental test for the existence of particles to the right of the step, but the tunnelling phenomenon illustrated in fig. 12.17(b) should allow some of the particles from the original beam to emerge, and be detected, beyond a *potential barrier* (a strong right–left force followed in a short distance by a strong left–right force) such as the one shown in fig. 15.2(b). This de Broglie wave tunnelling effect has been used to explain why a current can pass through an insulating oxide layer between two dirty copper wires, how a gas is ionized by a strong electric field, why charged particles of relatively low energy can penetrate a nucleus in spite of the strong repulsive electric force, and other phenomena for which classical physics has no satisfactory explanation.

The probability of tunnelling through a potential barrier depends very sensitively on the 'height' (in energy units) and the width of the barrier. We can see why if we consider a simple rectangular barrier of uniform height V_0 and width L, like the one shown in fig. 15.2(b). The ratio of the complex amplitude of the de Broglie wave downstream to that upstream will be roughly equal to $\exp(-\kappa L)$. We have

$$\hbar\kappa = -\mathrm{i}(\hbar k) = -\mathrm{i}p = -\mathrm{i}[2m(W - V_0)]^{1/2}$$

$$= [2m(V_0 - W)]^{1/2}$$

and so the negative exponent in the amplitude ratio will be proportional to $L(V_0 - W)^{1/2}$. The efficiency with which the beam can penetrate the barrier can therefore be expected to increase very rapidly (because of the exponential dependence) as either $V_0 - W$ or L is decreased.

A tunnelling phenomenon which exhibits this sort of dependence to a spectacular degree is alpha decay. The kinetic energies of the α-particles emitted by naturally occurring radioactive nuclei are all fairly similar (between 4 MeV and 8 MeV). The rates at which nuclei emit α-particles, which depends on a tunnelling efficiency, increase systematically with the kinetic energies of the emitted particles (a measure of $V_0 - W$ for the barriers involved). What is striking is the enormous variety of observed rates. These span a wider range of values than almost any other known physical quantity, the highest being more than 10^{20} times greater than the lowest. In this case the shape of the potential barrier is such as to amplify the dependence which we saw to be exponential in the case of a rectangular barrier.

Summary. The behaviour of de Broglie waves is similar to that of other waves. The fact that $\Delta k\,\Delta z \sim 1$ for a travelling wave group implies that $\Delta p\,\Delta z \sim h$ for a free particle (the uncertainty principle). Quantized energy levels of a particle confined to a restricted region of space can be understood in terms of standing de Broglie waves. The tunnelling of de Broglie waves leads to the penetration of particles through regions from which they are classically excluded by the forces acting.

Problems

15.1 Estimate the energy in eV of (a) an electron, (b) a neutron, (c) a helium atom, and (d) a uranium atom having a de Broglie wavelength of 0.1 nm.

15.2 Estimate the de Broglie wavelength of an oxygen molecule in air at room temperature.

15.3 If the kinetic energy of an electron, known to be about 1 eV, must be measured to within 0.0001 eV, to what accuracy can its position be measured simultaneously?

15.4 Estimate the diameter of an atom within which the minimum possible kinetic energy of an electron would be equal to 10 per cent of its rest energy $m_e c^2 \approx 500 \text{ keV}$.

15.5 For the travelling de Broglie wave

$$\psi = D \, e^{i(kz - \omega t)}$$

express $\omega \psi$ and $k^2 \psi$ in terms of partial differential coefficients of ψ. With the help of the dispersion relation (15.3) show that the wave equation

$$i\hbar \left(\frac{\partial \psi}{\partial t} \right) = -\frac{\hbar^2}{2m} \left(\frac{\partial^2 \psi}{\partial z^2} \right) + V\psi$$

is satisfied. (This is the *Schrödinger equation*. It is a postulate of wave mechanics that *all* de Broglie waves satisfy the Schrödinger equation, provided relativity effects can be neglected. None of its predictions has ever found to be at variance with experiment.)

16

Solitary waves

Our examination of various waves in the previous three chapters has highlighted a fact of great practical significance: *most wave systems are dispersive*. The chief characteristic of a dispersive system is that it distorts any travelling waves that are not sinusoidal (section 12.1). Sinusoidal waves are not very useful, since they cannot carry any information other than their frequency and amplitude: one is usually interested in sending signals as some kind of modulated wave, and in particular as a train of pulses. Such an enterprise might appear doomed to failure if the chosen wave system is a dispersive one.

The remarkable fact is, however, that dispersive systems can after all support the undegraded propagation of stable pulses of a special kind, known as solitary waves. The effective cancellation of the dispersion has a surprising source in another property of most real wave systems, namely *non-linearity*.

To see how this comes about, we must first learn how non-linearity on its own affects a wave system. To simplify the discussion, we shall consider in this chapter only travelling plane waves moving in the positive z-direction, and we shall also neglect all non-conservative processes like friction and viscosity, which dissipate energy.

16.1. Non-linear wave systems

In chapter 7 we examined free and forced vibrations of non-linear systems. In a non-linear system the return force is not exactly proportional to the displacement, and so the stiffness is not a constant. In a wave-carrying medium the stiffness and the mass, both of which are distributed continuously throughout the whole medium, fix the characteristic wave speed of the system; other things being equal, stiffer systems have greater wave speeds. If the system is non-linear, the wave speed will depend on the displacement ψ, which of course varies with both position and time in a wave. To explore the nature of this dependence, we shall base our discussion on two particular examples of non-linear wave systems.

Longitudinal waves. Before discussing non-linearity arising from the physics of particular systems, we consider a purely kinematic effect which is present in all *longitudinal* waves.

Figure 16.1 shows what part of a longitudinal travelling waveform might look like at some given moment t. We wish to follow the progress

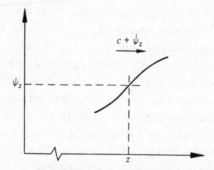

Fig. 16.1 Part of a non-linear longitudinal wave. The waveform is a plot of ψ_z against z, but the physical displacement is in the z-direction. The signal travels with speed c relative to the medium, but its absolute speed contains a correction for the longitudinal motion of the medium due to the wave itself.

of a point on the waveform where the longitudinal displacement ψ_z has some particular value. If we calculate the wave speed c for the system (by the methods of chapter 10), that will tell us how fast our point travels *relative to the medium*. But a longitudinal wave, as it passes, makes the medium itself move back and forth along the direction of travel with instantaneous velocity $\dot{\psi}_z$. What concerns us is the absolute speed with which the 'signal' (the particular ψ_z-value) travels through space, and we find this by adding $\dot{\psi}_z$ to the wave speed c. We shall call the result

$$v = c + \dot{\psi}_z \tag{16.1}$$

the *local signal speed*. It varies from point to point along the z-axis, and also fluctuates with time, because $\dot{\psi}_z$ does so.

Of course, if the velocities produced in the medium by the passing wave are always negligible compared with the wave speed ($\dot{\psi}_z \ll c$), then we can ignore the effect, and $v \approx c$ as linear theory predicts. For a longitudinal wave in a solid rod or wire, for example, the instantaneous value of $\dot{\psi}_z/c$ at any point is equal to the local strain. (See problem 10.3). Ordinary metals reach their yield points at strains well below 0.01, so the difference between c and v will be much smaller than 1 per cent under all normal conditions. The physics correction to the local wave speed is generally even smaller than this kinematic correction.

Acoustic waves. We now consider the particular case of an acoustic wave. We know that small pressure fluctuations in a gas can be propagated as a wave governed by the travelling wave equation (9.12), with speed

$$c = (\gamma p/\rho)^{1/2} \tag{16.2}$$

where p and ρ are the equilibrium pressure and density (10.12). This is true as long as the acoustic pressure ψ_p is much smaller than p.

When the pressure fluctuations do not meet this 'small displacement' requirement (10.20), we may consider instead a series of small acoustic pressure *changes* $\Delta\psi_p$, and find the non-linear behaviour by simple integration. Any such $\Delta\psi_p$, propagating through the already disturbed gas, will satisfy the wave equation; its propagation speed will be c, given by (16.2) but now calculated from the temporary and local values of p and ρ brought about by the wave. Since ρ is proportional to $p^{1/\gamma}$ (for adiabatic changes), c varies with p according to

$$c \propto p^{\frac{1}{2}(1-1/\gamma)} \tag{16.3}$$

and therefore fluctuates as the wave changes the local pressure.

In order to compare the size of this physical effect with that of the kinematic one (16.1), we wish to know how c varies with $\dot{\psi}_z$. The change in $\dot{\psi}_z$ resulting from the small pressure change $\Delta\psi_p$ is determined by the characteristic impedance (10.15), and is

$$\Delta\dot{\psi}_z = (c/\gamma p)\Delta\psi_p \tag{16.4}$$

Since the dependence of c on p is a simple power law (16.3) we may immediately write

$$\frac{\Delta c}{c} = \frac{1}{2}\left(1-\frac{1}{\gamma}\right)\frac{\Delta p}{p} \tag{16.5}$$

Now $\Delta\psi_p$ and Δp are actually the same thing, so we can combine (16.5) with (16.4) to obtain

$$\Delta c = \tfrac{1}{2}(\gamma-1)\Delta\dot{\psi}_z$$

This tells us that each small increase in $\dot{\psi}_z$ leads to a proportionate increase in c. Thus an overall increase from zero to $\dot{\psi}_z$ makes the wave speed increase from $c = c_0$ (the values corresponding to equilibrium values of p and ρ) to

$$c = c_0 + \tfrac{1}{2}(\gamma - 1)\dot{\psi}_z \qquad (16.6)$$

The pressure dependence of the wave speed is not a strong one: for air at STP ($\gamma = 1.40$), the value of the exponent in (16.3) is only 0.143. Thus it requires a doubling of the pressure (from $p = 1$ atm to $p = 2$ atm, say) to increase c by 10 per cent. An ordinary sound wave, as we saw (p. 179), produces pressure changes of order 10^{-4} atm at most. Thus observable non-linearity in an acoustic wave will require something extreme as the sound source, such as an explosion. Of course $c \approx c_0$ if $\dot{\psi}_z \ll c_0$.

For all physically possible values of γ the factor $\tfrac{1}{2}(\gamma - 1)$ is smaller than 1. Thus the kinematic correction (16.1) always has the greater effect. For $\gamma = 1.40$ the increase in the signal speed due to the physics of the gas is a fifth of the kinematic increase. Combining the two effects gives, for the acoustic wave,

$$v = c_0 + \tfrac{1}{2}(\gamma + 1)\dot{\psi}_z \qquad (16.7)$$

Shallow-water waves. Our second example is the case of gravity waves in shallow water. We saw in section 13.1 that the motion of the water at any point in such a wave is approximately longitudinal, as long as the water depth h is small compared with the wavelength. We consider an extra bump $\Delta\psi$ on top of a high wave (large ψ). Associated with the bump is a longitudinal velocity increment

$$\Delta\dot{\psi}_z = (g/h)^{1/2}\,\Delta\psi \qquad (16.8)$$

given by (13.29). Since the wave speed, according to (13.28), increases as $h^{1/2}$, we have

$$\frac{\Delta c}{c} = \frac{1}{2}\frac{\Delta h}{h}$$

Combining this with (16.8) and remembering that $\Delta\psi = \Delta h$, we obtain

$$\Delta c = \tfrac{1}{2}(g/h)^{1/2}\Delta h = \tfrac{1}{2}\Delta\dot{\psi}_z$$

which is easily integrated to give

$$c = c_0 + \tfrac{1}{2}\dot{\psi}_z$$

This time the physics contribution to the local signal speed is exactly half of that due to the sloshing motion of the water (16.1). The two effects combine to give

$$v = c_0 + \tfrac{3}{2}\dot{\psi}_z \qquad (16.9)$$

Non-linear waves in general. In each of these examples the overall change in the local signal speed is proportional to $\dot{\psi}_z$, the local longitudinal velocity of the medium itself, produced by the wave. Instead of using $\dot{\psi}_z$, we might be able to find v in terms of the 'natural' variable for the particular system under consideration. Thus, for acoustic waves we would use the acoustic pressure ψ_p which, conveniently, is in phase with $\dot{\psi}_z$ (10.13). The relationship

$$\dot{\psi}_z \approx (c_0/\gamma p_0)\psi_p$$

then leads to

$$v \approx c_0(1 + b\psi_p) \qquad (16.10)$$

where

$$b \equiv (\gamma + 1)/2\gamma p_0 \qquad (16.11)$$

and p_0 is the equilibrium pressure.†

For the second example (shallow-water waves) the natural variable is the height ψ of the water surface above its undisturbed position: this is also in phase with $\dot{\psi}_z$. We have, from (13.29),

$$\dot{\psi}_z \approx (c_0/h_0)\psi$$

and so

$$v \approx c_0(1 + b\psi)$$

where

$$b \equiv 3/2h_0 \qquad (16.12)$$

Here h_0 is the depth of the water before the wave arrives.

These two examples suggest a standard form for the local signal speed in any suitable non-linear wave, namely

$$v = c_0(1 + b\psi) \qquad (16.13)$$

---

† Equation (16.10) is an approximation because, instead of correctly integrating (16.4), we have used the characteristic impedance for small disturbances to connect large values of $\dot{\psi}_z$ and ψ_p, which are not strictly proportional to each other. The approximation is a good one if second and higher powers of ψ_p/p_0 can be neglected. This is of course less restrictive than the linear condition (10.20). (See problem 16.2.)

In this expression ψ is the chosen displacement variable, and b contains both the kinematic effect (for a longitudinal wave) and the relevant physics. Presumably there are systems in which $v(\psi)$ is not such a simple function. If the non-linearity is slight, however, (16.13) will be an acceptable approximation in any case for which $(v - c_0)$ is an odd function of ψ, as in fig. 16.2(a). This will still leave some systems out of account: for transverse waves on a string, for example, $v(\psi)$ must be even, as in fig. 16.2(b), since there is no kinematic contribution and any ψ-dependence of c must obviously be the same for equal upward and downward displacements. We shall, however, assume that any 'slightly non-linear' wave we discuss obeys (16.13), with $b > 0$.

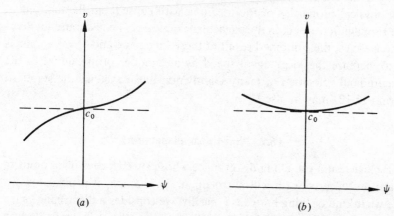

(a)   (b)

Fig. 16.2 Graphs characterizing non-linear wave systems. In each diagram the local signal speed v is plotted against the displacement ψ, and c_0 represents the wave speed for small displacements (the linear approximation). The broken lines describe truly linear systems.

(a) If $(v - c_0)$ is an odd function of ψ, v may be represented by a polynomial in odd powers,

$$v = c_0(1 + b\psi + b'\psi^3 + \ldots)$$

When the displacements we are dealing with are not too great, we may neglect the curvature of the graph and write $v \approx c_0(1 + b\psi)$.

(b) If $(v - c_0)$ is an even function of ψ, we require a polynomial in even powers,

$$v = c_0(1 + a\psi^2 + a'\psi^4 + \ldots)$$

Non-linear travelling wave equation. For any disturbance travelling in the positive z-direction, the displacement $\psi(z, t)$ must be a function of $vt - z$, where v is the signal speed. If v varies as in (16.13) we can easily show, by the argument which led to (9.12), that ψ will then satisfy

$$\frac{\partial \psi}{\partial t} + c_0(1 + b\psi)\frac{\partial \psi}{\partial z} = 0 \qquad (16.14)$$

This is a non-linear equation. (The *product* of ψ and $\partial\psi/\partial z$ is involved). It reverts as it should to the familiar travelling wave equation

$$\frac{\partial\psi}{\partial t} + c_0\left(\frac{\partial\psi}{\partial z}\right) = 0$$

when the medium is linear ($b = 0$), and comes very close to it for small disturbances in a non-linear medium ($b\psi \ll 1$).

Summary. In longitudinal travelling waves of large amplitude, the signal speed is affected by the to-and-fro motion given to the medium by the wave itself. Large disturbances can also cause significant fluctuations in the physical properties of the medium, with consequent fluctuations in the propagation speed. In the particular examples of acoustic and shallow-water waves, the combined result of these kinematic and physical effects is to increase the local signal speed by an amount proportional to the longitudinal velocity. For many slightly non-linear systems the standard form (16.13) may be used.

16.2. Non-linear dispersion

It is clear that a travelling disturbance whose speed varies from point to point on the waveform will not maintain its shape as it propagates. To see what kind of thing happens typically, we consider a disturbance such as the one illustrated in fig. 16.3, and assume that it is controlled by (16.13).

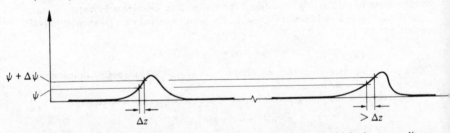

Fig. 16.3 Progressive distortion of a waveform travelling to the right in a non-linear medium. Points with larger displacements travel faster, with the result that the leading edge grows steeper and the trailing edge grows less steep. (The original waveform plotted here is the one given in problem 16.6.)

In a time interval t, points on the waveform at which the displacement is ψ will move a distance $c_0(1 + b\psi)t$ to the right. In the same time, however, neighbouring points where the displacement is $\psi + \Delta\psi$ will move *farther* to the right, by an amount $(c_0 bt)\Delta\psi$. Thus the gradient

$\partial\psi/\partial z$ everywhere must decrease as t increases: parts where the slope is positive become progressively less steep, while negatively sloping parts grow steeper. The whole waveform becomes more and more distorted as it propagates.

This non-linear effect differs from dispersion, which also causes progressive distortion of travelling waveforms, in one particularly important way: sinusoidal travelling waves are not immune. For a system obeying (16.13), it is the *trailing* edges of the negative-going parts of the waveform that steepen with time, so that the sine wave tends to turn into a sawtooth waveform (fig. 16.4).

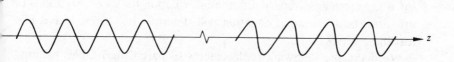

Fig. 16.4 Distortion of a sinusoidal wave travelling to the right in a non-linear medium.

If this process went on for ever, the peak of any positive-going waveform, where the signal speed is greatest, would eventually overtake all other points on the pulse, which would develop an overhang. Effects which we have neglected, particularly energy-dissipating processes, become crucially important as the wave is about to 'break', and so this does not happen in practice. Instead a limiting shape of pulse is formed, with a very steep leading edge known as a *shock front.*

Dispersion in a non-linear system. We now introduce dispersion. A waveform travelling in a medium which is both non-linear and dispersive might appear to have a double risk of progressive deterioration. But we shall find that the two effects can sometimes compensate each other in a very remarkable way, and that certain stable waveforms can then propagate indefinitely.

Our starting point is the travelling wave equation for a slightly dispersive medium (12.3) with an added term describing the non-linearity, taken from (16.14). The resulting equation

$$\frac{\partial\psi}{\partial t} + c_0(1 + b\psi)\frac{\partial\psi}{\partial z} + d\left(\frac{\partial^3\psi}{\partial z^3}\right) = 0 \qquad (16.15)$$

will approximately describe a travelling wave in almost any system which is slightly dispersive and slightly non-linear. The complete solution is beyond the scope of this book, but we can show that it may be satisfied by one particularly interesting travelling wave, discover some of the properties of that wave, and see how these properties depend on the non-linearity and the dispersion. The waveform in question is actually the one shown in fig. 16.3: a smooth, localized hump, with ψ going smoothly to zero on both sides.

For any disturbance $\psi(z, t)$ to be propagated (to the right) without change of form, we know that ψ must be some function of the single variable

$$Z \equiv vt - z$$

where v is a constant. We can turn the partial differential equation (16.15) for $\psi(z, t)$ into an ordinary differential equation for $\psi(Z)$. Solutions of this ordinary differential equation will describe the shapes of possible travelling waves.

We make the following replacements of partial derivatives by total derivatives:

$$\frac{\partial \psi}{\partial t} = v\left(\frac{d\psi}{dZ}\right) \equiv v\psi'$$

$$\frac{\partial \psi}{\partial z} = -\frac{d\psi}{dZ} \equiv -\psi'$$

$$\frac{\partial^3 \psi}{\partial z^3} = -\frac{d^3 \psi}{dZ^3} \equiv -\psi'''$$

We now have

$$(v - c_0)\psi' - bc_0\psi\psi' - d\psi''' = 0$$

and this may be directly integrated to yield

$$(v - c_0)\psi - \tfrac{1}{2}bc_0\psi^2 - d\psi'' + C = 0$$

where C is a constant of integration. We now multiply throughout by ψ' and integrate once more, obtaining

$$\tfrac{1}{2}(v - c_0)\psi^2 - \tfrac{1}{6}bc_0\psi^3 - \tfrac{1}{2}d\psi'^2 + C\psi + D = 0 \qquad (16.16)$$

in which a second integration constant D appears.

If a solution like the one in fig. 16.3 is to be permitted, we must have $C = D = 0$, since both ψ and ψ' have to go to zero at $Z = \pm\infty$. This allows us to rewrite (16.16) in the form

$$d\psi'^2 = (v - c_0 - \tfrac{1}{3}bc_0\psi)\psi^2 \qquad (16.17)$$

The condition for a maximum ($\psi' = 0$) then fixes the wave speed

$$v = c_0(1 + \tfrac{1}{3}b\psi_{max}) \tag{16.18}$$

Solitary waves. We have found that the undispersed propagation of a pulse (of one particular shape†) is indeed possible in a dispersive and non-linear medium. This is clearly a most valuable attribute, since it opens up the possibility of sending *large* pulses through real media without degradation. Pulses showing this kind of behaviour are known as *solitary waves*, or *solitons*.‡

What (16.18) shows is that large solitary waves (with large values of ψ_{max}) travel faster than small ones: a clue to their non-linear origin. The dispersion has no direct influence on the propagation speed, since d does not appear, but it does affect the *shape* of the pulse. Putting (16.18) into (16.17) gives

$$d(\psi'/\psi)^2 = \tfrac{1}{3}bc_0(\psi_{max} - \psi) \tag{16.19}$$

To interpret this result, it is helpful to remember that the quantity in brackets on the left is

$$\frac{1}{\psi}\frac{d\psi}{dZ} = \frac{d(\ln \psi)}{dZ}$$

Now we can see that, as we go off the peak (increasing values of $\psi_{max} - \psi$), the steepness of the ln ψ graph, and therefore of the waveform, increases more rapidly for small values of d than for large ones. That is, *the more dispersive the medium, the broader the pulse.* (The degree of non-linearity, through the size of b, also affects the width, but in the opposite sense.)

It is worth noting also that only 'normal' dispersion can give a solitary wave: equation (16.19) shows that d must be positive if ψ'/ψ is to be real, since b and c_0 are both positive, and ψ is never larger than ψ_{max}. This is quite understandable from a physical point of view. The effect of uncompensated non-linearity would be to steepen the leading edge of the pulse and flatten out the trailing edge. In Fourier theory terms this can be thought of as the continuous injection of short-wavelength components into the waveform at the front. The dispersion works by carrying these off towards the back of the pulse, and for this to happen the high-frequency waves must travel more slowly. Thus solitary waves can

† A formula for the particular waveform is presented in problem 16.6.
‡ The term soliton is usually reserved for certain solitary waves which have another remarkable property: when two of them 'collide', they emerge with their identities intact, as if they had undergone superposition in a linear system.

be found on shallow water (13.27), but not on a string. (Shallow-water solitary waves were actually the first to be observed: see problem 16.5.)

There are other non-linear and dispersive systems whose equations of motion have solitary wave solutions. (Of course, they also have periodic solutions analogous to the sinusoidal waves of non-dispersive systems, and more general solutions which are not simple travelling disturbances.) A notable feature of all solitary waves is their stability: they are one of the most indestructible ways in which a system can transmit or store energy. They have been invoked to explain an enormous variety of physical phenomena and processes, from the transmission of voltage pulses along nerves to the spiral structure of galaxies.

Summary. Since the signal speed varies with the position on the waveform, a wave travelling in a non-linear medium suffers progressive waveform distortion. This tendency can sometimes be balanced exactly by dispersion. We examined the travelling wave equation for a slightly non-linear, slightly dispersive system, and found that a solitary wave of special shape can be propagated without change of form. Increasing the dispersion broadens the pulse; increasing the non-linearity narrows it and makes it travel faster. Large solitary waves travel faster than small ones.

Problems

16.1   Show that, if pressure changes in an acoustic wave were isothermal, the local signal speed would be $c_0 + \dot{\psi}_z$.

16.2   Show that, in an acoustic wave,

$$\dot{\psi}_z \approx (c_0/\gamma p_0)\dot{\psi}_p$$

is a good approximation if second and higher powers of ψ_p/p_0 can be neglected.

16.3   Information is to be transmitted over a distance L by means of triangular pulses of height h and width w (fig. 16.5). If the medium is non-linear the pulses will arrive distorted. Show that, if (16.13) is obeyed, the condition for the change of pulse *shape* to be small is $b \ll w/hL$.

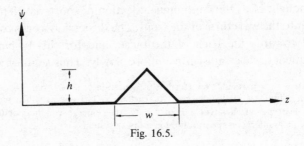

Fig. 16.5.

16.4   Calculate the speeds of solitary waves of height (a) 32 mm, (b) 160 mm, and (c) 320 mm, on water of depth 1.6 m. Compare the results with the speeds calculated in problem 13.2.

16.5   Solitary waves were first noticed in 1834 by John Scott Russell, when he was experimenting with high-speed horse-drawn boats on a canal in Scotland. In the course of subsequent laboratory experiments he found an empirical relationship

$$v = (gH)^{1/2}$$

between the speed v of a solitary wave and its height H, measured from the canal bottom. Show that this relationship is consistent with (16.18) for small waves.

16.6   Equation (16.17) may be simplified to

$$(\psi')^2 = (A - B\psi)\psi^2$$

where

$$A = (v - c_0)/d$$
$$B = bc_0/3d$$

(a) Confirm by direct differentiation and substitution that

$$\psi = C \operatorname{sech}^2 DZ$$

is a solution of the equation, with

$$C = A/B$$
$$D = \tfrac{1}{2}A^{1/2}$$

This is the solitary wave solution, and is the shape plotted in fig. 16.3.

(b) Confirm our qualitative finding in the text that increasing the dispersion broadens the pulse.

17

Plane waves at boundaries

The reflection and transmission processes that occur when a travelling wave meets a discontinuity in a string, and the standing waves set up when incident and reflected waves are superposed, were analyzed in chapter 9. We were able to adapt the results to other situations, such as the reflection of electromagnetic plane waves meeting a dielectric surface at right angles, or the formation of acoustic standing waves in a pipe.

Some of the most interesting ways in which the presence of a boundary between two media can affect the behaviour of waves in those media become apparent only when the waves are travelling in directions other than normal to the boundary. Likewise, standing waves in a region of three-dimensional space have characteristics which we cannot learn about by studying standing waves on strings. In this chapter, therefore, we think about plane waves meeting plane boundaries, not in general at normal incidence.

As a preliminary step, we learn how to handle plane waves in three-dimensional problems.

17.1. Reflection and refraction

Although we began by discussing transverse waves on strings, we have seen already that the disturbance in a one-dimensional wave need not be concentrated along a line. In a sinusoidal travelling wave $\psi(z, t)$ for which ψ is a quantity like acoustic pressure or the strength of an electric field, which can have a value at any point in space, the phase angle $\omega t - kz$ has the same value at all points which have the same value of z; these points lie on a surface perpendicular to the direction of propagation. A surface of constant phase (which need not be plane) is called a *wavefront*.

Adjacent wavefronts of given phase are separated by one wavelength. If the amplitude as well as the phase has the same value all over a plane wavefront,† we call the wave a *plane wave*.

Plane waves are essentially one-dimensional, but we can meet them in three-dimensional problems. It will not always be possible to choose coordinates such that the wave travels along an axis, as in fig. 17.1(a). We must therefore be able to express the wave in a form which does not depend on any particular orientation of the axes.

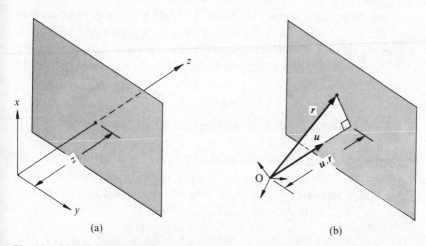

(a)   (b)

Fig. 17.1 (a) Plane wave propagating along the z axis. (b) Plane wave propagating in a direction indicated by the unit vector u, which does not in general point along any coordinate axis. The vector r measures the position of a point on the wavefront.

We can indicate the direction of propagation of the wave by means of a unit vector u, drawn at right angles to the wavefronts and pointing in the direction of travel. Points in space can be specified by a position vector r. These two vectors are shown in fig. 17.1(b). In a specific problem we would have to know their components with respect to the axes in use, but that is not necessary for a general discussion.

Now the scalar product $u \cdot r$ has the same value for every point r in a plane at right angles to u: that is, for every point on a plane wavefront. In fig. 17.1(a) the vector u would point along the z axis and $u \cdot r$ would be equal to z; in more general cases $u \cdot r$ is the distance along the propagation direction from the origin to the wavefront. Therefore, in any expression for a sinusoidal travelling plane wave (9.7) the term kz may be replaced by $k(u \cdot r)$.

† Later in this section we shall meet a case in which this is not true.

It is convenient to define the *wavevector*

$$k \equiv |k|u$$

and write the plane wave as

$$\psi(r, t) = A \cos{(\omega t - k \cdot r)} \tag{17.1}$$

with the usual alternatives in forms B, C and D.

We have thus turned the one-dimensional quantity k into a vector suitable for use in three dimensions; k could indicate (by its sign) only the *sense* of travel along a given direction, whereas k can give the direction also. The magnitude of k is $2\pi/\lambda$ where λ is the wavelength (9.10). Any dispersion relation derived in one dimension, such as (12.2), can be used in three dimensions if we first obtain the vector components of k and write

$$k^2 = k_x^2 + k_y^2 + k_z^2$$

or the equivalent expression in another coordinate system.

Reflection and refraction: directions. We are now in a position to consider the general case of a plane wave meeting a plane boundary. There will presumably be, in addition to the incident wave, a reflected wave and a transmitted wave, and we may write

$$\psi_i = A_i \cos{(\omega t - k_i \cdot r)}$$

$$\psi_r = A_r \cos{(\omega t - k_r \cdot r)}$$

$$\psi = A \cos{(\omega t - k \cdot r)}$$

To simplify later expressions we have put no suffixes on the transmitted wave, which is the one we shall be most interested in. The frequencies have been made equal since we know that reflected and transmitted waves are a consequence of some steady-state forced vibration: re-radiation by electrons shaken by the incident wave, for example (section 14.2).

The displacement at any point in the medium containing the incident wave is $\psi_i + \psi_r$. At points on the boundary there will be certain continuity conditions to be satisfied. In a simple case we might have

$$\psi_i + \psi_r = \psi$$

This would apply for acoustic waves, for example: the acoustic pressure just on one side of the boundary must be the same as that just on the other side. For transverse waves the conditions may be less simple; the electric field in an electromagnetic field, for example, must have the same tangential *components* on both sides of the boundary. But in either case the three displacements would have to be in phase (or in antiphase) with each other at the boundary for the continuity conditions to be capable of fulfilment at all times.

We therefore write

$$k_i \cdot r = k_r \cdot r = k \cdot r$$

where r now denotes a point *lying in the boundary* (fig. 17.2). Expanding these scalar products in terms of Cartesian components of the four vectors involved, we find

$$k_{ix}x + k_{iy}y + k_{iz}z = k_{rx}x + k_{ry}y + k_{rz}z = k_xx + k_yy + k_zz \quad (17.2)$$

where x, y and z are the components of r, and the notation for the wavevector components is obvious.

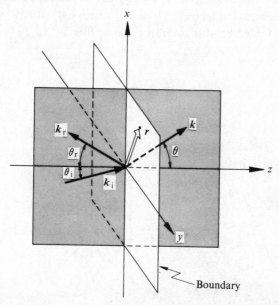

Fig. 17.2 Plane wave at a plane boundary. The axes are chosen so that the boundary coincides with the xy plane, and the incident wavevector k_i lies in the xz plane (shaded). It is shown in the text that the reflected wavevector k_r and the transmitted wavevector k also lie in the xz plane. The position vector r lies in the boundary.

With the axes chosen in fig. 17.2, the boundary coincides with the xy plane, and the incident wavevector k_i is drawn in the xz plane. These choices mean that

$$z = 0$$

$$k_{iy} = 0$$

and so (17.2) becomes

$$k_{ix}x = k_{rx}x + k_{ry}y = k_x x + k_y y$$

Every point lying in the boundary must satisfy these conditions; this will be possible only if

$$k_{ry} = k_y = 0$$
$$k_{ix} = k_{rx} = k_x \tag{17.3}$$

The first line of (17.3) expresses the fact that k_r and k may, like k_i, be drawn in the xz plane (as they were in fig. 17.2). The second line says that wavefronts of given phase (wave crests, for example) must have the same spacing, measured along the boundary, for all three waves. Since the three displacements must be in phase or in antiphase at any point on the boundary, we must be able to draw pictures like fig. 17.3.

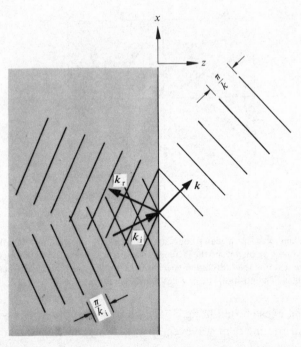

Fig. 17.3 Wave crests of incident, reflected and transmitted waves must always meet at the boundary. Wave crests of both signs (separated by half wavelengths) are drawn: there is no necessity for wave crests *of the same sign* to meet. For clarity only a limited section of each wavefront is shown.

In practice the directions of the waves will be indicated by the angles that their wavevectors make with some fixed direction. There is only one fixed direction in this problem: the normal to the boundary. In terms of the angles defined in fig. 17.2 the second line of (17.3) may be written

$$|k_i| \sin \theta_i = |k_r| \sin \theta_r = |k| \sin \theta$$

Since the incident and reflected waves travel in the same medium and have the same frequencies, k_i and k_r have equal magnitudes. Thus we have finally

$$\theta_r = \theta_i$$

$$|\mathbf{k}| \sin \theta = |\mathbf{k}_i| \sin \theta_i \tag{17.4}$$

The first of these conditions is the well-known law of specular (mirror-like) reflection. The second is more familiar in the form

$$n \sin \theta = n_i \sin \theta_i \tag{17.5}$$

involving the refractive indexes of the media (14.10), and is usually known as Snell's law.

The change in the direction of the transmitted wave relative to that of the incident wave is called *refraction*. Refraction is essentially a consequence of the differing phase velocities in the two media. If at least one of the media is dispersive, the amount of refraction will vary with the wave frequency; a mixture of waves of different frequencies (white light, for example) will be 'dispersed', or refracted through a range of different angles. This is, of course, the origin of the term 'dispersion'.

Refraction is observed only when the incident wavefronts are inclined to the boundary: putting $\theta_i = 0$ in (17.5) gives $\theta = \theta_i$.

Reflection and refraction: amplitudes and phases. These results (17.4) concerning the directions of the reflected and refracted waves hold for all plane waves and plane boundaries. In fact, they can be used to describe the *local* behaviour of any wave at any boundary, since we can always take a small enough section of a curved wavefront or a curved boundary to enable us to treat it as plane.

To find the amplitudes of ψ_r and ψ, or to be able to say whether they are in phase or in antiphase with ψ_i, we require specific information about the physics of the particular waves involved. We shall not pursue any examples in detail; the calculation of reflection and transmission coefficients is similar in principle to their derivation for stretched strings (section 9.2). As in that case the use of characteristic impedances is helpful.

An extra consideration arises in the case of a transverse wave, since we would need to know in which plane the displacement takes place. This information is called the *state of polarization* of the wave. In an electromagnetic plane wave, for example, the matching conditions to be satisfied at the boundary when E vibrates in the shaded plane in fig. 17.2 will clearly differ from those for E vibrating at right angles to that plane, or in some other direction between these two extremes.

Total internal reflection. We shall examine in more detail the case in which the medium carrying the incident and reflected waves has a lower phase velocity (a higher refractive index) than the medium carrying the refracted wave. This can describe a number of situations: acoustic waves passing from cold air to warm air, for example. The discussion is simpler, however, if we take a concrete example. We therefore assume that the 'dense' medium is glass of refractive index n, and consider what happens when a plane light wave comes up to a boundary behind which is a vacuum.

We ask how the dispersion relation for a vacuum

$$\omega^2 = c^2 k^2 = c^2(k_x^2 + k_z^2) \tag{17.6}$$

is affected by the presence of the glass. We must look at the x and z components separately.

In terms of the angles defined in fig. 17.2, k_x may be written as $|\mathbf{k}| \sin \theta$. Thus, by the refraction law (17.4) and the definition of the refractive index (14.10), we have

$$k_x = |k_i| \sin \theta_i = (n\omega/c) \sin \theta_i \tag{17.7}$$

From this we obtain the x part of the dispersion relation,

$$\omega = \frac{ck_x}{n \sin \theta_i}$$

Putting (17.7) into (17.6) yields the z part,

$$\omega^2 = \frac{c^2 k_z^2}{1 - n^2 \sin^2 \theta_i} \tag{17.8}$$

These relations tell us about wave propagation in the vacuum, but they contain the refractive index of the glass and the angle of incidence. The behaviour of the wave in the vacuum is thus affected by the presence of the glass.

We shall explore the influence of the angle of incidence θ_i more closely. If θ_i is not too large, $n \sin \theta_i$ will be smaller than 1, and both k_x and k_z will be real: we then have a non-dispersive plane wave travelling with phase velocity c, just as in an unobstructed vacuum.

If $n \sin \theta_i$ is made larger than 1, however, k_z^2 will become negative and so k_z must be imaginary. To find out what this does to the wave, we replace the imaginary k_z by $-i\kappa_z$ in the form D version of (17.1), written out in (x, z) coordinates. We obtain

$$\psi(x, z, t) = \text{Re} \{D \exp [i(\omega t - k_x x - k_z z)]\}$$

$$= \exp (-\kappa_z z) \, \text{Re} \{D \exp [i(\omega t - k_x x)]\}$$

This is a wave travelling in the x direction (along the boundary). Since the phase angle depends on x only, the wavefronts are plane (fig. 17.4).

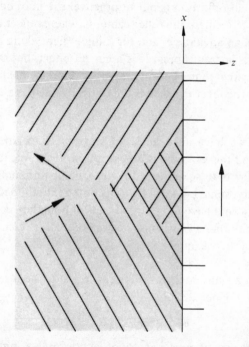

Fig. 17.4 Total internal reflection. The wavefronts of the refracted wave are schematically shown extending about one wavelength into the vacuum; in fact they extend to infinity but the amplitude falls off exponentially with z, decreasing by a factor $1/e$ in a distance of the order of a wavelength. For clarity only a small region of overlap between the incident and reflected waves is shown.

But the wave *is not a plane wave*, because the amplitude varies from point to point on a given wavefront, falling off exponentially with increasing depth z into the vacuum.

The degree of penetration can be judged by substituting $-i\kappa_z$ for k_z in the dispersion relation (17.8). This gives

$$\kappa_z = (n^2 \sin^2 \theta_i - 1)^{1/2} \omega / c$$

$$= (\sin^2 \theta_i - 1/n^2)^{1/2} |k_i|$$

We shall nearly always have $\kappa_z \sim |k_i|$, and so we can say that the order of magnitude of $1/\kappa_z$ is about a wavelength of the light in the glass.

In fact the z component of the transmitted wave has the characteristics of an evanescent wave: the vibrating electric field has the same phase at all depths, and its amplitude decreases exponentially with increasing depth. Hence no energy can be transported across the boundary into the vacuum, and we have *total internal reflection*.

Total internal reflection is put to practical use in optical instruments such as prism binoculars and reflex cameras, where reflection without loss of intensity is an advantage, and for 'piping' light along curved rods or bundles of transparent fibres. It is a cut-off effect, but one involving a cut-off angle rather than a cut-off frequency. The cut-off affects light of *all* frequencies, whenever θ_i exceeds the critical value θ_c given by

$$n \sin \theta_c = 1 \tag{17.9}$$

Wherever we have an evanescent wave we have the possibility of tunnelling, and this example is no exception. If the vacuum extends for only a wavelength or so in the z direction, being terminated by a second glass boundary, some light will be transmitted into the second dielectric. As usual we would have to apply two sets of boundary conditions to get the exact solution, which must restore a real z component to the wavevector in the vacuum.

Summary. In a sinusoidal travelling plane wave, the instantaneous displacement at any point r is

$$\psi(r, t) = A \cos (\omega t - k \cdot r)$$

where k is the wavevector: a vector of length $2\pi/\lambda$ at right angles to the wavefronts, pointing in the propagation direction.

When any such wave meets a plane boundary between two media, the directions of the reflected and refracted waves are fixed by (17.4), but their amplitudes and relative phases depend on the physics of the particular waves involved.

Total internal reflection occurs when a wave travelling in one medium is incident on a boundary with a less dense medium at an angle greater than the critical value given by (17.9).

17.2. Standing waves in an enclosure

If a sinusoidal travelling plane wave falls normally on a plane, totally reflecting surface, the incident and reflected waves will combine to produce a *standing plane wave*

$$\psi(r, t) = A \sin (k \cdot r) \cos (\omega t + \phi) \qquad (17.10)$$

This is merely the plane-wave equivalent of the standing wave produced by reflection at one end of a long string (9.24). The form of the space factor $\sin (k \cdot r)$ assumes that we have placed the origin in the reflecting plane, and that the reflection is accompanied by a phase change of 180°. The standing wave has stationary wavefronts. The wavevector k is still perpendicular to these wavefronts, but whether it points one way or the other is now of no significance.

In a standing wave formed between *two* reflecting surfaces, parallel to each other and to the wavefronts, the magnitude of k will be quantized just as $|k|$ was quantized in the case of a string fixed at both ends (9.26). The standing wave may now be written

$$\psi_n(r, t) = A_n u_n(r) \cos (\omega_n t + \phi_n)$$

in analogy with (9.30). The factors of interest are the eigenfunctions

$$u_n(r) = \sin (k_n \cdot r)$$

It is not difficult to extend these ideas to problems involving plane standing waves inside a region of three-dimensional space which is completely surrounded by plane, totally reflecting surfaces. We shall refer to such a box of waves as an *enclosure*. A specific example would be a concert hall in which a number of acoustic standing waves have been excited.

We shall discuss a rectangular enclosure, one of whose corners can be chosen as the origin of (x, y, z) coordinates lined up along three of the edges (fig. 17.5). For any standing wave in this enclosure, all three components of its wavevector k must be zero at any boundary. Following the argument which led to (9.26) for the string, we obtain†

$$k_x = l\pi/a$$

$$k_y = m\pi/b \qquad (17.11)$$

$$k_z = n\pi/c$$

---

† We shall not overburden the notation by writing k_{xl}, k_{ym} and k_{zn} in strict analogy with the previous k_n.

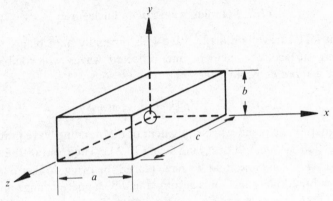

Fig. 17.5 A rectangular enclosure.

where a, b and c denote the dimensions of the enclosure in the x, y and z directions.

A standing plane wave with wavevector components (k_x, k_y, k_z) can be written

$$\psi(x, y, z, t) = A \cos(k_x x + k_y y + k_z z + \alpha) \cos(\omega t + \phi)$$

The value of the spatial phase constant α cannot be made equal to $\frac{1}{2}\pi$ as it was in (17.10), since that would necessitate a fresh choice of origin which is not open to us.

If we now use standard identities involving sines and cosines of sums of angles, and apply the conditions (17.11) which forbid terms containing $\cos k_x x$ and similar factors, we find†

$$\psi_{lmn}(x, y, z, t) = A \sin k_x x \sin k_y y \sin k_z z \cos(\omega t + \phi)$$

Each of the eigenfunctions

$$u_{lmn}(x, y, z) = \sin k_x x \sin k_y y \sin k_z z \tag{17.12}$$

carries three identifying labels l, m and n, instead of the single label n that sufficed in one dimension. These labels need not all have the same value, but can be chosen quite independently of each other. It should be noted that setting any one of them equal to zero causes the eigenfunction, and hence ψ, to vanish.

The wavevectors of the various standing waves have magnitudes given by

$$k^2 = k_x^2 + k_y^2 + k_z^2$$
$$= \pi^2[(l/a)^2 + (m/b)^2 + (n/c)^2] \tag{17.13}$$

† We have suppressed an uninteresting factor of -1.

These do not lie in a simple sequence such as we found in one dimension. It is even possible for two or more standing waves to have the same value of k, and hence the same frequency and the same energy.

We suppose the enclosure to have some degree of symmetry: we can, by way of illustration, choose one with $b = 2a$. Then (17.13) becomes

$$k^2 = \pi^2[(l/a)^2 + (m/2a)^2 + (n/c)^2]$$

It can be seen that, for all values of n, the eigenfunctions $u_{22n}(x, y, z)$ now have the same value of k^2 as the eigenfunctions $u_{14n}(x, y, z)$. They correspond to quite different standing waves, however: their wavefronts lie in different directions. Mathematically their distinctness is demonstrated by the fact that their eigenfunctions (17.12) are different. We could find many other such examples.

Two or more distinct standing waves with a common frequency are said to be *degenerate*. A stretched string has no degenerate standing waves. In an enclosure, the degeneracy increases as the enclosure becomes more symmetric. In the present example it is greatest for a cubic enclosure $(a = b = c)$.

Counting the standing waves. A box of standing waves (or modes) may not sound like a useful concept, but in fact such a picture is basic to our understanding of several phenomena which played crucial parts in the development of modern physics. In these problems the important quantity is the number of standing waves which have frequencies within a specified range.

We consider an enclosure of cubic shape $(a = b = c)$ and ask: how many of the standing waves which can occur in this box have wavevectors whose magnitudes do not exceed some given value k? The condition which must be satisfied for the occurrence of a mode (17.11) is, for a cubic box of side a, that the length of each component of the wavevector must be an integral number of units π/a. We can depict the modes by plotting the points (k_x, k_y, k_z) in three dimensions. From what we have said above about the allowed values of k_x, k_y and k_z, we deduce that the points representing the modes form a cubic lattice (fig. 17.6). The number of modes with wavevectors whose magnitudes do not exceed k is simply the number of lattice points contained by a spherical surface of radius k.

Unless we are specifically interested in those few modes whose wavelengths are comparable with the enclosure dimensions, k will be much larger than the lattice spacing π/a. Then the sphere will enclose a very large number of lattice points, and we can count them by a method which ignores the graininess of the lattice.

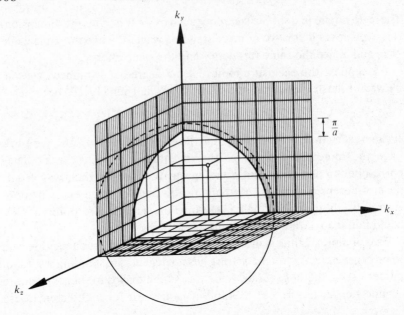

Fig. 17.6 Counting the modes in a cubic enclosure of side a. The points (k_x, k_y, k_z) representing the modes form a cubic lattice of characteristic spacing π/a. The point $(2, 3, 1)$ representing a mode of wavelength $0.53a$, is shown for illustration. The number of modes whose wavevectors have magnitudes not exceeding k is equal to the number of lattice points within a sphere of radius k. In the example shown there are approximately 50 modes; when k is much larger than the lattice spacing π/a the graininess of the lattice can be ignored and the number of modes calculated from the volume of the sphere.

We recognize that each lattice point takes up a 'volume' $(\pi/a)^3$. Then, since the 'volume' of the sphere of radius k is $4\pi k^3/3$, and since lattice points representing modes fall only within the octant for which k_x, k_y and k_z are all positive, we can say that the required number of modes is

$$N(k) \equiv \frac{1}{8} \times \frac{4\pi k^3}{3} \times \left(\frac{a}{\pi}\right)^3$$

$$= \left(\frac{a^3}{6\pi^2}\right)k^3$$

If we denote the volume of the enclosure a^3 by V, this result can be written

$$N(k) = \left(\frac{V}{6\pi^2}\right)k^3 \tag{17.14}$$

It is possible to generalize the method to deal with enclosures of any shape; it is found that the shape does not in fact affect the result (17.14), as long as k corresponds to a wavelength which is much smaller than the enclosure dimensions.

Having counted the modes with all sizes of wavevector up to k, we can find the number in the range from k to $k + dk$ by differentiating (17.14). If that number is denoted by $g(k) \, dk$ we have

$$g(k) \equiv \frac{dN}{dk} = \left(\frac{V}{2\pi^2}\right) k^2 \tag{17.15}$$

We shall call $g(k)$ the *density of modes function* (fig. 17.7). The term refers, of course, to the density with which the modes are packed per unit range of k, and not to any density in space. Another way of describing the density of modes function is to say that the average increase in k when we go from one mode to the next is $1/g(k)$.

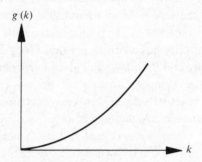

Fig. 17.7 Density of modes function for a cubic enclosure increases as k^2.

In real physics problems the interest usually lies in the way in which the various modes contribute to the energy of a system. To be able to discuss such questions we shall first have to convert (17.14) or (17.15) into a corresponding function of the frequency, and to do that we need to know what dispersion relation is obeyed. We shall consider two different examples.

Heat capacity of a solid. Much of our knowledge of how the detailed structure of a substance enables it to store energy has been obtained from measurements of heat capacities, and in particular their dependence on temperature. In a solid, by far the largest contribution comes from vibrations of the lattice into which the atoms (or molecules, or ions) of the substance are bound together, as if by interconnecting springs.

The total number of modes available for energy storage by the lattice is $3N$, where N is the number of atoms in the piece of solid. (The number is $3N$ rather than N because we must count both longitudinal vibrations and transverse vibrations, and the latter can occur in two independent directions.)

The theorem of the equipartition of energy states that the mean energy of a system at absolute temperature T is $\frac{1}{2}k_B T$ per degree of freedom, where k_B is the Boltzmann constant. By a degree of freedom we mean a quadratic term in the expression for the total energy. Each mode of the solid lattice has two degrees of freedom (its kinetic energy and its potential energy) and so the lattice as a whole has $6N$ degrees of freedom, giving a total energy W of $3Nk_B T$.

The heat capacity is defined as

$$C_V \equiv \frac{\partial W}{\partial T}$$

We assume that the volume is kept constant so that we do not alter the lattice spacings (by thermal expansion) and thus the inter-atomic forces which help to determine the mode frequencies. With a total energy of $3Nk_B T$, the heat capacity has a constant value $3Nk_B$. This prediction is known as the Dulong and Petit law, and gives fairly good agreement for most solids at ordinary temperatures.

The result does not actually depend on a wave picture, since all we have assumed is that there are $3N$ degrees of freedom, each storing energy $\frac{1}{2}k_B T$. Even if the atoms were considered as separate entities, each vibrating independently about its own lattice site, the result would be the same since each vibrating atom has six degrees of freedom: kinetic and potential energies in three independent directions. At very low temperatures, however, the standing-waves picture and the independent-vibrators picture lead to quite different results.

An essential ingredient in the explanation of the low-temperature behaviour is quantum theory. If equipartition of energy occurred under all conditions, every mode (or vibrator) would contribute equally to the heat capacity, whatever the temperature; the value $3Nk_B$ would then be obtained at all temperatures. The conflicting observation that C_V tends to vanish altogether as the temperature approaches absolute zero was one of the awkward facts which had no satisfactory classical explanation. Einstein (using an independent-vibrators picture) injected the essential idea that a vibrator of angular frequency ω cannot take up just any amount of extra energy, but must accept it in small but finite quanta $\hbar\omega$, and showed how this explained the disappearance of the heat capacity at $T = 0$. To obtain the correct temperature dependence of C_V in the low-temperature region, however, it is necessary to combine the quantum concept with a standing-waves picture, and this was done by Debye.

The essential feature of an energy quantum is that it is *proportional to the frequency*: the high-frequency modes can increase their energy only in relatively large steps. The probability that a mode of frequency ω is excited at all is given by the Boltzmann factor $\exp(-\hbar\omega/k_B T)$, and so modes with $\hbar\omega \gg k_B T$ will make a negligible contribution to energy storage. Now we can begin to understand the significance of $N(\omega)$: as the temperature is raised above absolute zero, the number of 'useful' modes (those with $\hbar\omega \lesssim k_B T$) will increase, and the temperature variation of the heat capacity will depend strongly on the manner in which new modes become available.

The complete derivation of C_V as a function of temperature is quite difficult, but the low-temperature behaviour can be understood in a simplified way. If we say that all modes with angular frequencies between zero and $k_B T/\hbar$ are excited, while modes of higher frequencies store no energy at all, then the number of contributing modes is

$$N(T) \approx \left(\frac{V}{6\pi^2 v_\phi^3}\right)\left(\frac{k_B}{\hbar}\right)^3 T^3$$

To obtain this expression we have used (17.14) for $N(k)$ and written

$$\omega \approx v_\phi k$$

which neglects dispersion. (We do not expect the important long-wavelength modes to be appreciably affected by the lumpiness of the crystal: compare section 12.2.)

If we further suppose that each active mode stores, on average, about $\tfrac{1}{2}k_B T$, then the total vibrational energy of the lattice will be

$$W \approx \tfrac{1}{2}k_B T \times N(T) \approx \left(\frac{V k_B^4}{12\pi^2 \hbar^3 v_\phi^3}\right) T^4$$

Thus the heat capacity is

$$C_V \equiv \frac{\partial W}{\partial T} \approx \left(\frac{V k_B^4}{3\pi^2 \hbar^3 v_\phi^3}\right) T^3 \tag{17.16}$$

The interesting feature of (17.16) is the cubic dependence on the absolute temperature, which is a consequence of the form of $N(\omega)$ in a non-dispersive enclosure. Einstein's independent-vibrators model gave slight but real differences from this behaviour. Figure 17.8 shows the experimental data for argon, which forms a particularly simple solid whose heat capacity can be attributed entirely to lattice vibrations.

Conduction electrons in a metal. In wave mechanics, a gas of weakly interacting particles in a container is viewed as an assembly of standing de Broglie waves in an enclosure. If we take the potential energy inside the

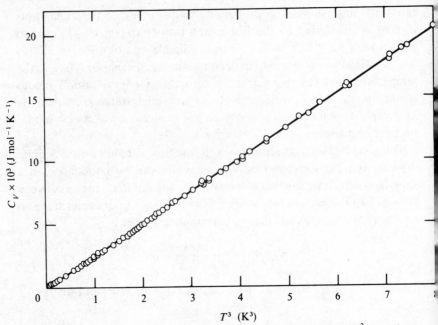

Fig. 17.8 Heat capacity of solid argon at temperatures below 2 K, showing T^3 dependence. (Based on data of L. Finegold and N. E. Phillips, *Phys. Rev.* **177** (1969) 1383.)

enclosure to be zero, the appropriate dispersion relation (15.3) is

$$\omega = \left(\frac{\hbar}{2m}\right) k^2$$

Substituting this in (17.14) gives

$$N(\omega) = \left(\frac{V}{6\pi^2}\right)\left(\frac{2m}{\hbar}\right)^{3/2} \omega^{3/2} \qquad (17.17)$$

and so the density of modes function in terms of angular frequency is given by

$$g(\omega)\, d\omega = \frac{3}{2}\left(\frac{V}{6\pi^2}\right)\left(\frac{2m}{\hbar}\right)^{3/2} \omega^{1/2}\, d\omega$$

Since the particle energy is proportional to the frequency (15.2) we can write

$$g(W)\, dW = \frac{3}{2}\left(\frac{V}{6\pi^2}\right)\left(\frac{2m}{\hbar^2}\right)^{3/2} W^{1/2}\, dW$$

The function $g(W)$, usually called the density of states function for the system, helps to determine the distribution of energy amongst the particles. It is not the only function involved, however. It tells us how the *available* standing waves are distributed in energy, but we must also ask: how many particles are actually represented by each given standing wave?

There is one important application in which this question has a particularly simple answer. We have seen that some properties of metals can be understood by regarding the conduction electrons as a gas of electrons moving freely through the positive-ion lattice (sections 2.3 and 14.3). In wave mechanics we can view the electrons as standing de Broglie waves in the enclosure formed by the boundaries of the piece of metal.

An electron, in common with other particles called fermions, must obey a rule of wave mechanics known as the Pauli exclusion principle. A consequence of this rule is that not more than two electrons can be represented by the same standing de Broglie wave.† For most metals at ordinary laboratory temperatures, the electrons effectively select the standing waves that make the total energy of the system as small as possible, while satisfying the exclusion principle. The result is two electrons per mode, starting with the mode of lowest frequency (kinetic energy) and using up the higher-frequency modes in succession until all the conduction electrons have been allocated.

The number of electrons $N_e(W)$ with energies in the range W to $W + dW$ is thus

$$N_e(W) \, dW = 2\left(\frac{V}{4\pi^2}\right)\left(\frac{2m_e}{\hbar}\right)^{3/2} W^{1/2} \, dW \quad (W \leqslant W_F)$$

$$N_e(W) \, dW = 0 \quad (W > W_F)$$

where W_F is the energy at which we run out of electrons (fig. 17.9). This quantity is known as the Fermi energy, and its value obviously depends on the number of conduction electrons present (problem 17.11).

An experimental test of this simple picture is possible with the aid of the photoelectric effect. The metal is irradiated by x-rays of a single wavelength, whose energy quanta are therefore all of the same size. The kinetic energies of the ejected photoelectrons are measured. Electrons whose kinetic energy in the metal is low will require more energy to remove them from the metal, and will thus emerge with less spare kinetic energy than those whose kinetic energy in the metal is high. The kinetic energy distribution of the photoelectrons should therefore mirror the energy distribution within the metal.

---

† The exclusion principle states that at most one electron can occupy a given 'state'. A standing de Broglie wave can be used to describe the motion in space of electrons with two different 'spin' directions, and these are counted as separate states.

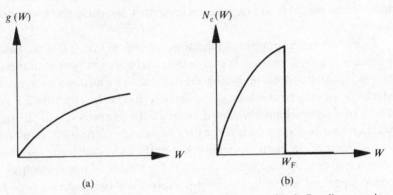

(a)   (b)

Fig. 17.9 (a) The density of modes as a function of energy, for de Broglie waves in an enclosure. (b) The electron energy distribution is twice the density of modes function up to energy W_F, and zero at higher energies.

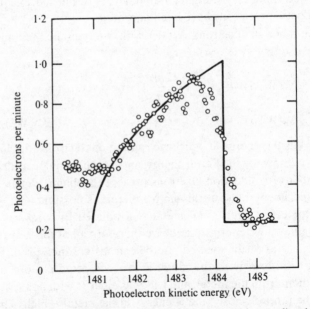

Fig. 17.10 Kinetic energy distribution of photoelectrons ejected from sodium by x-rays of energy 1486.6 eV. The smooth curve has the form shown for $N_e(W)$ in fig. 17.9. The finite energy resolution of the electron spectrometer causes the sharp drop in count rate at the Fermi energy to appear as a finite slope. These photoelectrons have a kinetic energy which is 2.4 eV smaller than the x-ray energy: this amount of energy (the *work function* of sodium) is required to remove an electron from the metal to rest at infinity. (After P. H. Citrin, *Phys. Rev. B* **8** (1973) 5545.)

Figure 17.10 shows the results of such an experiment with sodium. It can be seen that the shape of the spectrum is broadly similar to the distribution of fig. 17.9(b).

Summary. The number of standing waves in an enclosure with wavevectors in the range k to $k + dk$ is proportional to the volume of the enclosure and to $k^2 dk$. If we know the dispersion relation we can calculate the number of standing waves in a given frequency range.

Problems

17.1 (a) Calculate the critical angle of incidence for total internal reflection in glass with a refractive index of 1.50. (Assume that the second medium is a vacuum.)

(b) Estimate, in vacuum wavelengths, the thickness ($\sim 1/\kappa_z$) of a vacuum layer which would permit appreciable tunnelling, for light incident at 42.0° and for light incident at 45.0°.

17.2 Shallow-water waves with straight wavefronts, travelling in water of depth 1 m, come to a step where the depth changes to 2 m. (a) If the incident waves meet the step at an angle of 30°, calculate the direction of the waves beyond the step. (b) Calculate the critical angle of incidence for total reflection at the step.

17.3 Electromagnetic plane waves are incident at angle θ_i on the plane surface of a plasma. (a) Show that total reflection will occur at wave frequencies lower than $\nu_p/\cos\theta_i$ where ν_p is the plasma frequency. Assume $\gamma \ll \omega_p$. (This result generalizes our conclusions in section 14.3 for normal incidence.) (b) Angles of incidence larger than about 75° cannot be used for radio propagation involving ionospheric reflection, because the reflected wave will not return to earth. Estimate the maximum wave frequency that could usefully be reflected from the F_2 layer ($\nu_p \approx 10\,\mathrm{MHz}$).

17.4 An electromagnetic wave travels in a long copper pipe (a waveguide) of rectangular cross-section. The electric field in the wave is transverse, and oscillates at right angles to two of the sides. The width of the pipe in the direction perpendicular to the electric field is a.

Because the walls are conducting, the tangential component of the electric field must be zero at any wall; thus in the direction perpendicular to the electric field (taken as the x direction) we have

$$k_x = l\pi/a \qquad (l = 1, 2, 3, \ldots)$$

(a) If the z axis lies along the pipe, find the z part of the dispersion relation for the wave with $l = 1$, on the assumption that the waveguide contains a vacuum. Show that low-frequency cut-off occurs, with $\omega_c = \pi c/a$. (b) Calculate the cut-off frequency for a waveguide with $a = 100\,\mathrm{mm}$.

17.5 Write down the values of k for all the modes of a cubic enclosure with l, m and n up to 3. (Give the answers in units of k_{111}, the wavevector for the lowest mode.)

17.6   In the earliest working laser, a burst of fluorescent light of mean wavelength 694.3 nm was produced within a cylindrical piece of ruby. The end faces of the cylinder were parallel to each other and perpendicular to the axis, and were polished flat and silvered. One or more of the longitudinal modes of the resulting enclosure could be used to select and amplify light of certain frequencies. Estimate the number of modes available for this purpose in a cylinder 20 mm long, if the light produced by the atomic process in the ruby has a wavelength spread of 0.1 nm.

17.7   Find the frequencies of notes which would be particularly rewarding to an amateur bass singer in a shower enclosure 1.0 m $\times$ 1.0 m $\times$ 2.0 m. (Assume his range is from 100 Hz to 300 Hz, and take the speed of sound in warm, moist air as 350 m s^{-1}.)

17.8   A steady note of frequency 440 Hz is played by loudspeaker in a room of interior dimensions 20.0 m $\times$ 10.0 m $\times$ 5.0 m, and the acoustic pressure amplitude at one position in the room is measured continuously. When the sound is abruptly stopped, the acoustic pressure takes 0.39 s to fall to $1/e$ of its previous value. The speed of sound in the room is 340 m s^{-1}.

Estimate the number of acoustic standing waves excited while the note was being played. (Assume that a mode is appreciably excited if its frequency lies within half a width of the driving frequency.)

17.9   The coefficient of T^3 in the expression for the lattice heat capacity at low temperatures (17.16) can be evaluated if we know v_ϕ, the speed of sound in the solid. To be more accurate we should use separate values v_l and v_t for the longitudinal and transverse modes. Show that the heat capacity is then given by

$$C_V \approx \left(\frac{Vk_B^4}{9\pi^2\hbar^3}\right)\left(\frac{1}{v_l^3}+\frac{2}{v_t^3}\right)T^3$$

17.10   Show that the Fermi energy for the conduction electrons in a metal is

$$W_F = \frac{h^2}{8m_e}\left(\frac{3N}{\pi V}\right)^{2/3}$$

where the symbols have the same meanings as in the text.

Calculate W_F in eV for copper. (Use the data given in section 2.3.)

17.11   Show that the average kinetic energy of the conduction electrons in a metal is $\frac{3}{5}W_F$, where W_F is the Fermi energy. Use only the fact that $g(W)$ is proportional to $W^{1/2}$.

18

Diffraction

The boundaries between different wave-carrying media discussed in the previous chapter were assumed to be extensive. Reflected and refracted waves generated at such boundaries have a family resemblance to the incident wave giving rise to them; plane waves arriving at plane boundaries, for example, result in plane reflected and refracted waves.

If the second medium occupies only a small pocket of space embedded within the first medium, we think of it less as a medium and more as an object obstructing the progress of the wave through the medium proper. A new type of behaviour, known as *diffraction*, is apparent in such cases. Particularly striking effects are obtained when large numbers of identical diffracting objects are arranged in some regular way. A familiar example is provided by a street lamp as seen through a fog of water particles: the halo round the lamp would not be there if the lamp were viewed through a continuous mass of water.

The most characteristic feature of diffraction is that incident waves appear to 'bend round' objects made of materials which, in the form of extended media, are quite opaque. In a similar way, waves passing through a transparent hole in an opaque object are 'spread out'. We shall refer to an obstacle in a transparent medium, or an aperture in an opaque object, as a *diffraction centre*.

We shall find that these effects are appreciable when the diffraction centres have dimensions which are comparable with the wavelength of the waves involved. Diffraction thus sets a limit to the precision with which light or other waves can be used to record fine detail in a physical object under examination. It also means that attempts to create an infinitely narrow *beam* of plane waves by blocking off parts of an extended plane wave are doomed to frustration; for beam widths of a few wavelengths or so, this method will actually make the width greater.

Diffraction had a historical role in providing incontrovertible evidence of the wave nature first of light and later of electrons and other material particles. It also has important practical applications, since by its means lengths of objects can be very precisely measured in terms of the wavelength of the diffracted waves: visible light, x-rays, or the de Broglie waves of electrons, for example.

Although diffraction produces new phenomena, no new physical processes are involved. All observed effects are due to forced vibration of parts of the diffraction centres at the wave frequency, the production of new waves by these shaken parts, and the superposition of the new waves and the incident wave to produce 'the diffracted wave'. But it is easy to see that the mathematical complexities are greater than in similar cases we have studied previously: in principle it is necessary to apply boundary conditions at the surfaces of all the diffraction centres, which may be irregular in shape.

All kinds of waves show diffraction, but to make the discussion concrete we shall conduct it mostly in terms of light waves.

18.1. Features due to the arrangement of the diffraction centres

The main principles can be illustrated by means of a simple example in which there are just two identical diffraction centres. Plane waves are incident normally on a screen with two apertures which take the form of long narrow slits, parallel to each other and perpendicular to the plane of the diagram (fig. 18.1). With such a system we can restrict the discussion to two dimensions. Apart from the slits, the screen is opaque (either totally reflecting or totally absorbing).

Points beyond the screen will receive waves originating in the general regions of the two illuminated slits. We do not know how the slits radiate. If they are very narrow, we might expect them to generate cylindrical waves such as a glowing filament would produce; but in general the waves will be more complicated than this.

In section 18.2 we shall discuss the principles by which the nature of the individual diffracted waves can be worked out. At present we shall merely assume that both slits send light forward in all directions, though not necessarily uniformly. In some of these directions the two waves will superpose constructively, and in others destructively; and *we can deduce these special directions without detailed knowledge about the forms of the waves coming from the two slits.*

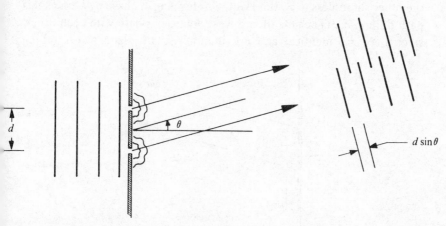

Fig. 18.1 Two-slit diffraction. Plane waves are incident from the left on a screen in which there are two identical long slits, perpendicular to the page. The slits give rise to identical waves; their form is unknown, but at large distances beyond the screen the diffracted wavefronts can be taken as plane. Disturbances are out of phase owing to the differing path lengths from the two slits.

In talking about the superposition of waves going in a certâin *direction* we are presupposing that the light is being observed very far downstream of the diffracting screen: many times the slit separation distance d, in fact. The conditions we have assumed (incident plane waves and observation in given directions) are met in most practical situations in which diffraction is observed; they are referred to as *Fraunhofer conditions*. If the incident wavefronts are appreciably curved, or if we observe the diffracted waves at points whose distance beyond the screen is comparable with d, we have *Fresnel conditions*. The physics is the same in both cases, of course, but its description is easier in the limiting Fraunhofer case.

We denote the direction of observation by θ as indicated in fig. 18.1; the symmetry of the set-up means that we need to consider positive values of θ only. Since we are observing a long way from the screen, we may take the two sets of wavefronts to be plane over the range of angles involved, which will usually be fairly small in practice. For the same reason we may take the *amplitudes* of the superposed disturbances as equal. Their values will in general depend on θ. We do not know the directional characteristics of the waves from a centre; at present we shall denote the amplitude of the disturbance from each centre by $A_c(\theta)$.

Small differences in the distances from the two slits must not be neglected when we are considering the *phase angles* of the two disturbances. The waves from one slit have travelled farther than those from the other slit (unless $\theta = 0$) and will therefore lag in phase by k times this path difference. (The slits themselves radiate in phase with each other, since they are identical and are illuminated by plane waves falling

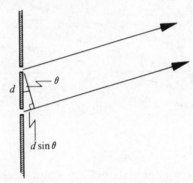

Fig. 18.2 The path difference in fig. 18.1 is $d \sin \theta$. The corresponding phase difference is $kd \sin \theta$ where k is the magnitude of the wavevector.

normally on the screen.) Simple geometry (fig. 18.2) gives the phase difference as

$$\alpha(\theta) = kd \sin \theta \qquad (18.1)$$

We can now calculate the amplitude $A(\theta)$ of the combined wave with the aid of the usual type of vector diagram (fig. 18.3). We find

$$A(\theta) = 2A_c(\theta)|\cos[\tfrac{1}{2}\alpha(\theta)]| \qquad (18.2)$$

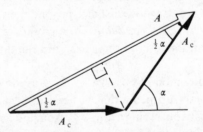

Fig. 18.3 Vector diagram for calculating the amplitude A, when disturbances of equal amplitude A_c and phase difference α are superposed. Both A_c and α vary with θ, the direction of observation indicated in fig. 18.1.

It can be seen that $A(\theta)$ is a product of two expressions: we may write

$$A(\theta) = A_c(\theta)\Phi(\theta)$$

where $A_c(\theta)$ depends on the diffracting properties of an individual slit, whereas $\Phi(\theta)$ depends only on the phase difference resulting from the different distances from the two slits. In this particular case we have

$$\Phi(\theta) = 2|\cos\left[\tfrac{1}{2}\alpha(\theta)\right]| = 2|\cos\left(\tfrac{1}{2}kd\sin\theta\right)| \qquad (18.3)$$

The intensity of the waves observed in direction θ will be proportional to

$$A^2(\theta) = A_c^2(\theta)\Phi^2(\theta)$$

We can expect that the amplitude of the wave radiated by a slit will vary relatively gently with θ, so that the intensity pattern will be dominated by the behaviour of $\Phi^2(\theta)$. For the two-slit system we have

$$\Phi^2(\theta) = 4\cos^2\left[\tfrac{1}{2}\alpha(\theta)\right] = 4\cos^2\left(\tfrac{1}{2}kd\sin\theta\right)$$

which has maxima in directions given by

$$\sin\theta = 2n\pi/kd = n\lambda/d \qquad (n = 0, 1, 2, \ldots)$$

If θ is small enough that we can write $\sin\theta \approx \theta$, the angular separation of the maxima is approximately uniform, and equal to λ/d. By measuring this angular separation (or the separation of the intervening directions in which the intensity is zero) we can find λ in terms of d, or *vice-versa*. *This does not require a knowledge of* $A_c(\theta)$.

Multiple-slit diffraction. The extension from 2 to N regularly spaced identical slits (fig. 18.4) is straightforward. The results form the theoretical basis of the spectroscopic device known as the *diffraction grating*.

The vector diagram of fig. 18.3 is replaced by the one in fig. 18.5. With the aid of the geometrical construction shown in the figure, we find

$$\Phi(\theta) \equiv \frac{A(\theta)}{A_c(\theta)} = \left|\frac{\sin\left[\tfrac{1}{2}N\alpha(\theta)\right]}{\sin\left[\tfrac{1}{2}\alpha(\theta)\right]}\right| \qquad (18.4)$$

where $\alpha(\theta)$ is the phase difference between the disturbances from two adjacent slits (18.1). It is easy to confirm that (18.4) embraces the particular case (18.3) of $N = 2$.

Because N is an integer, $\Phi^2(\theta)$ will start to repeat itself every time $\tfrac{1}{2}\alpha(\theta)$ increases by π, or $\sin\theta$ increases by λ/d. Whenever we have

$$\alpha(\theta) = 2n\pi \qquad (n = 0, 1, 2, \ldots) \qquad (18.5)$$

the numerator and the denominator on the right of (18.4) are both zero. As these values of $\alpha(\theta)$ are approached, however, we know that $\sin\left[\tfrac{1}{2}N\alpha(\theta)\right]$ and $\sin\left[\tfrac{1}{2}\alpha(\theta)\right]$ can be replaced by $\tfrac{1}{2}N\alpha(\theta)$ and $\tfrac{1}{2}\alpha(\theta)$ respectively. Thus, at angles satisfying (18.5), we have

$$\Phi(\theta) = N$$

$$A(\theta) = NA_c(\theta)$$

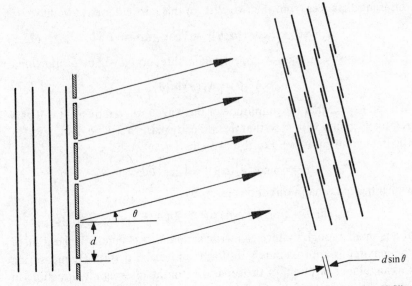

Fig. 18.4 Multiple-slit diffraction: as fig. 18.1, but the number of slits is increased to *N*. Slits are identical and uniformly spaced. Waves from adjacent slits have phase difference $kd \sin \theta$.

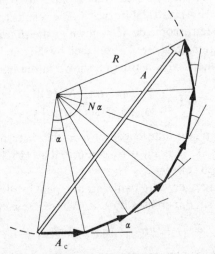

Fig. 18.5 Vector diagram for calculating the amplitude *A* in the *N*-slit case. Each component disturbance has amplitude A_c and phase difference α relative to the next. Each component vector subtends an angle α at the centre of the circle drawn through the ends. The radius *R* is given by

$$\tfrac{1}{2}A = R \sin \tfrac{1}{2}N\alpha$$

$$\tfrac{1}{2}A_c = R \sin \tfrac{1}{2}\alpha$$

Eliminating *R* gives *A* in terms of A_c and α.

In these directions lie the *principal maxima* of intensity, characterized by complete constructive superposition of the contributing disturbances. The angular separation of the principal maxima is the same as the separation of the maxima for $N = 2$ (about λ/d if θ is small). The integer n in (18.5) is known as the *order*. The zero-order principal maximum is formed by waves going in the forward direction ($\theta = 0$), and higher orders occur at larger angles.

The numerator $\sin\left[\frac{1}{2}N\alpha(\theta)\right]$ also goes through zero when $\frac{1}{2}N\alpha(\theta)$ is an integral multiple of π; this will happen $N-1$ times between adjacent principal maxima, giving rise to directions of zero intensity which are not present when $N = 2$ (fig. 18.6). Between these directions there must be

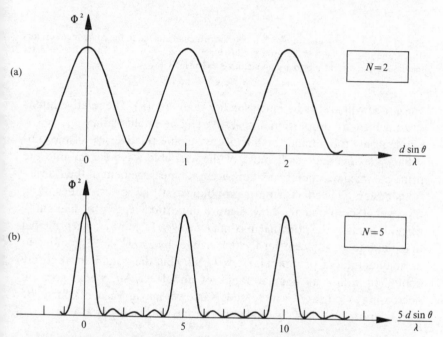

Fig. 18.6 Plots of Φ^2 against α for (a) two slits, and (b) five slits. The intensity of the light is proportional to $A_c^2\Phi^2$; the graphs approximately represent the intensity variation in cases where A_c varies slowly with θ. (The vertical scales are adjusted to show the *relative* variation in the two cases.)

subsidiary maxima, but these are small compared with the principal maxima. This can best be understood by sketching vector diagrams (fig. 18.7). As θ increases, making α increase, the chain of vectors tends to curl round until a closed figure is formed; when that happens we have a zero of intensity. A subsidiary maximum will always correspond to a chain of vectors which has curled through more than 360°, and the resultant

(a)

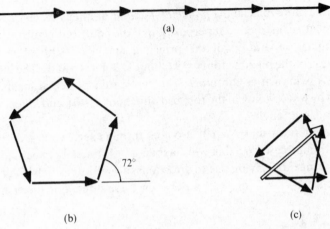

(b)   (c)

Fig. 18.7 Vector diagrams for $N = 5$. (a) At the principal maximum the component vectors are in line. (b) At the first zero the vectors form a closed figure. (c) The first subsidiary maximum; the second subsidiary maximum is even smaller.

amplitude will always be much smaller than $NA_c(\theta)$. The relative intensities, which are proportional to $A^2(\theta)$, will be smaller still.

Increasing the number of slits decreases the relative size of the subsidiary maxima and sends more of the available wave energy into the principal maxima, thereby making accurate measurement of their directions easier. This effect is most pronounced at the lowest values of N, however. (See problem 18.6.) A more important reason for making N larger is to make the principal maxima *narrower*. In going from a principal maximum to an adjacent zero, $\alpha(\theta)$ changes by $2\pi/N$, corresponding to an angular interval of approximately λ/Nd. Thus the maximum is entirely contained within an angular range of about $\pm\lambda/Nd$, which goes on decreasing as long as N is increased. Narrower maxima can, of course, be located more precisely, leading to better measurements of λ or d.

Diffraction by crystals. Most of our knowledge of the structure of crystalline solids has been obtained from experiments involving diffraction. The waves most commonly used are x-rays, whose wavelengths are comparable with the distances between atoms in crystals. Diffraction of neutron de Broglie waves is also valuable, because neutrons interact with the magnetic moments of the diffracting atoms and so help to elucidate the magnetic structure of the solid.

A crystal consists of atoms (or ions) stacked in a regular array. In every crystal there can be found a basic atomic group (the *unit cell*) which is endlessly repeated in a geometric pattern known as the *lattice*. Only a limited number of distinct types of lattice can be formed in three dimensions, but unit cells can be constructed in countless ways.

The unit cell is the diffraction centre, corresponding to the slit in a diffraction grating. To predict the diffraction pattern for the crystal in complete detail, we need to know how the atoms are arranged (more exactly, how the scattering electrons are distributed) in the unit cell. The *directions* of the diffraction maxima are, however, fixed by the spatial arrangement of the diffraction centres, that is by the lattice, just as the directions of the principal maxima from a diffraction grating are fixed by the slit spacing d. The crystal, in fact, acts as a three-dimensional diffraction grating.

To discover the possible directions, we consider the scattering of x-rays from two unit cells (fig. 18.8). We take the position of one cell (the lower

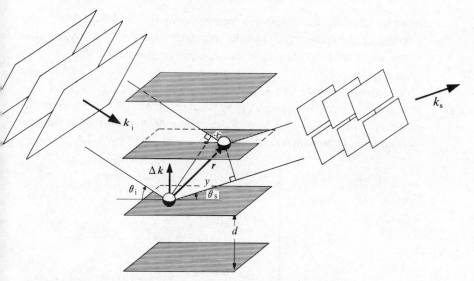

Fig. 18.8 Diffraction of x-rays by two unit cells. The lower cell in the figure is taken as the origin, and r measures the relative position of the other cell. Diffracted waves are out of phase unless the second cell lies in one of the planes perpendicular to Δk, with separation d satisfying Bragg's law. (For simplicity the cells are shown as single atoms.)

The phase difference due to path difference x is $k_i \cdot r$; that due to path difference y is $k_s \cdot r$.

In x-ray diffraction it is conventional to specify directions by giving angles (such as θ_i and θ_s) between the wave and the plane rather than those between the wave and the normal.

one in the picture) as the origin; the position of the second cell, relative to the first, is indicated by the position vector r. The incident x-rays have wavevector k_i; each cell produces scattered waves of wavevector k_s, and $|k_s| = |k_i|$. We wish to compute the phase difference between the two sets of scattered waves when they reach a distant detector.

From the geometry of the figure we can see that the incident disturbance at the upper cell *lags* that at the lower cell by an amount $k_i \cdot r$; but a
phase *lead* of $k_s \cdot r$ is introduced by the path difference between the cells
and the detector. The x-rays from the upper cell thus have an over-all
phase lead of

$$(k_s - k_i) \cdot r \equiv \Delta k \cdot r$$

where we have defined a *scattering vector* Δk.

Because the number of diffraction centres in the whole crystal will be
very large indeed, the amplitudes of the subsidiary maxima will be quite
negligible. We thus consider only the principal maxima, formed by
complete constructive superposition of the diffracted x-rays. For complete constructive superposition the condition

$$\Delta k \cdot r = 2n\pi \qquad (n = 0, \pm 1, \pm 2, \ldots) \qquad (18.6)$$

must be satisfied. For $n = 0$ this means simply that r (which defines the
position of the second cell) should be perpendicular to Δk. Thus x-rays
scattered by all cells lying in a plane perpendicular to Δk will reinforce
each other.

Moreover, since k_i and k_s have equal magnitudes, the angle of constructive scattering is equal to the angle of incidence (fig. 18.9). Thus *any
plane of cells acts as if it were a partially reflecting mirror obeying the
normal law of specular reflection* (17.4).

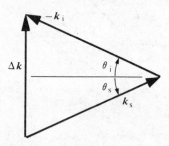

Fig. 18.9 The magnitudes of k_i and k_s are equal (k, say); thus $\theta_i = \theta_s \equiv \theta$ and $\Delta k = 2k \sin \theta$.

For non-zero values of n, equation (18.6) defines planes of cells parallel
to the one with $n = 0$. (Compare the discussion at the beginning of section
17.1.) The perpendicular distance between adjacent planes is d, given by

$$\Delta k \cdot r = \Delta k \, d = 2kd \sin \theta$$

Equation (18.6) states that *there will be constructive superposition of the
x-rays 'reflected' from these planes, provided the condition*

$$2kd \sin \theta = 2n\pi$$

is fulfilled. In terms of the x-ray wavelength λ, the condition takes the form

$$2d \sin \theta = n\lambda \qquad (18.7)$$

known as *Bragg's law*.

A crystal, by reason of its regular construction, contains many sets of parallel and equally spaced planes which are rich in diffracting cells. For each set of planes a few values of θ will be allowed by Bragg's law; these will depend on the spacing d of the particular planes involved, and on the x-ray wavelength. By measuring these angles we can work out the geometry of the lattice.

Summary. We have examined three diffracting systems, each consisting of regularly spaced, identical diffraction centres.

In the two-slit example, directions of zero and maximum intensities alternate regularly, the angular interval between adjacent maxima being approximately λ/d.

Increasing the number of slits while maintaining their spacing at the value d gives a pattern which still repeats at angular intervals ≈λ/d, but the principal intensity maxima become increasingly narrow, and increasingly strong relative to the subsidiary maxima.

A crystal contains identical building blocks known as unit cells, arranged in a periodic lattice. The unit cells act as diffraction centres for x-rays. An isolated plane of cells partially 'reflects' x-rays incident at any angle, like a mirror; but x-rays reflected from parallel planes reinforce each other to make a diffraction maximum only if the x-ray direction and wavelength, and the spacing of the planes, satisfy Bragg's law (18.7).

18.2. Features due to the nature of the diffraction centres

We have found that the amplitude of a disturbance at any point in the diffracted wave can be written as a product of two expressions: a factor $A_c(\theta)$ which depends on the diffracting characteristics of a centre (an aperture, an obstacle or a group of atoms in a crystal) and a factor $\Phi(\theta)$ which depends on how these centres are arranged in space. Finding $\Phi(\theta)$ in any particular example is a matter of superposing disturbances due to the different centres, taking account of their relative phases.†

---

† Phenomena involving overlapping waves from different sources are often classified under the heading of *interference*. An absolutely rigid distinction between diffraction and interference is not always possible, however.

In this section we turn to the problem of finding $A_c(\theta)$ for a given type of diffraction centre. At the outset it can be stated that it is quite impossible to calculate it from first principles, except in a very few rather artificial cases. Fortunately there is available an approximate method which, under certain conditions which are usually fulfilled in practice, gives results which agree fairly closely with observation.

Huygens's principle. We consider plane waves falling normally on a single plane diffraction centre: an aperture in a screen, or an opaque obstacle. We are interested in the disturbance ψ produced at some fixed observation point downstream (fig. 18.10). To find ψ from first principles, we should calculate the forced vibrations of all the moving parts (namely electrons, if we are dealing with light waves), work out what new waves

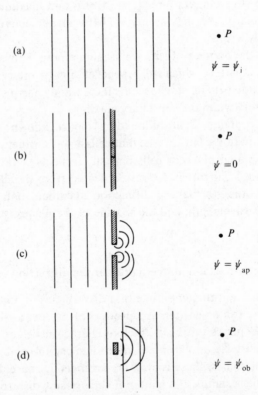

Fig. 18.10 (a) A plane wave causes a disturbance $\psi = \psi_i$ at an observation point P. (b) When P lies beyond an opaque screen the disturbance is zero. (c) With an aperture in the screen, the disturbance ψ_{ap} can be calculated approximately with the aid of Huygens's principle. (d) With an opaque diffracting obstacle, the disturbance ψ_{ob} can be found from ψ_{ap} for an aperture of the same shape and size as the obstacle.

are thereby generated, and superpose these with the incident wave. This is, as we have said, an impossible task; but the following argument recognizes that these are the physical processes involved when diffraction takes place.

First we consider the situation shown in fig. 18.10(b), in which the screen has as yet no aperture. An opaque screen is one giving $\psi = 0$; this implies that the disturbances at P cancel each other. There are two sources of disturbance at P: the incident wave and the wave radiated by the irradiated screen. If we divide the latter disturbance into two parts ψ_s and ψ_p we may write

$$\psi_i + \psi_s + \psi_p = 0 \tag{18.8}$$

By ψ_s we mean the disturbance due to waves reaching P from all parts of the screen except the plug of material which will shortly be removed to create an aperture, and ψ_p arises from that plug itself.

In fig. 18.10(c) the aperture has been opened, giving rise to a disturbance ψ_{ap} at P; this disturbance is the one we wish to calculate. Proceeding as before, we write

$$\psi_i + \psi_s' = \psi_{ap} \tag{18.9}$$

We have used ψ_s' instead of ψ_s because the electrons in the screen will not have exactly the same forced vibration as before: they now experience no driving force from the wave radiated by the plug. Our method for calculating ψ_{ap} rests, however, on the assumption that

$$\psi_s' \approx \psi_s \tag{18.10}$$

Any difference between ψ_s and ψ_s' will be due mainly to screen electrons within a few wavelengths of the aperture, and the approximation will therefore be better for large apertures than for small ones since a smaller fraction of the total disturbance will come from these electrons at the edges.

Combining (18.8), (18.9) and (18.10) gives

$$\psi_{ap} \approx -\psi_p \tag{18.11}$$

Now ψ_p was the disturbance produced by the plug *when it was in place*, and driven by a plane wave. It must then have been uniformly covered with electrons vibrating in phase with each other and with equal amplitudes. Thus we can calculate ψ_{ap} (approximately) by replacing the entire system (screen plus plane wave) by an imaginary object uniformly covered with point wave sources of equal strength, all radiating in phase.

By means of these arguments we have bypassed the complicated details of the re-radiation processes that go on in the screen. It is important to realize that it is only at points *beyond* the screen that ψ produced by the real system is (approximately) equal to ψ produced by the replacement system of imaginary wave sources spread across the aperture. If we try to work out the disturbance at a point in front of the screen in this way, we shall get the wrong answer. This cannot be otherwise, since we have merely described the screen as 'opaque'; conditions in front of the screen will obviously depend on whether opaque means reflecting or absorbing.

The idea of replacing a section of a wave spanning an aperture by a set of imaginary wave sources is known as *Huygens's principle*. Because most diffracting apertures are many wavelengths wide, and because observations are usually made at distant points, far from the edges where approximation (18.10) is least reliable, nearly all diffraction calculations use this procedure.

We can make a similar argument for the diffracting obstacle in fig. 18.10(d). If that obstacle is identical in shape to the plug in fig. 18.10(b), we can write

$$\psi_i + \psi_p' = \psi_{ob}$$

where ψ_p' is the disturbance due to radiation from the *isolated* plug, acting as an obstacle. In this case we make the approximation

$$\psi_p' \approx \psi_p$$

and find

$$\psi_{ob} \approx \psi_i - \psi_{ap}$$

which uses (18.11). Thus the diffracting effect of an obstacle can be found if we know the diffracting effect of an aperture of the same shape. Again, the result will be more accurate for large obstacles than for small ones.

Single-slit diffraction. As in section 18.1, we shall deal with apertures rather than opaque obstacles, to avoid the slight extra complication of the incident wave. We illustrate the method by finding $A_c(\theta)$ for a single uniform slit of width a (fig. 18.11).

The aperture in this example is to be replaced by a strip uniformly covered with line sources of light, radiating in phase. Disturbances at the observation point have phase differences resulting from path differences from these elementary sources, and the calculation follows the derivation of $\Phi(\theta)$ for N slits.

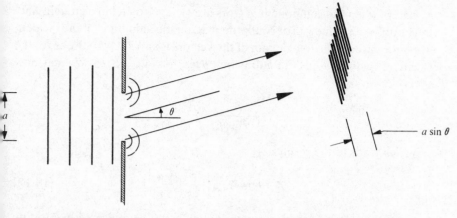

Fig. 18.11 Diffraction by a uniform slit. The waves from the imaginary sources spread over the aperture have a continuous spread of path differences spanning the range $a \sin \theta$.

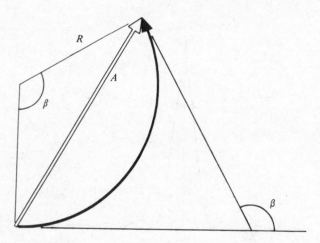

Fig. 18.12 Vector diagram for calculating the amplitude A, when the component vectors in fig. 18.5 become very small and very numerous. The phase difference between the first and last vectors is $\beta = ka \sin \theta$. (See fig. 18.11.) The length of the curved chain of vectors is A_{max}.

The component vectors in fig. 18.5 now become very short and very numerous, and so form the arc of a circle (fig. 18.12). The angle subtended at the centre by this arc is

$$\beta(\theta) = ka \sin \theta$$

This is just the phase difference between the disturbances originating from the two edges of the slit.

The maximum amplitude

$$A_{\text{max}} \equiv A_{\text{c}}(0)$$

occurs when the component vectors are set end to end in a straight line. (Its absolute value will of course depend on the amplitude of the incident plane waves.) The total length of the vectors is also equal to the arc of the circle,† whose radius (for a direction θ) is thus $A_{max}/\beta(\theta)$. We may now write

$$\frac{A_c}{|2 \sin \frac{1}{2}\beta|} = \frac{A_{max}}{|\beta|}$$

which leads to the result

$$A_c(\theta) = A_{max} \left| \frac{\sin\left[\frac{1}{2}\beta(\theta)\right]}{\frac{1}{2}\beta(\theta)} \right| \qquad (18.12)$$

The variation of intensity with direction can be indicated by plotting $(A_c/A_{max})^2$. Figure 18.13 shows that most of the wave energy is concentrated into a forward maximum at $\theta = 0$, though there are other small

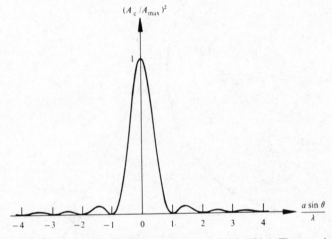

Fig. 18.13 Intensity pattern for diffraction by a uniform slit of width a. The quantity plotted is $[A_c(\theta)/A_{max}]^2$ as given by equation (18.12).

maxima farther out. The intervening zeros appear when

$$\beta(\theta) = 2\pi, 4\pi, 6\pi, \ldots$$

The first of these zeros is at

$$\theta \approx 2\pi/ka = \lambda/a$$

---

† This assumes that the imaginary sources radiate uniformly in all directions. In the mathematical derivation of Huygens's principle the amplitude is found to contain a factor $1 + \cos\theta$, whose effect we can neglect as long as θ is small. It is interesting to note that this factor also leads to a disturbance of zero when $\theta = \pi$, i.e. in front of the screen.

Thus λ/a may be taken as a measure of the angular width of the forward maximum. For a given wavelength λ, we find that the waves are 'spread out' in inverse proportion to the slit width a. The result for a two-dimensional aperture will be more complicated, but we may expect an angular spread *of the order of* λ/a with any aperture whose dimensions are $\sim a$.

Multiple-slit diffraction. We are now in a position to write down the amplitude $A(\theta)$ for diffraction by N slits of width a, spaced at intervals d. We merely combine $\Phi(\theta)$ from (18.4) with $A_c(\theta)$ from (18.12) to obtain

$$A(\theta) = A_{\max} \left| \frac{\sin\left[\frac{1}{2}N\alpha(\theta)\right]}{\sin\frac{1}{2}\alpha(\theta)} \right| \cdot \left| \frac{\sin\left[\frac{1}{2}\beta(\theta)\right]}{\frac{1}{2}\beta(\theta)} \right| \qquad (18.13)$$

The phase differences involved are, as before,

$$\alpha(\theta) = kd \sin\theta$$

$$\beta(\theta) = ka \sin\theta$$

Equation (18.13) allows for the diffracting characteristics of the particular slits involved (calculated with the aid of Huygens's principle, and therefore a good approximation if a is not too small) as well as their arrangement.

The intensity in direction θ is proportional to $[A(\theta)/A_{\max}]^2$. This quantity is plotted in fig. 18.14 for a five-slit system. We notice two important features.

The first is that *the spacing of the principal maxima is fixed by the* $\Phi(\theta)$ *factor alone.* This is because $A_c(\theta)$ varies relatively slowly with θ, as we assumed in section 18.1. In this particular case the $A_c(\theta)$ factor will always vary more slowly than the $\Phi(\theta)$ factor, because d must be larger than a.

The second feature is that *the intensities of the principal maxima depend on the $A_c(\theta)$ factor*, which modulates the more rapid fluctuations produced by $\Phi(\theta)$. The $[\Phi(\theta)]^2$ pattern is in fact formed within an envelope whose shape is just the intensity pattern for a single slit. Other slit shapes, with different diffraction patterns, would modulate the five-slit pattern in different ways, but would leave the spacing unchanged as long as the slit separation d was kept constant.

Without going into details, we can make analogous statements about diffraction by a crystal. In that case the *directions* of the diffracted intensity maxima are fixed by the geometrical arrangement of the diffraction centres (the unit cells) within the crystal. By measuring these

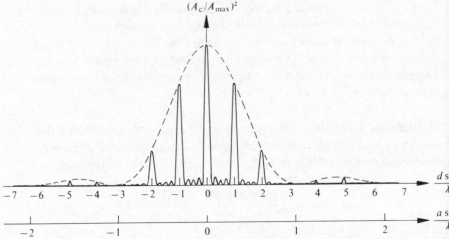

Fig. 18.14 Intensity pattern for diffraction by five uniform slits of width a, uniformly spaced at intervals 3.3a (solid curve). The broken curve represents the intensity pattern for one of the slits on its own, and has the same shape as the curve shown in fig. 18.13.

directions we can learn the basic *symmetry* of a crystal, just as we could find the spacing d for a diffraction grating by measuring the directions of the principal maxima.

The *intensities* of the diffracted beams, on the other hand, depend on the diffraction characteristics of a unit cell. By measuring the intensities we can learn the crystal *structure*. In practice this step can be a major undertaking. Finding the structure from the measured intensities is analogous to deducing the nature of the slits in a diffraction grating from the shape of the modulating factor $[A_c(\theta)]^2$; we worked the other way round, in one simple case only.

Fourier theory and diffraction. We end by indicating how diffraction patterns produced by objects of a more general kind are calculated. We

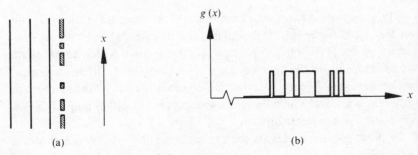

Fig. 18.15 (a) Plane waves incident on a screen with parallel slits of arbitrary width and spacing. (b) The aperture function.

still discuss plane waves incident normally on a plane screen containing long, parallel slits, but now we allow the widths and spacings of the slits to be arbitrary, and not necessarily regular (fig. 18.15).

The screen may be described by means of an *aperture function* $g(x)$, which indicates the amplitude of the imaginary wave source at a distance x along the screen perpendicular to the slits. The value of $g(x)$ will be zero where the screen is opaque; at a slit it will have some constant value which will depend on the amplitude of the incident waves.† Figure 18.15(b) shows the aperture function for the screen in fig. 18.15(a). If we take the disturbance from the strip at $x = 0$ to have zero phase constant, then the disturbance in direction θ due to wave sources in the strip of screen between x and $x + dx$ is

$$\psi(x, \theta, t)\,dx \equiv \mathrm{Re}\,\{g(x)\exp[i(\omega t - kx\sin\theta)]\}\,dx$$

To find the disturbance $\psi(\theta, t)$ produced by the entire screen, we must superpose the disturbances from all strips; this means integrating $\psi(x, \theta, t)$ over all x (which is why we have written it in form D). The answer may be expressed in the form

$$\psi(\theta, t) = \mathrm{Re}\,[D(\theta)\exp(i\omega t)]$$

where

$$D(\theta) = \int_{-\infty}^{+\infty} g(x)\exp(-ikx\sin\theta)\,dx$$

The significance of this result may be seen more clearly if we indicate the observation direction by means of the x component of the wavevector

$$k_x = k\sin\theta$$

instead of by the angle θ itself. We then have

$$D(k_x) = \int_{-\infty}^{+\infty} g(x)\exp(-ik_x x)\,dx$$

Comparing this with (11.16) we see that *the complex amplitude of the diffracted disturbance is proportional to the Fourier transform of the aperture function.*

---

† The theory can also cope with more complicated cases. The transparency of the slits may be variable, giving a $g(x)$ which takes on intermediate values. We can also envisage screens (a non-uniform glass plate, for example) with complex aperture functions which indicate phase differences between the waves emitted from different points.

The simplest possible aperture function is one describing a single uniform slit. For a slit of width a, situated between $x = -\frac{1}{2}a$ and $x = +\frac{1}{2}a$, we have

$$g(x) = A_0/a \qquad (|x| < \tfrac{1}{2}a)$$

$$g(x) = 0 \qquad (|x| > \tfrac{1}{2}a)$$

where A_0 is a constant proportional to the amplitude of the incident wave. We find immediately

$$D(k_x) = \frac{A_0}{a} \int\limits_{-a/2}^{+a/2} \exp(-ik_x x)\, dx = A_0 \left(\frac{\sin \frac{1}{2}k_x a}{\frac{1}{2}k_x a} \right) \qquad (18.14)$$

Simple substitutions show that $|D(k_x)|$ is equal to $A_c(\theta)$ as given by (18.12) and that the constant A_0 in (18.14) is just the forward amplitude A_{max}.

This method of computing diffraction patterns under Fraunhofer conditions is seen at its most effective in more complicated situations which involve two-dimensional apertures in two-dimensional arrangements. It is another indication of the power and elegance of Fourier theory, and a further reason for the central place that it occupies in the more advanced study of vibrations and waves.

Summary. The diffraction pattern produced by plane waves falling normally on a plane aperture can be calculated approximately by imagining the aperture to be covered uniformly with wave sources of equal amplitude, radiating in phase (Huygens's principle).

The angular divergence of waves diffracted by any aperture of width a is of the order of λ/a.

In the intensity pattern produced by a regular array of identical diffraction centres, the spacing of the maxima is characteristic of the array geometry, while their relative intensities depend on the detailed form of the individual centres.

Problems

18.1   Ultrasonic acoustic waves are diffracted by a pair of parallel slits 100 mm apart. The zeros in the diffraction pattern are spaced at angular intervals of 8.3°. Calculate the wavelength.

18.2   In the two-slit system of fig. 18.1, the slits can be made to radiate out of phase with each other by making the incident plane waves arrive at an angle $\theta_i \neq 0$. (a) Show that the phase difference between the two disturbances observed in direction θ is then $kd(\sin \theta_i + \sin \theta)$. (b) Show that the minimum value of θ_i which will lead to *destructive* superposition in the forward direction $\theta = 0$ is approximately $\lambda/2d$.

18.3 The two-slit system is the basis of modern versions of Young's historic experiment demonstrating the wave nature of light. The incident waves are derived from a single illuminated slit of width a, parallel to the diffracting slits and a distance L in front of them (fig. 18.16).

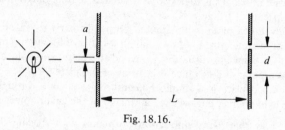

Fig. 18.16.

Use the result of problem 18.2 to show that the condition

$$a \ll L\lambda/d$$

must be satisfied if visible 'interference fringes' are to be observed.

18.4 For multiple-slit diffraction, show that the number of principal maxima cannot exceed d/λ in practice. For light of wavelength 500 nm, evaluate the number of orders obtainable with gratings of spacing (a) 0.1 mm, (b) 0.01 mm, and (c) 0.001 mm.

18.5 For the N-slit system of fig. 18.4, show that the phase difference between the total disturbance and the disturbance due to slit 1 (or slit N) is $\frac{1}{2}(N-1)kd \sin \theta$.

18.6 For N-slit diffraction, show that the subsidiary maximum adjacent to a principal maximum has an amplitude proportional to $1/\sin(3\pi/2N)$. Find the ratio of this amplitude to the amplitude of the principal maximum, for (a) $N = 5$, (b) $N = 10$, and (c) $N = 1000$. (Assume that A_c is independent of θ.)

18.7 A mixture of light of wavelengths λ and $\lambda + \Delta\lambda$ falls on an N-slit diffraction grating with spacing d. Show that the angular separation of the two principal maxima of order n is

$$\Delta\theta = \frac{n\,\Delta\lambda}{d\cos\theta}$$

For a grating with $d = 2.5\ \mu$m, evaluate $\Delta\theta$ for the sodium D lines ($\lambda = 589.0$ nm, $\lambda + \Delta\lambda = 589.6$ nm) with $n = 1$.

18.8 A radio aerial (antenna) consists of a horizontal array of uniformly spaced vertical dipoles, excited in phase. It is required to concentrate as much as possible of the radiation along a single axis (forwards and backwards) within a horizontal angular range $\pm 3°$. Find the minimum number of dipoles required, and their maximum spacing in units of the wavelength.

18.9 (a) A photographic plate is illuminated perpendicularly by plane monochromatic light of wavelength λ. A very small object placed at a perpendicular distance b from the surface of the plate acts as a scattering point. When the plate is developed it shows a series of concentric circular fringes. Show that the circles of maximum density have radii r_n given by

$$r_n^2 = n^2\lambda^2 + 2nb\lambda \qquad (n = 1, 2, 3, \ldots)$$

(b) If the set of fringes is regarded as a circular diffraction grating, show that the grating spacing at fringe n is $(n\lambda^2 + b\lambda)/r_n$.

(c) The developed plate is illuminated perpendicularly by plane monochromatic light of wavelength λ. Show that first-order diffraction produces light converging to a point at a distance b beyond the plate, and light appearing to diverge from a point at a distance b in front of the plate.

(The developed plate is a simple *hologram*: if you look through the plate, towards the light source, you will see an image of the small object.)

18.10 (a) Show that x-ray diffraction by planes with spacing d requires a wavelength which does not exceed $2d$. (This is why visible light cannot be used.)

18.11 Figure 18.17 shows a layer of Na^+ and Cl^- ions in a sodium chloride crystal. The shaded region indicates the basic repeating group (the unit cell), which has a side of 0.563 nm.

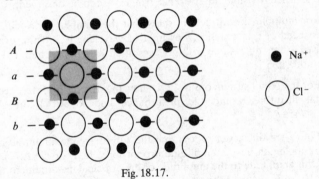

Fig. 18.17.

(a) Calculate the first five Bragg diffraction angles for x-rays of wavelength 0.120 nm, associated with the planes $A, B, \ldots$ (b) Show that the x-rays 'reflected' from planes $a, b, \ldots$ will in fact cancel those from planes $A, B, \ldots$, and that diffracted beams with odd n will have zero intensity for this crystal structure. These are analogous to the 'missing orders' produced by zeros in $A_c(\theta)$ for a multiple-slit system. (See problem 18.18.)

18.12 (a) Show that the intensities of the subsidiary maxima in the single-slit diffraction pattern, relative to the intensity of the central maximum, are approximately given by $1/(m + \frac{1}{2})^2\pi^2$ where $m = 1, 2, 3, \ldots$

(b) Calculate the relative intensities of the first three subsidiary maxima.

18.13 If you stand vertically and form slits of width 1 mm with your eyelids, what approximate angular separation is necessary between two horizontal illuminated slits for you to be able to resolve them? (Assume that the limit is imposed by diffraction. Use the *Rayleigh criterion*, according to which overlapping diffraction patterns can just be distinguished if the central maximum of one lies over the first zero of the other.)

18.14   The depth of field of a camera lens (the range of object distances giving an acceptably sharp image) can be increased by reducing the aperture. This will, however, increase the effect of diffraction, which tends to blur the image. Estimate the aperture (as a fraction of the focal length) at which diffraction will become appreciable, for a film which can resolve 60 lines per millimetre. (Assume that there are no significant lens defects.)

18.15   We have seen that waves diffracted by a slit of width a have a spread of directions $\Delta\theta \sim \lambda/a$. Find the corresponding spread Δk_x in the component of k at right angles to the slit, and show that

$$\Delta k_x \, \Delta x \sim 2\pi$$

where Δx is the spread of x values in the beam. (Compare this with the analogous result (11.13) relating spreads in the propagation direction.)

18.16   (a) Show that the amplitude of a single-slit diffraction pattern in the direction $\theta = \lambda/2a$ is approximately $2/\pi$ times the amplitude in the forward direction.

(b) Ocean waves of wavelength 5.0 m are diffracted by a breach of width 10.0 m in a harbour wall. Over what angular range will the amplitude of the waves within the harbour exceed $2/\pi$ times the maximum value?

18.17   In an electron microscope a beam of electrons is used to cast magnified shadows of thin specimens. If the 'apertures' in the specimens are typically 0.1 nm across, estimate how high the kinetic energy of the electrons has to be to give sharp images. (Assume that the limiting effect is diffraction; in practice the diffraction limit cannot be achieved.)

18.18   In a multiple-slit diffraction pattern, principal maxima predicted by $\Phi(\theta)$ may be absent because $A_c(\theta)$ happens to be zero in these directions. (a) For uniform slits of width a and separation d, show that 'missing orders' occur if the ratio d/a is an integer. (b) If $d = 2a$, corresponding to a screen with alternating opaque and transparent strips of equal width, show that all even orders (except order zero) are missing.

Answers to problems
and hints for solution

Chapter 1

1.1   (a) $60\,\mathrm{s}^{-1}$.
     (b) 9.5 Hz.
     (c) 0.10 s.

1.2   (a) 57 mm.
     (b) $30°$.
     (c) 0.059 J.

1.3   (a) 28 mm.
     (b) $3.0\ \mathrm{m\ s}^{-1}$ to the left.
     (c) $t = 35$ ms.

1.4   (a) $p_1/m\omega_0$.
     (b) $+\frac{1}{2}\pi$.

1.5   (a) 20 mm.
     (b) $-60°$.
     (c) $10\ \mathrm{mm} - (17\ \mathrm{mm})\,\mathrm{i}$.

1.6   Work in form B.
     $30\ \mathrm{mm} + (17.5\ \mathrm{mm})\,\mathrm{i}$.

1.7 $10\,000g$.

Chapter 2

2.1 $5 \times 10^{-4}\,\mathrm{N\ m\ rad}^{-1}$.

2.3   15 per cent lower.

2.4   Frequency falls by 3.3 per cent, corresponding to a pitch drop of just over half a semitone.

2.5   Frequency is determined by mass of cone (area a) and stiffness $a^2\gamma p/V$ of air in cabinet (volume V).
     52 Hz.

2.7 (a) CV_1.
  (b) 0.
2.9 $1.6 \, \text{m s}^{-2}$.
2.10 49 g, 45 mm.
2.11 (a) $x = 0$.

Chapter 3

3.1 (a) $0.02 \, \text{kg s}^{-1}$.
  (b) 30.
  (c) $1.2 \, \text{kg s}^{-1}$.
3.7 $30 \, \text{kg s}^{-1}$.

Chapter 4

4.3 Use equation (4.8).
  35°.
4.4 Use equation (4.5).
  0.94 s.
4.5 (a) 0.12 N.
  (b) $0.55 \, \text{N} < F_{\text{st}} < 0.79 \, \text{N}$.

Chapter 5

5.2 (a) Motion is stiffness controlled.
  100 mm, −6.5°.
  (b) 74 mm, −125°.
  (c) Motion is mass controlled.
  0.56 mm, −176°.
5.3 Make the denominator of $R(\omega)$ a minimum.
5.5 (a) $314.16 \, \text{s}^{-1}$.
  (b) $313.77 \, \text{s}^{-1}$.
  (c) $313.37 \, \text{s}^{-1}$.
  (d) $314.94 \, \text{s}^{-1}$.
5.8 (a) $600 \, \text{s}^{-1}$.
  (b) $38 \, \text{s}^{-1}$.
  (c) 16.
  (d) 2.5.
5.9 Use the value of $\langle P \rangle$ at resonance.
  3.9 N.
5.10 $\langle P \rangle = \frac{1}{2} m \gamma (\omega A)^2$ from (5.11). Use $B_{\text{q}} = -A \sin \phi$ and find $\sin \phi$ from fig. 5.2.
5.12 (a) 4.4 mm.
  (b) −77°.

5.15 (a) 0.99 Hz.

    (b) 79 s.

5.16 133°.

5.17 (a) $\gamma\dot{\psi} + \omega_0^2\psi = \ddot{\psi} + \ddot{y}$ where y is ground displacement; solve as (5.3).

    (b) $A/H \approx 1$ when $\omega \gg \omega_0$.

5.18 Maximum amplitude occurs half a beat period after switching on. At that time, damping factor in (5.29) must be 0.1 or smaller. $Q \lesssim 70$.

Chapter 6

6.2 (a) $-i/\omega C$.

    (b) R.

    (c) $i\omega L$.

6.4 For x-rays $\omega \gg \omega_0$ and the motion is mass controlled.

6.5 Use an approximation for the compliance, valid when the damping is very heavy and $\omega \ll \omega_0$.

Chapter 7

7.1 (a) Yes.

    (b) No.

7.2 (a) $0.961\nu_0$.

    (b) 23°.

    (c) Stored energy is approximately proportional to A^2. (Neglect very small effect due to changing frequency). Thus A^2 decreases as $\exp(-\gamma t)$.

       56 s.

7.3 (a) Return force is proportional to $(1 - a_0/a')\psi$ where a' is length of string when mass has displacement ψ. Find a' in terms of a and ψ, and expand a/a' as a power series in ψ/a. Compare force with (7.1) to find $\alpha \approx 2a_0/a^2(a - a_0)$.

    (b) 0.14 m.

7.4 Compare ψ given by (7.14) at $t = 0$ and $t = \frac{1}{2}\tau$.

7.5 (a) -6.0.

    (b) -7.0.

    These differ by only 8 per cent from the value for $p = 9$.

7.6 2×10^{-21} J $= 0.01$ eV.

7.7 Solve equation (7.12) for ψ and use binomial series up to F^2 term.

7.8 0, 10, 50, 60, 100, 110, 120 Hz.

7.9 30, 40, 60, 70 Hz.

Chapter 8

8.2 $\psi_1 = 8.5$ mm, $\psi_2 = 5.1$ mm.

8.4 For amplitude ratios use equations (8.25) and (8.27).
  (a) 0.46.
  (b) 6.5.
  (c) 1.9.

8.5 $q_1 = (m^{1/2}/2)(3^{1/2}\psi_1 + \psi_2)$, $q_2 = (m^{1/2}/2)(-\psi_1 + 3^{1/2}\psi_2)$.

8.6 1.23 s, 1.27 s.

8.7 (a) $q_1 = 0.41x_1$, $q_2 = 0.41x_2$.
  (b) 2.0 Hz, 4.4 Hz.

8.9 (a) $\omega_1 = \omega_2 = (1/LC)^{1/2}$.
  (b) $\gamma_1 = 2R/L$, $\gamma_2 = 0$.
  (c) In the 'in-phase' mode the currents through R always cancel, and so there is no damping.

8.10 170 Hz and 180 Hz.

Chapter 9

9.1 Compare problem 1.5.
  (a) 20 mm.
  (b) $-60°$.
  (c) 10 mm $- (17$ mm$)$i.

9.2 (a) 2.4 Hz.
  (b) 0.84 m.
  (c) -2.0 m s^{-1}.
  (d) 0.10 m.

9.3 (a) 15 mm s^{-1}.
  (b) 0.0028 per cent.

9.4 A upwards, B upwards (faster than A), C stationary, D downwards, E stationary.

9.5 (a) 90 N.
  (b) 0.10 kg m^{-1}.

9.6 Reduce tension by a factor of 9.

9.7 Points A, D and E were moving in the same direction, which was opposite to that of C. Points A and D had approximately the same velocity. Points C and E had approximately the same speed. (Alternatively, all points might have been stationary.)

9.10 Downwards, $3h$, $4l$.

9.13 76 N.

9.15 Use conservation of energy.

9.16 0.10 s^{-1}.

9.17 Solve equations (9.41).

9.18 (a) $\Gamma/\omega = 1/10\pi$.

40 m.

(b) Use (9.47).

0.0080, 90° advance.

Chapter 10

10.1 (i) and (iv).

10.2 8.3×10^{10} N m^{-2}.

10.3 Stress is $\rho c\dot{\psi}$, and $E = \rho c^2$ from (10.5).

10.4 15 per cent lower. (See problem 2.3.)

10.5 4.4 m.

10.6 (a) 1.4 km s^{-1}.

(b) 1.4×10^6 kg s^{-1} m^{-2}.

(c) 60:1.

(d) Characteristic impedance mismatch makes transmission coefficient small ($T \approx 6 \times 10^{-4}$).

10.8 77 μV.

10.9 2.8 W.

10.10 1.4 dB km^{-1}.

Chapter 11

11.1 Spectrum contains components at $n = \pm9$ and $n = \pm11$, all of the same amplitude. Note that the fundamental has zero amplitude in this example.

11.3 Spectrum contains components at $n = \pm N$ (amplitude $\frac{1}{2}A$) and $n = \pm(N \pm 1)$ (amplitude $\frac{1}{4}B$).

11.5 (b) $c_0 = 0$, $c_n = i(A/2n\pi) \cos n\pi$ $(n > 0)$.

11.6 (a) $\psi = d(1 + 2z/L)$ $(z < 0)$.

$\psi = d(1 - 2z/L)$ $(z > 0)$.

(b) $c_n = 0$ (n even),

$c_n = 4d/n^2\pi^2$ (n odd).

(d) $\psi(z, t) = \sum_{n \text{ odd}} (8d/n^2\pi^2) \cos(n\pi z/L) \cos(n\pi ct/L)$.

11.7 (a) 5 ms.

(b) 0.6 s.

11.8 63 ± 6 m^{-1}.

Chapter 12

12.4 Relative increase is $-(d/c)(n\pi/L)^2$.

10 per cent.

12.5 (a) 0.24.

(b) 1.8 mm.

12.6 Use the fact that $v_\phi = 2v_g$.

12.7 Yes. The transverse force $-T(\partial\psi/\partial z)$ is out of phase with the displacement, and so not in quadrature with $\partial\psi/\partial t$. (This is the kind of wave that exists in a cut-off region of finite length, allowing tunnelling to occur.)

12.9 100 Hz.

Chapter 13

13.1 (a) $0.21\,\mathrm{m\,s^{-1}}$, $0.32\,\mathrm{m\,s^{-1}}$.
(b) $1.2\,\mathrm{m\,s^{-1}}$, $0.62\,\mathrm{m\,s^{-1}}$.
(c) $7.0\,\mathrm{m\,s^{-1}}$, $7.0\,\mathrm{m\,s^{-1}}$.

13.2 Use (9.8), (12.2) and (13.27).
(a) $3.3\,\mathrm{m\,s^{-1}}$.
(b) $3.8\,\mathrm{m\,s^{-1}}$.
(c) $4.0\,\mathrm{m\,s^{-1}}$.

13.3 Use group velocity from (13.23).

13.4 Use (12.2), (12.4) and (13.27).

13.5 $230\,\mathrm{mm\,s^{-1}}$.

13.6 Use result of problem 13.5.

13.7 1.6 m.

13.8 (a) Use (13.19) and (13.20), and make coefficient of k^4 zero.
(b) 4.7 mm.

13.9 (a) 5.1 mm, 58 mm.
(b) $0.15\,\mathrm{m\,s^{-1}}$, $0.45\,\mathrm{m\,s^{-1}}$. Gravity group is slower than stream, surface tension group is faster than stream.

Chapter 14

14.2 (a) $238\,\Omega$.
(b) $0.53\,\mu\mathrm{T}$.

14.3 0.040.

14.4 Show $R < 0$.

14.5 0.77.
See section 6.4.

14.6 710 m.

14.8 $5 \times 10^{10}\,\mathrm{m^{-3}}$.

Chapter 15

15.1 (a) 150 eV.
(b) 0.08 eV.
(c) 0.02 eV.
(d) 0.0003 eV.

15.2 Take the kinetic energy as $kT \approx 4 \times 10^{-21}$ J.
      30 pm.
15.3 To within about 4 μm.
15.4 ~5 pm.

Chapter 16

16.1 Use (16.2). For isothermal changes ρ is proportional to p; thus c
      is independent of p.
16.2 Integral from 0 to ψ_p is integral from p_0 to p. By binomial expansion

$$p^n - p_0^n \approx n p_0^{n-1}$$

16.3 Peak shifts by bhL relative to pulse centre. Shape change will be
      small if this shift is much smaller than pulse width w.
16.4 Use (16.12) and (16.18).
      (a) 4.0 m s^{-1}.
      (b) 4.2 m s^{-1}.
      (c) 4.4 m s^{-1}.
      Small solitary waves have the same speed as long sinusoidal waves.
16.5 Use (13.28), (16.12) and binomial expansion of $(gH)^{1/2}$ with
      $H = h_0 + \psi$.
16.6 (a) Use $d(\text{sech } x)/dx = -\text{sech } x \tanh x$ and $\tanh x = (1 -$
      $\text{sech}^2 x)^{1/2}$.
      (b) D varies as $d^{-1/2}$. Thus increasing d decreases D and makes
      $\text{sech}^2 DZ$ curve broader.

Chapter 17

17.1 (a) 41.8°.
      (b) 1.85λ, 0.45λ.
17.2 (a) 45°.
      (b) 45°.
17.3 (b) $\approx$40 MHz.
17.4 (a) $\omega^2 = c^2 k_z^2 + (\pi c/a)^2$.
      (b) 1.5 GHz.
17.5 1, $\sqrt{2}$ (3 degenerate modes), $\sqrt{3}(3)$, 2, $\sqrt{(17/3)}(3)$, $\sqrt{(22/3)}(3)$, 3.
17.6 Use (17.13) with $m = n = 0$.
      $2a\Delta\lambda/\lambda^2 \approx 8$.
17.7 175, 196, 247, 262.5 Hz.
17.8 Use (17.15).
      50.
17.10 Substitute $\omega = W_F/\hbar$ in (17.17) and equate to $\frac{1}{2}N$.
      7.0 eV.

17.11 Average is integral of $Wg(W)$ divided by integral of $g(W)$, both integrals taken from 0 to W_F.

Chapter 18

18.1   14.5 mm.

18.4   (a) 200.

     (b) 20.

     (c) 2.

18.5   Use fig. 18.5. Phase lead is $(90° - \frac{1}{2}\alpha) - (90° - \frac{1}{2}N\alpha)$.

18.6   (a) 0.25.

     (b) 0.22.

     (c) 0.21.

18.7   0.014°.

18.8   Use (18.4).

     20, $19\lambda/20$.

18.11 (a) 6.8°, 13.7°, 20.7°, 28.0°, 36.8°.

18.12 (a) Assume that the maxima of $(\sin x)/x$ occur close to the maxima of $\sin x$.

     (b) 0.045, 0.016, 0.008.

18.13 ≈ 0.5 mrad $= 0.03°$.

18.14 Assume image is formed in focal plane. Then angular spread due to diffraction is $\sim \lambda D/f$ if aperture diameter is f/D.

     $f/32$.

18.16 (b) ±14°.

18.17 For $\lambda \ll 0.1$ nm we must have $T \gg 150$ eV.

Constants and units

Physical constants

electron charge	e	1.60×10^{-19} C
electron mass	m_e	9.11×10^{-31} kg
proton mass	m_p	1.67×10^{-27} kg
permittivity of vacuum	ε_0	8.85×10^{-12} F m^{-1}
permeability of vacuum	μ_0	$4\pi \times 10^{-7}$ H m^{-1}
speed of light in vacuum	c	3.00×10^8 m s^{-1}
Avogadro constant	N_A	6.02×10^{23} mol^{-1}
Boltzmann constant	k or k_B	1.38×10^{-23} J K^{-1}
Planck constant	h	6.63×10^{-34} J s
	$\hbar$	1.05×10^{-34} J s

Mathematical constants

π	3.142
e	2.718
ln 10	2.303

Conversion factors

atmosphere   1 atm $= 1.01 \times 10^5$ N m^{-2}

electronvolt   1 eV $= (1.60 \times 10^{-19}$ C$) \times (1$ J C$^{-1}) = 1.60 \times 10^{-19}$ J

Decimal prefixes for SI units

m (milli-)	10^{-3}		k (kilo-)	10^3
μ (micro-)	10^{-6}		M (mega-)	10^6
n (nano-)	10^{-9}		G (giga-)	10^9
p (pico-)	10^{-12}			

Index